D1072497

Local Organizations

Local Organizations

INTERMEDIARIES IN RURAL DEVELOPMENT

Milton J. Esman AND
Norman T. Uphoff

Cornell University Press
ITHACA AND LONDON

334.683
E76L

Copyright © 1984 by Cornell University Press

All rights reserved. Except for brief quotations in a review, this book, or parts thereof, must not be reproduced in any form without permission in writing from the publisher. For information, address Cornell University Press, 124 Roberts Place, Ithaca, New York 14850.

First published 1984 by Cornell University Press.
Published in the United Kingdom by Cornell University Press Ltd., London.

International Standard Book Number 0-8014-1665-5
Library of Congress Catalog Card Number 83-73340
Printed in the United States of America

Librarians: Library of Congress cataloging information appears on the last page of the book.

The paper in this book is acid-free and meets the guidelines for permanence and durability of the Committee on Production Guidelines for Book Longevity of the Council on Library Resources.

Contents

UNIVERSITY LIBRARIES
CARNEGIE-MELLON UNIVERSITY
PITTSBURGH, PENNSYLVANIA 15213

Preface

Interest in local organizations operating in the rural areas of less developed countries has grown in recent years as experience with them accumulates. When they "work," their accomplishments in the midst of poverty and disarray prompt praise from a variety of sources. Their "failures," of course, are cited as one more piece of evidence that development will be slow in coming or, if it comes at all, must be a top-down process.

In the following analysis of experience with local organizations in rural development, the discussion sometimes becomes abstract. To show that the subject matter remains utterly concrete, however, we shall begin with the evidence of some diverse observers—a business executive, a famed writer, and two social scientists.

The first is Peter T. Jones, senior vice-president and general counsel for the manufacturing firm of Levi Strauss. In his other role as chairman of the board of directors of the Inter-American Foundation, he has had occasion to observe grassroots development in Latin America in recent years. After a trip in 1981 to the Dominican Republic, he wrote:

> What struck me most on this visit was the contrast—not between the rich and poor—but between the hope and accomplishments of poor people who have effectively organized themselves, and the fatalism and difficulties of those struggling alone. I talked to small landowners producing coffee and rice, to small entrepreneurs who make inexpensive clothing and shoes, and to fruit and vegetable venders who, with Foundation help, have acquired their own tricycle pushcarts which they formerly rented at exorbitant rates.
>
> By any standards, these are poor people. But they are successfully managing their own farms and businesses, and they have formed cooperative organizations to achieve what they are unable to do alone. Through their organizations, they have gained access to credit; jointly purchased tractors and facilities for drying their produce; reduced the costs of raw materials; and expanded their markets.
>
> In 1979, farmers in the Dominican Republic saw their homes, land and equipment devastated by Hurricanes David and Frederick. [Some] were able to rebuild quickly and efficiently because their cooperative organizations were in place. These small landowners show pride and self-confidence in managing

their own affairs, and they are determined in their joint efforts to create better lives for their families and their community. They know they are doing it for themselves, but they also acknowledge the importance of the credit that was made available to them through their federation's loan fund.

The IAF's contribution to that loan fund helped the federation to mobilize local resources four times the amount of our grant. . . .

The contrast with another farming community we visited was stark. This community, which had also suffered severe hurricane damage in 1979, lacked any strong local organization that could stimulate and reinforce individual efforts. The existing farmers' association was loosely structured and required little financial commitment or other involvement from its members.

According to the farmers themselves, hundreds of thousands of dollars had been made available in the aftermath of the hurricanes, but local participation and initiative were not encouraged in rehabilitation activities. Loans were not paid back, nor was there much expectation by donor agencies that they would be. The money provided, whether in loans or gifts, was largely squandered. [IAF, 1982:4–5][1]

In background and vocation, V. S. Naipaul differs greatly from Jones. Descended from sugar plantation workers brought to Trinidad from India, he is a writer who has traveled widely throughout the Third World. In his book *India: A Wounded Civilization,* he tells of visiting a squatters' settlement in the slums of Bombay to observe the work of the Shiv Sena organization, which sought to improve the lives of thousands otherwise bereft of opportunity and security:

A squatters' settlement, a low huddle of mud and tin and tile and old boards, might suggest a random drift of human debris in a vacant city space; but the changes now were that it would be tightly organized. The settlement in which the engineer [Naipaul's guide] had stayed, and where we were going that morning, was full of Sena "committees," and these committees were dedicated as much to municipal self-regulation as to the Sena's politics. . . .

The bus stopped and we were just outside the settlement. The entrance lane was deliberately narrow, to keep out carts and cars. And, within, space was suddenly scarce. . . . But the lane was paved, with concrete gutters on either side; without that paving—which was also new—the lane, twisting down the hillside, would have remained an excremental ravine. [1977:63–64]

Naipaul meets with one of the committee groups in its "stuffy little shed with a corrugated iron roof; but the floor, which the engineer remembered as being earth, was now of concrete," and then walks through the slum. The crowded conditions are not pleasant, as there are few facilities for hygiene or sanitation. But the committees have been able to build washing sheds

[1]Complete references to studies identified in the text by author and year are to be found in the Bibliography.

and latrine blocks, to install dustbins and get them used, and to establish some open spaces for community use and aesthetic purposes. They also help residents find employment. It is a continuing struggle to bring some order and progress out of the filth and despair. But Naipaul is able to tell by sight the areas of the slum in which the committees are operating. Where they are not, he reports some miserable scenes.

At the end of his rounds, Naipaul writes: "We were now back where we had started, at the foot of the hill, at the entrance, with the washing sheds full of women and girls, and the latrine blocks full of children: slum life from the outside, from the wide main road, but, approached from the other side . . . evidence of what was possible" (1977:69). The committee system had not been able to organize and improve the whole slum—owing to "perhaps internal political reasons, perhaps a clash of personalities, or perhaps simply an absence of concerned young men." Local organization is not always possible or always effective. But this "evidence of what was possible" is one of the few notes of optimism in Naipaul's survey of contemporary India.

Moving to a pair of communities in the Philippines, we see the following contrast in the provision of water supply, as documented by Gerald Hickey and Robert Flammang:

> The importance of strong local participation in AID-assisted projects is underlined very well in the cases of two Frankel-type water purification plants near Naga City, Camarines Sur. The first, near Milaor, just south of Naga, is a great success. There is a constant stream of people coming to use it. Some come from as far as 6 or 7 kilometers away. The local people marvel at how this plant can extract the filthy, stinking water from the watercourse next to it and convert it into clear, sweet-tasting drinking water.
>
> The barangay chief was a young dynamo of a person who was clearly just as poor as the other clients of the project; he showed us how the barangay financed the operations of the plant with a 1.50 peso assessment per month. There were a few families in arrears, but some had paid as far ahead as December. People came out of their homes to tell us what they thought of *their* water system, and it was uniformly favorable. They said they no longer have to haul water so far, or buy it from vendors along the highway. They don't have to worry about getting sick from drinking the water. Their lives are better than what they used to be.
>
> A twin water purification plant was constructed just north of Naga City in Magarao. It is standing unused, rusting. Children are swimming in a nearby creek that looked just about as dirty as the one at Milaor, but they told us that they did not get their drinking water from the creek. No, they said, they bought their water from a nearby owner of a private well (who had a nice house and car) for 2 pesos a month. The children said that there was not always enough money to buy gas for the engine powering the (Frankel) unit; while it was shut down, apparently, the spark plug rusted and the local people couldn't get it started again.

> At any rate, there was no great urge on anybody's part to fix it as long as the private well water was available. No one came out to talk with us as we inspected the unit except the children. With no apparent local support, the project appears to be a failure, benefiting no one except the firms and workers which built it. We conclude again that the grassroots rural poor can make or break an AID-supported project intended for their benefit, depending on how much they feel themselves to be a part of the project. [1977:59]

We do not believe that local organization is sufficient in itself to eliminate the burdens of poverty or the deprivations of status and opportunity. But these observations make vivid the possibilities for development from below. These possibilities can in many circumstances, and with circumspection as well as sympathy, be facilitated by government or private agencies. The extent of such possibilities, the problems impeding them, and the means for taking advantage of this potential are the concerns of this book.

Our earlier comparative analysis of the role of local organization in rural development dealt with systems of local organization in Asia (Uphoff and Esman, 1974). In this work we extend our analysis to all parts of the Third World and focus on the performance of specific local organizations to determine what features and approaches are likely to be productive under varying conditions in establishing and sustaining grassroots institutions for development.

Because different readers will have different interests and needs, our analysis was prepared in two forms. First, we worked on this book, including an examination of the data and relationships found in a sample of 150 rural local organizations. They range from the most to the least successful, and we explore specific problems and alternative remedies in some detail. Recognizing that many persons might prefer a more succinct treatment, we also prepared a shorter state-of-the-art paper (Esman and Uphoff, 1982). This made the focus sharper, in much the same way that reducing a photograph in size makes the objects and contrasts stand out more clearly. The effort to present such a summary statement in turn pressed us to revise and elaborate the more complete exposition. We hope that each of these versions will find its way to the appropriate readership.

This book is the culmination of five years of work at Cornell and overseas by faculty members, graduate students, and researchers associated with the Rural Development Committee under a cooperative agreement with the Office of Rural Development and Development Administration (now Office of Multi-Sectoral Development) of the United States Agency for International Development (USAID). The Rural Development Participation Project (RDPP) carried out under this agreement was, we think, an unusual example of collaboration between an academic group and a government agency. Its purpose was to increase the effectiveness of "participatory" approaches to rural development. It mobilized the efforts of 30 Cornell faculty members, 45 graduate students, and 40 associates from many disciplines and eight countries.

This analysis is the fourth under the RDPP. It was preceded by analyses of rural development participation (Uphoff, Cohen, and Goldsmith, 1979), of the use of paraprofessionals in rural development (Esman et al., 1980), and of participatory approaches to agricultural research and development (Whyte, 1981). More than 60 other research publications were issued under the project's sponsorship, many of which are listed in the Bibliography.

We acknowledge and thank the research assistants who participated actively in our Working Group on Local Organization during its two years. Forrest Colburn was the first coordinator of the group; Virginia Haufler took over this responsibility when Colburn left for dissertation research in Nicaragua; Farhat Haq, Cynthia Moore, and Nancy St. Julien, together with Colburn and Haufler, did the data searches and drafted the case analyses; Katrina Eadie and Darryl Roberts served as statistical consultants and helped with the computer data processing; Chris Brown participated in the early stages of deliberation and assisted us in extracting the state-of-the-art paper from the larger manuscript.

We also thank our colleagues who provided helpful criticism of drafts of the analysis: Harvey Blustain, John M. Cohen, Merilee Grindle, Bruce F. Johnston, David Leonard, John D. Montgomery, Elliott Morss, Emery Roe, Mitchell Seligson, William F. Whyte, and Frank W. Young. We also wish to express our appreciation to John Harbeson, AID's project manager for the RDPP (1979–82) while on leave from the University of Wisconsin, Parkside, with whose cooperation and encouragement our work was carried out. We thank Barbara Williams Smith, Debi Ostrander, Krishna Mukherjee, Jane Kellogg, and Susan Garey for their indispensable help in the production of the manuscript.

MILTON J. ESMAN *and* NORMAN T. UPHOFF

Ithaca, New York

Local Organizations

CHAPTER 1

Local Organizations as Intermediaries

When looking at the rural areas of the developing countries in Asia, Africa, and Latin America, one finds significant differences both between and within countries. There are, naturally, differences in soils, rainfall, and other agronomic conditions and in population density and social structures. One finds more complex differences in land tenure, in the commercialization of production and marketing, in occupational structure and income distribution, in public services and access to these services, in political regimes, and in government policies relevant to the rural economy.

Among the most important differences are the opportunities available to rural people to manage their own affairs, to influence public decisions, and otherwise to participate in activities that affect their economic productivity and quality of life. We shall argue that (1) such involvement is essential for accomplishing broad-based rural development; (2) eliciting and sustaining such participation will require some configuration of organizations that are accountable and responsive to their members; (3) the variety of interests and needs among rural people requires a variety of organizations even in the same area; and (4) the particular forms of organization that are likely to work will vary with specific local experience, the tasks to be performed, and the political-administrative environment.

Any concrete measures to facilitate rural development through local organization must take into account the realities distinctive to each society. This does not mean, however, that environmental factors determine local organization or rural development outcomes. It is possible to make some general statements that address trends and delineate the political economy of rural areas in developing countries. The most urgent general observation is that the populations of rural areas, with some exceptions in Latin America, continue to increase in absolute numbers—despite massive migration to urban areas, often caused more by intolerable rural poverty than by attractive opportunities in the cities. These growing numbers of rural people must somehow earn their livelihood on a land base that is not expanding; indeed, in many cases intensive exploitation is destroying the physical environment on which the livelihood of future generations depends. Though migration has transferred a portion of rural poverty to the shanty towns of urban areas, the greatest concentrations of severe poverty remain in rural areas.

In contrast to the evident poverty and deprivation in urban centers, the rural poor tend to be invisible to national planners and to foreign visitors alike (Chambers, 1980). This situation is exacerbated by the well-documented phenomenon of "urban bias," the tendency of governments to favor urban areas and urban publics in the provision of services, in the investment of public funds, and in incentives for private investment (Lipton, 1977). The discovery of "small farmers" by donor agencies has not yet greatly altered patterns of public expenditure. Even the more substantial landed farmers must often struggle for public policies that serve their economic needs. Small and marginal farmers, tenants at will, landless workers, and such weak and disadvantaged groups as female-headed households, who together constitute a majority in most rural areas, have limited access to public services and tend to be overlooked by governments (Esman, 1978a).

During the past decade it has become increasingly clear that macroeconomic growth alone does not eliminate and may not even reduce rural poverty. In many of the countries that have enjoyed rapid and sustained economic growth, widespread disabilities continue to afflict the rural population. Members of the growing rural labor force in most developing countries have experienced little improvement in their productivity or quality of life. Thus, attention has begun to turn to strategies that benefit larger numbers of people, that emphasize increased employment and productivity, and that improve health and welfare for the disadvantaged majority, most of whom continue to live and to earn their livelihood in rural areas. Such broad-based development is the policy objective of this study.

Elements of Rural Development Strategy

There are a number of essential components of development strategies that focus on assisting the majority of the rural population. Certain situations may require basic, even radical changes in the ownership of productive assets and in the distribution of political and economic power. Some people who have advocated radical structural changes believe that any lesser measures can be only palliatives and are doomed to failure. Our working hypothesis, however, is that real economic and institutional improvements are possible in most developing countries even without revolutionary change, which is in most cases unlikely. We consider that such improvements can contribute to higher productivity and a better quality of life for rural people and can, over time, enhance their ability to influence their future.

We should make clear that our concern with local organizations does not presume that they are all that is needed to promote rural development. Broad-based strategies of rural development, which can improve productivity and distribute its benefits more equitably among the rural majority, require a range of supports:

(1) *Public investments* in physical and social infrastructure. This includes roads, water supply for irrigation and household consumption, and such facilities as schools and health clinics. Infrastructural investments sufficient to deal with rural needs usually require a substantial shift of priorities from urban to rural areas and include not only the construction of facilities but also their operation and maintenance. Though such costs can and indeed should be shared with organized local publics, major and continuing governmental expenditures are required.

(2) A *policy environment* that is supportive of and responsive to the interests of rural constituencies, including the poor. This orientation can be expressed, for example, in market prices that provide production incentives to small farmers and small entrepreneurs, in import duties and tax and credit policies that discourage premature labor-displacing mechanization in areas of high underemployment, or in programs aimed directly at creating employment among landless laborers and marginal farmers.

(3) *Technologies* suitable to the circumstances and the capabilities of small farmers and other rural producers. Farming practices that foster improved productivity should be within the financial means of small operators. Similar technological improvements should increase the productivity of craftspersons and small rural industries. These are important because an increasing proportion of rural employment and income will have to come from off-farm and nonagricultural activities.

(4) *Effective institutions* operating at various levels and enhancing capacities in the public and private sectors. Three sets of institutions are required for rural development:

   (a) The first is *networks of government agencies* providing the public services that have come to be associated with higher productivity and an improved quality of life in all countries which have experienced successful rural development. These range from production-oriented assistance (agricultural research, extension, credit and marketing agencies, small industry advisory services, and irrigation departments) to social services (school systems, public health and sanitation facilities, and family planning agencies).

   (b) Some of these activities can be handled by *private enterprises* through market channels, and others may be undertaken by *private voluntary agencies* acting on philanthropic principles. The main burden of providing services is likely to fall on governments acting through their administrative networks. But depending on the tasks and circumstances, private-sector and market-based institutions may be better equipped to carry out such functions, provided they disperse rather than concentrate opportunities and benefits.

(c) The third set consists of *local institutions* ranging from local government bodies to rural enterprises. In Chapter 3 we outline a typology which orders the diverse array of local institutions that are not part of government or of the profit-oriented private sector. Potentially the most interesting and dynamic part of this continuum is the variety of locally based membership organizations—cooperatives, farmers' associations, mothers' clubs, health committees, water users' groups, ethnic unions, tenant leagues, and the like—which we designate collectively as *local organizations*.

All four of these components—public investments and expenditures, government policies, appropriate technologies, and effective institutions—are necessary for rural development. Ideally, the management of rural development would make them mutually reinforcing, so that one activity does not cancel out another. Local organizations of marginal farmers and laborers cannot offset the adverse effects, for example, of a government policy that subsidizes capital and fosters labor-displacing mechanization.

By and large, the previous academic literature on rural development and the preoccupations of government planners and development assistance organizations have focused on the first three components. The institutional aspects of rural development have received less attention than they deserve. Yet without institutions—bureaucratic, commercial, or participatory—infrastructure will not be built or maintained, public services will not be provided or well utilized, available technology will not be put to its best use on a wide scale, and governments will not be able to maintain satisfactory information exchanges with the publics that must be served if broad-based rural development is to become a reality rather than a slogan.

Within the set of institutions undergirding development, we would emphasize the importance of *local* institutions. These can act as intermediaries between rural residents and both government agencies and private commercial firms. They amount to a third sector, distinguishable from the public and private sectors as conventionally defined. This sector can affect the performance of the whole institutional framework of a country, serving to make central institutions more effective and responsive and filling an operational gap between the state and private enterprise.

Within the set of local institutions that can contribute to rural development, our focus is on local organizations (LOs). These we define as organizations which act on behalf of and are accountable to their membership and which are involved in development activities. This distinguishes them from organs of the state and also from more purely social or cultural associations. Examples of LOs are given in (4c) above. Not all are accountable to their members in the same way and not all are equally involved in rural development work. But as a set they have these basic characteristics. We wish to examine the contributions that such organizations can make to the goals of increased productivity and well-being, particularly for the poor rural majority.

We are dealing with "organization" for rural development not in a generic sense, but in a more limited and focused manner. Membership organizations have a logic and natural history that set them apart from bureaucratic structures and from market or communication networks. The term "local organization" has received increasing attention,[1] and its definition has gained greater consensus over the last decade, though some other terms have been used for the same set of organizations.[2] We find the other terms less serviceable than the one used here, which has wider usage and seems appropriate for most purposes.

Our focus on local organizations as a third sector parallels the distinctions offered by Hunter (1969) among administrative, economic, and political approaches to rural development. The first relies on bureaucratic mechanisms and seeks enforced compliance with government decisions in a regulatory mode. Decisions are made by experts according to technical principles and criteria, following policy objectives set by top officials. The second relies on market mechanisms and seeks to promote desired behavior through price incentives. Decisions are left to individuals, who calculate private advantage without explicit reference to broader interests or the public good. The third alternative is "political" in the way an Indian editorial in the *Economic Weekly* [Bombay] made the distinction: "The clue to the failure of rural development [in India] lies in this, that it cannot be administered, it must be organized. While administration is something which the civil service can take care of, rural development is a political task which the administration cannot undertake" (cited in S. Huntington, 1968:395). The third approach relies more on voluntaristic mechanisms, appealing to people's sense of interest and values. Cooperation is sought through processes of bargaining, discussion, accom-

[1]For example, Owens and Shaw (1972); Uphoff and Esman (1974); Johnston and Kilby (1975); Hunter (1978); Gow et al. (1979); Griffen and Ghose (1979); Bottrall (1980), Oxby (1980), and others with the Overseas Development Institute, London; Johnston and Clark (1982); and Leonard and Marshall (1982). The number of persons working collaboratively on the subject perhaps reflects its complex as well as collective nature.

[2]The Rural Organizations Action Project of FAO (van Heck, 1979) used the term "rural organizations" (also "people's organizations"), but this suggests that they are unique to rural areas or function differently there. The study by Nelson (1979) of similar groups in urban areas suggests the inadvisability of defining them geographically. D. Korten (1980) calls them "community organizations," but this suggests that they include the whole community, when in fact they may incorporate only a part of it or may involve persons from a number of communities. The term "participatory organizations" is used in the studies in Galjart and Buijs (1982), but this may limit the range of cases unduly or convert differences in degree into contestable differences in kind. Our work has benefited from the analyses and observations of these writers even though our nomenclature differs.

Table 1-1. Alternative approaches to rural development

	I	II	III
Principal mechanism	Bureaucratic structures	Market interactions	Voluntary associations
Decision-makers	Administrators and experts	Individual producers, consumers, and investors	Leaders and members
Guides for behavior	Regulations	Price signals	Agreements
Criteria for decisions	Policy—technically best means to implement it	Efficiency—best way to maximize profit and/or utility	Interests of members
Sanctions	State authority	Financial loss	Social pressure
Mode of operation	Top-down	Individualistic	Bottom-up

modation, and persuasion. Decisions are taken with reference to both group and individual interests; neither state authority nor rules of profit maximization determine choices, though both regulatory and price inducements may be invoked through members' decisions. The three approaches are contrasted in summary form in Table 1-1.

These alternatives can be illustrated with reference to achieving a goal such as range management, where the number of cattle grazed should be controlled so that water and pasture resources are not reduced through overuse. A pure public-sector approach would rely on expert advice to determine the "carrying capacity" of the area and would proceed to bring herd size into conformity with this determination. If the number of cattle was too large, regulations would be issued commanding herd reduction, perhaps apportioning authorized herd sizes to specific sub-areas. Persons or groups holding cattle in excess of these levels would be fined; if that did not reduce the numbers enough, the government might confiscate or even destroy cattle to bring the herds down to allowable levels. A pure market approach would manipulate price to induce owners to sell off any "excess" cattle. A pure approach of the third kind would seek to deal with herders in groups (on a tribal, community, or economic-interest basis), explaining the dangers and costs (to them) of overstocking and seeking agreement and compliance based on common understanding and social controls.

These examples are somewhat stereotyped to bring out the essential differences, but we can see that each approach relies on a different mechanism of compliance.[3] They do not constitute a pure continuum, since there are

[3]Students of organization theory will recognize that these three correspond to the ideal types of organization which Etzioni (1960) characterized as coercive, utilitarian (or remunerative), and normative. Boulding (1963) has written in parallel terms of threat systems, exchange

some important qualitative differences among them. In practice, no rural development effort could succeed by following only one approach, as Hunter points out. A well-rounded strategy would draw on a combination of these mechanisms, since each by itself produces diminishing returns. There are some things that each one can accomplish better (more cheaply, more reliably) than the others, such as amassing resources for investments having long-term payoff, promoting innovation to raise productivity levels, or modifying programs to meet local conditions.

A pure market approach in the example just given, subsidizing the price paid to producers, might be very costly, but perhaps no more so than mounting an administrative campaign to destock the area. To be sure, in some situations cattle owners might be indifferent to price incentives, making the market approach ineffective. The third approach might be attractive on ethical grounds, but the difficulties encountered in establishing or working with local groups might be formidable and the results slow or unreliable. This approach might be less attractive to the extent the environment was imminently endangered. Administrative methods are generally preferred by governments because they seem more uniform and reliable, and this belief commonly outweighs cost or normative considerations. Still, bureaucratic capacity is limited and uneven in most developing countries, and it seldom achieves the efficiency or effectiveness expected. It commonly needs the supplementation of either or both of the other approaches.

The third approach has some characteristics of each of the other two. The first and third are similar in that they would solve problems through collective rather than individual action, making efforts to achieve objectives that transcend individual self-interest. In this regard, the third sector is somewhat like the public sector. On the other hand, the second and third approaches do not wield state authority to gain their objectives (though both can influence it). Being less able to coerce, they are more likely to be flexible and accommodating in seeking solutions. In a way that is usually more effective than individuals can achieve, the third sector represents private interests vis-à-vis the state and its bureaucracies. Local organizations thus may be seen as "intermediate" in the sense that Berger (1977) characterized them as standing between the individual and the state.

Each approach has strengths as well as weaknesses. The temptation to justify one by pointing to evident deficiencies in the others should be resisted. It is wiser to look for complementarities among them, to take advantage of what each can do best. In the past there may have been too much

---

systems, and integrative systems. The first two correspond, respectively, to the techniques of hierarchy (bureaucracy) and exchange (market) analyzed in Johnston and Clark (1982:158–60), whereas the third combines aspects of the other two techniques they analyze—polyarchy and bargaining. These three approaches were analyzed as alternative channels for service delivery in Esman and Montgomery (1980).

readiness to assume that governmental solutions would be most effective for promoting rural development. At present we sense an inclination to see the private sector as having more beneficent effects than can realistically be expected. Particularly where one wishes to raise the productive capacities of poorer households, communities, and regions, reliance only on private enterprise is unlikely to produce such results. Profit-oriented activities cluster where there are the greatest and most immediate opportunities for returns. Thus, although there is an important and often neglected role for private-sector initiative in rural development, we do not see the market as the solution to all or even most of the problems. True, the rural majority would generally benefit from fewer distortions of the market, but market "imperfections" come from private as well as public sources.

In any case, it is spurious to try to choose between public and private approaches, since both are needed and since local organizations as a third sector can serve to make the other two more effective. The development of vigorous local organizations has important implications for extending the outreach of public administration and for improving the performance of government agencies and personnel. Local organizations, by aggregating the demands and resources of private citizens, can also supplement and make more effective the efforts of individuals in the private sector, representing their needs more persuasively and helping to solve local problems in more appropriate ways.

Our purpose in focusing on LOs is to increase knowledge of how their potential contributions can be tapped. LOs have too often been overlooked or neglected, or the efforts to work with them poorly informed or misdirected. The tendency of governments to ignore LOs is reported vividly by a sociologist on the staff of the World Bank:

> I talked recently to an economist who had just completed an in-depth study of the agricultural sector of two developing countries. He told me that in order to increase resource mobilization, the two governments have used almost all the conventional economic remedies: they have attempted to raise various agricultural levies; they have expanded the commercial banking and credit network; they have subsidized certain inputs (e.g., fertilizers) so as to encourage their usage; they have invested in road infrastructure; and so on.
>
> But in spite of all these efforts, the agricultural economy has remained sluggish and obstinately refuses to develop at a steady rate. The only thing the governments have not paid attention to, my colleague commented sadly, was institution building in rural areas. They did not at all perceive the urgency of providing the peasants with more adequate social institutional structures for increased economic effectiveness, so that the farmers themselves could achieve a self-sustaining and durable growth. [Cernea, 1982:137]

Fortunately, such neglect is not universal. The past decade has witnessed growing recognition among scholars and government officials of the importance of local organization and popular participation in any broad-based strategy of rural development. One example is this passage from the draft of India's sixth Five-Year Plan:

> Critical for the success of all redistributive laws, policies and programs [proclaimed in this plan] is that the poor be organized and made conscious of the benefits intended for them. Organized tenants have to see that the tenancy laws are implemented. Organizations of the landless have to see that surplus lands are identified and distributed to them in accordance with the law within five years. . . .
>
> The general lesson of the experience so far is that because of leakages in delivery systems and ineffective administration, rural programs fail to improve the distribution of income. The Planning Commission is proposing a massive shift of resources in favor of rural areas with an inbuilt redistributive character in almost every program. But whether these [programs] . . . will have the desired equalizing effect will depend on the extent to which the organized pressure of the beneficiaries counteracts the weaknesses of the administration and the opposition of vested interests. [Government of India, 1978:1.98]

The final report of the 1979 United Nations' Food and Agriculture Organization (FAO) declared that "rural development strategies can realize their full potential only through the motivation, active involvement and organization at the grass-roots level of rural people, with special emphasis on the least advantaged, in conceptualising and designing policies and programmes and in creating administrative, social and economic institutions, including cooperative and other voluntary forms of organization for implementing and evaluating them (FAO, 1979:8)." This theme is now reflected in the writings of a growing number of academic students of economic development (e.g., Johnston and Clark, 1982; Cummings, 1982; Chambers, 1983).[4] One finds it also in the policy pronouncements of such development assistance agencies as USAID (1975), the British Overseas Development Ministry (ODM, 1975), and the World Bank (1975:35–38). It does not yet represent a firm consensus, but it does constitute an increasingly influential stream of thought.

Analytical and Policy Considerations

The contributions that local organizations can make to rural development can be analyzed in different ways. In Chapter 3 we deal with eight basic

[4]Johnston and Kilby wrote previously, "There is an almost universal need to encourage the creation of local organizations of farmers—to facilitate the dissemination of technical information, to reduce the administrative costs of distributing credit to small farmers, to manage the distribution of irrigation water, and the various other activities that may require group action. But the instrument of choice will depend on local circumstances." (1975:445–46)

functions that LOs can perform. Here we address such contributions schematically under three headings (which, conveniently, all begin with the same letter): *efficiency, equity,* and *empowerment*. Not only are these analytically distinguishable, but in practice an LO may further only one or two if not all three. To be sure, ineffective LOs may accomplish none of these objectives, and some LOs may diminish them.[5] As discussed below, some connection and even progression may be drawn from one objective to the other, so they are not unrelated outcomes. This makes LOs more interesting than if their contributions were quite discrete and not, at least potentially, cumulative.

We begin by recognizing that LOs may produce few or no positive benefits, in which case they are unlikely to last very long or to be more than nominal entities. Gow and his associates with Development Alternatives, Inc. (DAI) describe the high failure rate of LOs as a "sobering fact" (1979, I:28). As discussed at length in Chapter 6, LOs are vulnerable to many "pathologies" that impair their performance on behalf of their members, but these can be counteracted in various ways (Chapter 7). When LOs perform up to the expectations of their members (and sympathetic outside observers), they can accomplish some remarkable things, as suggested in the Preface and shown in the chapters that follow.

In the resource-poor environments of most developing countries, governments lack the funds, personnel, and administrative capacity to provide services directly to the masses of small farmers and other marginal publics in scattered and sometimes isolated rural areas. Frequently, the services that governments attempt to provide prove to be irrelevant and even detrimental to the circumstances of these people. There is commonly a real information gap between governments and the majority of their rural constituents, and an incapacity on the part of the latter to articulate their common needs and preferences and to make credible claims for effective assistance.

To the extent that LOs are able to increase the *efficiency* of resource use, they are likely to be widely welcomed. Even though LOs usually entail costs to their members and leaders, as well as to the government and other agencies, there are a number of ways in which net productive benefits can be achieved.

(1) The provision of public services by government agencies can be facilitated by more *accurate and representative information* about the needs, priorities, and

[5]Gow and associates (1979, I:2) suggest that LOs can be vehicles for a two-way flow of information to provide technical knowledge and to reinforce individuals trying new approaches, for minimizing risk and achieving economies of scale, for achieving greater political and economic independence, and for attempting to spread the benefits of assistance to a large number of the rural poor. The first and second are "efficiency" functions, the third is "empowerment," and the fourth "equity," as conceived here. They also note that LOs can be vehicles for perpetuating inequitable social systems (counter-equity) and for controlling the rural population (counter-empowerment), important observations to bear in mind.

capabilities of local publics, and more reliable feedback on the impact of government initiatives and services. Public services often fail to be effective because program designers or donor agencies are poorly informed as to the circumstances of the people they intend to assist. Through organization, these often locally distinctive needs can be summarized and expressed. Listening only to individuals makes it possible, indeed likely, that planners and administrators will hear unrepresentative views. These views may even discourage agencies from listening to members of the publics they are to serve, so that in the absence of representative channels, a certain "deafness" may result.

(2) Given the high degree of natural and social variability in rural areas, there is a great need for *adaptation of programs* to meet local conditions if the resources employed are to be efficiently used.[6] "Standard packages" enhance central control and seem to save on administrative costs, but if they are locally ineffective, the funds and staff time involved are essentially wasted. Local organizations can share the responsibility and cost of making activities serve priority needs more appropriately.

(3) Local organizations can provide opportunities for *group communication* where reaching rural publics is costly because of the great number and dispersion of persons to be contacted—for agricultural extension advice, nutrition education, immunizations, supervised credit, or family planning information, for example. Communicating with intended beneficiaries individually is prohibitively expensive, as is making individual small loans. LOs, on the other hand, can extend services directly to their members, "retailing" to individual households the information, credit, and material resources that governments provide on a "wholesale" basis, thereby reducing administrative costs.

(4) Local *resource mobilization* on a self-help or matching grant basis helps stretch the resources that the government can contribute to rural improvements. Even if communities are generally poor, they usually have some resources that can be applied to meeting priority needs. Rural people's labor is seldom as unemployed as urbanites surmise, but labor is a valuable resource that they can contribute, along with materials, reliable information, and managerial skills.[7] Resource-scarce governments can hardly afford to

[6]A vivid demonstration can be seen in the case study by King (1981) of the introduction of cooperatives in northern Nigeria. He found the potentials for group action different in villages only a few miles apart—because of different distributions of village power and different agricultural opportunities, credit institutions, and physical environments. The standard agricultural credit and input program being introduced within the area was mostly unsuccessful, since it was not attuned to local variations.

[7]The building of bridges through local committees in the Baglung district of Nepal is one of the more dramatic examples (P. Pradhan, 1980). Within five years, 62 suspended bridges, some of them as much as 300 feet in length, were built with local labor and mostly local materials—establishing a network across the whole district—much faster than the government could have done and at one-fourth to one-eighth the cost. The government provided only steel cables and some modest funds for skilled labor. The conventional use of cement and steel rods, which would have been both expensive and difficult to transport into the remote areas of the district, was avoided by utilizing the indigenous technology of masons and blacksmiths—which relates to the next point.

forgo such opportunities for meeting more needs than their budgets would otherwise allow.

(5) Although the *technical knowledge* of local people may not be highly sophisticated, it is commonly quite appropriate, based on long experience and intimate practical knowledge of their environment. Through LOs, such knowledge can be tapped and utilized to produce desired results and avoid costly technical mistakes in government investments.[8]

(6) It has been observed that *utilization and maintenance* of facilities and services is better with programs grounded on local organization and participation.[9] A World Bank study of rural water supply systems found that these were better maintained and less abused and had a higher level of financial performance to the extent that there was village participation in decisions about installing the systems, plus local contribution to construction costs and payment of water fees once the systems were in operation (1976:63). This means that the benefit from government investments can be greater and resources more efficiently used.

(7) Finally, *cooperation* in new programs involving economic, social, or technological change is more likely to occur when a local organization enjoying the confidence of rural people shares responsibility for the innovation.[10]

There is not likely to be much disagreement on the desirability of increasing the efficiency of development programs through local organization. There may be less consensus on the second objective to which LOs contribute: promoting equity of opportunities and benefits. However, national policy-makers even of a somewhat conservative disposition (e.g., in Taiwan, Malawi, South Korea) may value relatively equitable economic and social relationships at the base, since this is likely to contribute to political stability and a better international image for the regime.

There is considerable debate in the literature as to whether LOs do or can improve the position of the poorer, weaker sectors of rural society (e.g., UNRISD, 1975; Gow et al., 1979, I:8–9), and we address this issue at various junctures in this book. The common experience is similar to that

[8]The study of a random sample of small-scale irrigation schemes in Nepal by the Agricultural Projects Service Centre (APROSC) found that "those projects in which the local people participated in the identification and design of the project proved to be technically more successful" (APROSC, 1980:7). We found three examples in the literature, from Nepal, the Philippines and Mexico, where local people warned engineers that the design of the proposed dams would make them inadequate to withstand peak flow; their warnings were vindicated when the dams washed out shortly after construction (Shrestha, 1980; Coward, 1979a; Cernea, 1983:16–17).

[9]Utilization of rural health facilities is also reported to be greater when local committees contribute to health education, scheduling visits, and patient follow-up (Isely et al., 1979).

[10]Making changes in public sanitation practices to reduce the burden of parasites is one of the more difficult undertakings in rural development, but Fountain (1973) reports that by working through committees of village leaders in the Bandungu province of Zaire, public health workers were able within one month to reduce the average ascaris infestation from 70 percent to less than 15 percent.

reported in one study of an agricultural cooperative in the Dominican Republic, where peasant coffee growers with middle-sized and large holdings benefited from the extension and marketing arrangements, while the majority of small-holders and, particularly, landless laborers received few or no benefits (Sharpe, 1977:214–15). Still, we find numerous cases where local organizations have benefited poor women, marginal farmers, disadvantaged ethnic groups, and other frequently bypassed categories.[11]

Since economic and social gains for the poor do not necessarily change the social rank-order in rural areas, more equitable distribution of benefits may be fairly widely accepted as a goal of local organization. To be sure, in most cases where disadvantaged sectors are assisted, there is some outside agency performing what we would characterize as a "catalyst" role (see Chapter 8).[12] That relatively equitable outcomes of development efforts can be achieved is itself worthy of note and should not be discounted simply because those efforts receive outside support. We wish to learn from such experiences, to know how marginal groups as well as mainstream ones can be benefited through local organization, since improvements for the poorly endowed constitute one criterion of development in many countries' or donor agencies' programs.

The third possible objective is likely to be more controversial. The contribution of LOs to the empowerment of rural people may or may not be appreciated by nonmembers, whether local notables already influential or those in positions of authority who must deal with demands concerning public policy and allocations. LOs do not invariably add to the influence of rural people, but they can equip local publics with the voice and capacity to make credible demands on government and on others who control resources. If they are able to influence the content of public policy and affect the terms of access to needed services, LOs can counteract the "urban bias" noted above. Actually, since this bias not only is a major impediment to developing agriculture and rural productive capacities but thereby also

[11]Examples of local organizations that improved the economic and social position of the poor in India can be found with the Bhoomi Sena (de Silva, 1979) and Kottar Social Service Society (Field, 1980). One of the KSSS projects involved digging ancillary ditches off main irrigation channels to serve the fields of small farmers: the 40 kilometers of ditches serve 1,600 acres of land and benefit 9,700 farmers, 61 percent of whom have less than one-quarter of an acre. KSSS cooperatives among poor fishermen and potters have assisted particularly disadvantaged social groups. Netmaking centers have employed young girls in fishing villages and at the same time upgraded the quality of the nets. In Latin America, cooperatives have benefited poor women in Bolivia (Wasserstrom, 1982) and Colombia and Panama (Hartfiel, 1982).

[12]Catholic priests and laypersons, who were supportive but not sufficient to achieve desired results in the Dominican case referred to above, were highly successful in the KSSS case. The Inter-American Foundation assisted the women's cooperatives referred to in footnote 11, as well as hundreds of other groups in Latin America and the Caribbean. Bhoomi Sena grew out of consciousness-raising work by the Communist Party of India in the 1940s, but the one episode of an outside agency's coming in from an urban area to assist that movement proved to be a debacle, and its successes have been largely self-directed.

holds back the expansion of national resources and capabilities, such redress can bring benefits beyond the rural sector. We will discuss below why political leaders are not necessarily adverse even to empowerment as a concomitant of local organizational development.

These three outcomes—efficiency, equity, and empowerment—represent something like steps in local development. The first is widely accepted and sought; the second somewhat less so; and the third still less, though it may ultimately be the most important for a country's progress. Although LOs can pursue, and to some extent achieve, these objectives separately, the three are connected. Making gains on the first front, efficiency, expands the volume of available goods and services and increases the scope for benefiting the disadvantaged by redistributing resources at the margin if not at the source. Without advances in efficiency it is difficult to promote equity, since the political power to reallocate a "pie" that has not been made larger is uncommon. In a similar way, advances in equity may support empowerment, by bringing more persons into the structure of economic and social benefits; these produce the greater security and confidence needed to foster participation in political processes beyond the household and community.

We have not been able to demonstrate empirically the progression that seems logical among these goals. We find some LOs that contribute only to the first but not the others, and others that enhance equity but have not empowered people beyond their self-reliant circle. Some philosophers and practitioners of development would see empowerment as the ultimate objective and test of development, but that becomes partly a matter of definition. We are satisfied that all three are, under a variety of circumstances and in a variety of ways, desirable policy objectives. If local organizations contribute only to the first at present, they may nevertheless further the others over time. Persons or agencies may be concerned only with efficiency goals in development and may choose to work along those lines, though they should know that LOs which benefit only or mostly the well endowed can impede subsequent progress for the poorly endowed. These larger issues are important for policy debates but cannot be resolved by our analysis. We have some evident normative concerns on this subject and will elaborate our views in Chapter 2. The main task of this study is to provide a broader and sounder empirical basis for assessing and promoting local organizations for rural development.

The case for local organization draws on the knowledge that most rural publics that have experienced significant improvement in their productivity and welfare, including those in the United States, have had some organizations at the local level that perform these functions and conduct these activities. By contrast, rural societies characterized by persistent or increasing poverty due to inequity, exploitation, or neglect tend not to have such networks of organizations. While the establishment and strengthening of active networks of local organizations is a feasible policy option for most governments, as discussed below, its implementation is by no means politically or financially costless. Nor should one lose sight of the fact that establishing and maintaining LOs may entail economic and psychological costs for their members and leaders.[13] The process of building such institutions can involve conflict and uncertainty; it can be facilitated only by measures appropriate to distinctive local needs and circumstances.

One of the first efforts to assess systematically the relationships of local organizations to rural development was our own comparative study of 16 countries' experience.[14] The findings demonstrated "a strong, empirical basis for concluding that local organization is a necessary if not sufficient condition for accelerated rural development, especially development which emphasizes improvement in the productivity and welfare of the majority of rural people" (Uphoff and Esman, 1974:xi). The analysis, taking a "macroperspective," dealt with whole systems of local organization and with their developmental consequences, rather than with the varieties, activities, and functions of specific local organizations or with their internal dynamics and governance. The latter require a "microperspective." In this study, while exploring the macroperspective somewhat further, we go more deeply into micro issues.

The main conclusions from the earlier study should be reviewed here, since they will be further tested and refined with more extensive and current data. We found that the most effective systems of local organization were structured to operate at various levels:

> Local institutions should have *more than one level of organization,* at least a two-tier pattern, in which the lower tier performs functions at the neighborhood or small-group level, while the other undertakes more complex business and governance activities that require relatively larger-scale operations. A multi-tiered approach to local organization can combine the benefits of solidarity and of scale, both for mobilizing resources and for organizing and implementing development projects.

[13]Johnston and Clark (1982:171) criticize a "more is better" view of participation "which flies in the face of both experience and common sense. Anyone who has been active in a local government council, a committee of concerned citizens, or a community cooperative knows a truth virtually ignored in the current development debate: effective local organization is expensive to those who choose to participate in it." They criticize the unreserved endorsement by the World Bank (1975) of more participation by the rural poor in the planning and implementation processes, and approve the more qualified approach in Uphoff, Cohen, and Goldsmith (1979).

[14]That analysis (Uphoff and Esman, 1974) was based on detailed studies we commissioned on Bangladesh, China, Egypt, India, Indonesia, Israel, Japan, Malaysia, Pakistan, the Philippines, South Korea, Sri Lanka, Taiwan, Thailand, Turkey, and Yugoslavia, examining a wide range of rural development experience geographically and ideologically. The South, East, and Southeast Asian cases, along with the comparative analysis, have been updated and published in three volumes (Uphoff, 1982–83).

Links with political and administrative centers that control information and other resources are indispensable to effective local organization.

> Organization for rural development should be seen as a *system* of institutions performing various functions in the rural sector of a particular country. We found no case where only one institution was carrying the full responsibility for rural development or where *complementarities* among institutions were not important.
>
> Of key significance was the extent and effectiveness of *linkages* between and among institutions, horizontally with other organizations at the same level and vertically between local organizations and higher-level organizations which set policy and allocate resources essential to success in rural development.
>
> Local institutions which are separated and isolated from other levels are likely to be impotent developmentally. Local autonomy by itself provides little leverage for development. What make the most difference are systems or networks of organization that make local development more than an enclave phenomenon.

Articulating local needs and gaining access to resources will be most effective if there are multiple channels:

> Local communities should be linked to higher-level decision centers by *multiple channels,* both to achieve the benefits of specialization in communication and to enjoy alternative avenues of influence. Because any single channel may at any time be blocked or monopolized, may fail to function, or may yield unsatisfactory results, it is important to have multiple channels which local leaders can resort to singly or in concert to meet their needs. At the same time, central governments do well to rely on more than a single channel for reaching out to the village.

We found that successful local organization seems more viable where there is relative equity in the ownership and access to productive assets, especially land. This does not necessarily mean equality in ownership or access, but rather the absence of severe inequality in socioeconomic stratification. We stated that the implications of this finding required further analysis, which we have attempted here.

We found, not to our surprise, that local organization cannot be treated as a politically neutral or technocratic phenomenon. Any membership organization is a theater for internal competition among individuals and among groups of its members. It can also be a target of influence or control by outsiders seeking to use it to their political advantage. It can be a force that asserts claims on government for scarce resources and for policy support; thus it has the potential to ally with or oppose political parties or regimes.

We made a number of observations about the importance of decentralizing administrative decision-making so that government agencies could be more responsive to the needs, preferences, and capabilities of organized local publics. We argued that the deconcentration of administrative decision-making or devolution to local organizations need not be uniform across all sectors. Rather the extent of decentralization should vary according to the subject at hand and according to the capacity and orientation of lower-level officials and institutions. A government that means to help the rural majority may find that turning authority over to local staff or organizations without any requirements or supervision can reinforce local elites or self-serving leadership.

We have since come to believe more firmly that effective local organization requires far-reaching changes in the overcentralized structures and rigid operating procedures of the agencies of public administration that have evolved in most developing countries. These must and can adjust to meet the requirements of more active and organized rural publics. The strengthening of local organizations must therefore be accompanied by changes in administrative structures and procedures and in the behavior of officials and technicians working in agencies that serve rural publics (D. Korten and Uphoff, 1981).

The Experience of Industrialized Countries

In industrialized countries, peasant and labor organizations have played an important role in modernizing agriculture, raising the economic level of rural people, and improving the quality of rural life.[15] Even prior to the Industrial Revolution, protests and uprisings erupted when the peasantry sensed that their traditional rights or their minimum subsistence was being threatened by governments or by members of the dominant class (Landsberger, 1973b; Hilton, 1973). More permanent organization among rural people to advance their interests by fostering collective self-help and demanding favorable treatment from government, however, has been a product of the last hundred years.

In most of western Europe, the emancipation of the peasantry began with the French Revolution in the late eighteenth century. The nineteenth century witnessed the emergence of organizations among peasant owner-cultivators, many of which contained cooperative features. These were mostly service cooperatives, emphasizing credit at reasonable interest rates, the purchase of commercial inputs, marketing, insurance, and education of

[15]The most extensive treatment of this subject is the historical analysis and contemporary survey conducted for the International Labour Office (ILO) by Flores (1970). Unfortunately this and other sources seldom deal with the variety and complexity of rural organizations or with their internal governance. Histories of European peasantry—e.g., Weber (1976)—rarely throw much light on the question of organization.

members and their families. The organizations demanded favorable public policies and established links with government agencies at a time when governments were beginning to expand their services to agriculture. Thus, they were able to articulate the needs and preferences of their members to agents of the state. The gradual extension of the franchise to peasants during the nineteenth century increased the bargaining power of their organizations at the national level and their influence and power over local affairs.[16] In Japan, accompanying the government's support for agriculture following the Meiji Restoration in 1868, farmers' groups became an active factor in technological progress, starting with seed-exchange societies (Aqua, 1974). Indeed, Westney (1983) argues that the earlier experience with local organization in Japan contributed importantly to economic and social transformation in the post-Restoration period. In the United States, the emergence of rural organization having economic and political functions was particularly associated with the Populist movement, which arose at the end of the nineteenth century and wielded influence but little power.[17]

By 1937 there were nearly 25,000 rural cooperative associations in Europe, most of them affiliated with national federations, with an estimated membership of 25 million farm families. This mass membership and the impressive scope and scale of their activities should not gloss over the fact that many millions of farm families did not belong and did not participate. While direct information on this subject is not available, it appears that a large proportion of the nonmembers worked on marginal, isolated, or insecure holdings. In the United States, where many owner-cultivators have belonged to their own organizations since the late nineteenth century and exercised a major influence on matters affecting the welfare of farmers, even the most progressive of the farm organizations, the National Farmers Union, has not succeeded in enrolling many of the poorer and less secure farmers in the areas where it is active (Crampton, 1965).[18] In Scandinavia the cooperative movements were relatively broad-based, though still not

[16]The role of such organizations in raising agricultural productivity and transforming the political system during the nineteenth century is noted on a comparative basis contrasting the experience of Denmark and Prussia, in Gerschenkron (1966:39–41). The two systems of landholding influenced the functioning or absence of rural local organizations, a finding consistent with our 1974 conclusions. See also Hyden (1980).

[17]The political successes of "populism" were most dramatic in the case of the Non-Partisan League centered in the state of North Dakota (Hicks, 1931; Morland, 1955). The roots of such organization were more diffuse than the emphasis some historians have given to key personalities as seen in the fact that already in 1890, without any prominent leadership and before the Populist Party was formed, the North Dakota Farmers Alliance had 40,000 members when the state population was only 141,000 (Conrad and Conrad, 1975:5).

[18]The extent and importance of farmers organizations in the U.S., moving from economic functions into political lobbying, should not be overlooked. The first cooperative, for dairy marketing, was formed in 1810, but a cooperative movement did not begin to emerge until the 1870s, spurred by the Grange movement and concentrated mainly in the Midwest. Grain and livestock marketing was the main cooperative activity, with consumer co-op and credit union

incorporating all of the lower rural strata. One of the most inclusive rural organizations among more developed countries was in Japan where, after land reform was imposed by the victorious Allies, the national federation of agricultural cooperatives became and remains an important economic and political force (Aqua, 1974).

In Europe, the United States, and Japan, marginal farmers have been gradually disappearing since World War II and have been absorbed for the most part into industry and the service sector. In most developing countries, however, the rural labor force continues to increase. Thus the experience of industrialized countries, demonstrating the utility of farmers' organizations particularly to middle- and large-scale farmers, cannot be a sufficient guide to the organizational future of small and marginal farmers in less developed countries. Methods will have to be found to extend the opportunities and benefits of organization to this sizable and, in many cases, growing constituency.

In most industrialized countries, the organization of landless agricultural workers was not attempted until the early twentieth century when labor unions finally gained legitimate status. The organization of farm workers lagged behind that of industrial workers because of the survival of patron-client relationships in rural areas, the bitter opposition of farm owners, and the absence of government support. The scattered and isolated situation of many rural workers, the seasonality of their employment, the fact that in some countries (including the United States) many were recent immigrants or members of racial minorities, and especially the excess of labor supply over demand all reduced the collective bargaining power of agricultural workers and their propensity to organize. Membership tended to be small, from 3 to no more than 10 percent of rural workers. Only after World War II, when the rapid expansion of industrial employment and the consequent decline in the supply of rural labor increased their bargaining power, was there notable growth in union membership among agricultural workers in European countries. Though conditions have greatly improved, governments have been slow to extend social security, minimum wages, and other benefits—equivalent to those enjoyed by industrial and office employees—to rural workers. Agricultural labor unions remain weak in the United States and in Japan (Flores, 1970:484).

In the socialist countries of eastern Europe, where agricultural land has been nationalized except in Poland and Yugoslavia, cultivators are enrolled

---

(savings and loan) activities increasing over time, especially once the federal government began to provide legal and financial assistance. Since 1935, rural electrification cooperatives have provided electricity to 99 percent of farm families. As of 1977, more than 50 million Americans and one out of three families, were members of some kind of cooperative (McGrath, 1978). See also Fite (1965) and Knapp (1969). On the application of cooperative approaches to contemporary rural poverty in the U.S., see Marshall and Godwin (1971).

either in government-sponsored production cooperatives or, if they work on state farms, in labor unions. While these forms of organization tend to be compulsory and to be dominated by the state and the official Communist Party, they have led to improved material conditions, especially for the lower strata of agricultural workers. The socialist pattern of collective agricultural production and its concomitant organizational structures present developing countries with an alternative to individual proprietorship, especially in areas of high population density or where communal ownership and group cultivation are culturally sanctioned.[19] In Poland and Yugoslavia, peasant proprietorship and collectives coexist, but in both systems the cultivators are highly organized and their organizations closely linked with the agencies of the state (Jackson, 1973; Galaj, 1973; R. Miller, 1977).

The lessons of these experiences for the contemporary Third World are that organizations can yield impressive benefits to those who earn their livelihood from farming, while those who are unable or fail to organize, or are prevented from organizing, have limited ability to assert and protect their group interests. They are dependent on the good will of patrons at a time when agriculture is becoming increasingly commercialized, or on market mechanisms that offer little advantage to those having limited economic resources. Usually they are effectively excluded from the benefits and protection of government services to agriculture. While the more secure owner-cultivators seem to find the means to organize, often with the assistance of governments, the economically and socially weaker and marginal rural groups in all the industrialized countries had difficulty forming and maintaining organizations for mutual assistance and collective advocacy.

In Europe and the United States, organization of rural workers was fostered only with changes in the supply and demand for rural labor or with strong governmental intervention, including radical land reform. Since in most developing countries agricultural laborers will constitute the majority of the rural households and labor force in years to come, methods of bringing them the indispensable benefits of collective organization will represent a major challenge to government officials, political leaders, and development assistance agencies interested in promoting broad-based patterns of rural development. Unfortunately, because there are relatively few published case studies of agricultural laborers' organizations in developing countries, our quantitative analysis does not cover them adequately. We were able to consider in some detail a broad range of existing LOs, many of them specifically of, by, and/or for the poorer sections of rural society. Few of the LOs studied are rural labor unions or peasant leagues, however. Most

[19] We note this without advocating it, since the record of collective agricultural production efforts in the Third World is on the whole quite unfavorable. Various forms of cooperative, group, or joint farming, on the other hand, appear to have some productive possibilities (Wong, 1979).

often they are organizations in which a majority of members have at least some assets besides their labor.[20]

Current Status

The reader may well ask, if local organizations are so important to rural development, why are they not more vigorous and conspicuous phenomena in most Third World countries? While there have been notable successes, there have been many more failures, and in many locations efforts to promote such organizations have not even begun. Why have significant successes been relatively uncommon? Why are rural local organizations so much weaker in most developing countries than they are in contemporary Europe, the United States, or Japan?

A first consideration is that forming and maintaining local organizations is no easy task. There is no reason to believe that LOs will necessarily come forth as the logic, need, or demand for them arises. Such a view has been criticized by Goodell (1982), who takes issue with the "induced institutional development" theory of Hayami and Ruttan (1971). She says they "substantiate their analysis by citing the institutional resilience and responsiveness of Japan and the United States which helped launch the development of those nations in the past 150 years—as though after centuries of harsh colonial rule, Bolivia, Zimbabwe or Bangladesh find themselves at a starting point for development comparable to two societies which virtually had always governed themselves" (1982:121). Her point is, however, diminished by the existence of significant, if not yet widespread, LOs in these very countries (e.g., Healy, 1980; Smith and Dock, 1981; Alam, 1979).

In fact, we find the organizational impulse quite widespread, with some roots in almost all cultures. Our review of LO experience worldwide suggests that the problem is more a matter of sustaining LOs and making them effective than of calling them into existence. Given our previous finding (supported by most other writers on the subject) that LO performance over time depends significantly upon supportive interaction with government agencies, we would begin looking for answers to the questions posed above by considering political factors.

One important reason why active LOs are not found more widely is the hostility of central governments. Some regimes regard any organized activity in rural areas as potentially subversive. Certain political elites regard themselves as modernizers and see local influentials as reactionaries who compete with them for power and who are bent on preserving their priv-

[20]In their study similar to this one, Gow et al. found that LOs for the poorest of the rural poor—those with little or no resource base, the land-poor, and the landless—"were few and far between" (1979, I:5).

ileges at the expense of national goals and larger publics. Some central governments accept the necessity and even the desirability of local organizations but only if such organizations can be effectively controlled by the state or the official party. They are not willing to tolerate associations that can operate independently of central tutelage, make claims for resources that may strain or embarrass the government, or become instruments of opposing political groups. To prevent such unwelcome contingencies, some governments effectively proscribe local organizations or suffocate them with surveillance and attention so paternalistic as to undermine all local initiative.

Even when elements of the political elite are willing to tolerate constituency organizations, the state's administrative organs may find ways to thwart or block them. The centralized, authoritarian, and formalistic nature of many Third World bureaucracies, particularly their extensions in rural areas, calls for a passive populace. Organized and vocal publics are likely to annoy and complicate the lives of bureaucrats, add unwelcome "political" dimensions to their work, threaten their technical and operational control of public services, and require changes in programming procedures and in methods of delivering services. To avert such unwelcome prospects—which they may also believe would attenuate the quality, impartiality, and predictability of services—and to maintain some social distance from poorly educated and low-status rural publics, officials and technicians may discourage, circumvent, or undercut local organizations in rural areas.

Local elites represent yet another problem, given their traditional status, their greater education, the contacts they enjoy with officialdom, and their control over land and other resources on which others depend for their livelihood. They tend to perceive organizations of the poor and the marginal as threats to their domination of rural society. In some countries they are even willing at times to employ organized violence to intimidate or destroy such efforts. Thus, they may be able to forestall the formation of local organizations, to dominate and control those that exist, or to preempt for themselves resources that were intended by governments to reach poor and disadvantaged people through cooperatives or other kinds of local organization. Though the commercialization of agriculture in various parts of the world is undermining established patron-client arrangements, these relationships survive in many areas. Where they survive, they inhibit the development of collective organizations among the rural poor because the latter often prefer particularistic security and material benefits, even when these are manifestly inequitable. The first impulse of rural people when familiar patron-client links are broken is usually to find a new patron rather than resort to the much more dangerous strategy of class organization.

In some societies, elite control may nevertheless represent a net, even if not equal, benefit to all classes in the community. Where communal solidarity persists along with a sharing ethic, the elites may provide continuing legitimacy for enforcing contracts and managing conflicts within traditional organizations, thus guaranteeing subsistence and other minimal rights to all members of the community.[21] They may be successful in attracting government funds that benefit all sections of the community, enhancing rather than weakening their dominant economic and social position. (The role of local elites in community and LO leadership is taken up in Chapter 8.)

Some of the weaknesses and failures in rural local organization can be attributed neither to the hostility of governments nor to the machinations of local elites, but to disabilities among the poor that prevent them from taking advantage of the opportunities open to them. These disabilities may arise from social cleavage and lack of trust—based on ethnic, kinship, political, or other differences—that undermine their ability to cooperate. Their inexperience in working within the framework of formal organizations and their inability to impose discipline on some members or to control unscrupulous leaders who plunder the assets of the organization often contribute to the breakdown of local organizations or to their domination by unrepresentative and corrupt minorities.[22] In many cases local people are not convinced that organizations would yield them benefits sufficient to justify the effort and the risks. Frequently, they do not command the resources, in the absence of outside assistance, to bring an organization into existence. Or they simply may not trust outsiders, including government officials, who attempt to enroll them in organizations.

Governments often contribute to the breakdown of local organization, not out of willfulness but because of their failure to deliver what they promise. Often this is the result of poor program design or administrative incompetence. Once expectations have been built up, the failure to deliver as promised weakens, sometimes fatally, the incentives of local people to participate in or even to continue affiliation with local organizations, because these have not provided benefits that justify the cost of participation. This is one reason that reforms and improvements in public administration must go hand in hand with successful local organization in rural areas.

Owner-cultivators seem to be best equipped to form and sustain associations and make an impact on local authorities, especially where there is some degree of equity in the distribution of assets. In such cases governments are likely to be sympathetic, and the power of local elites is relatively modest. Organization is more problematical among those who need organization

[21]This is argued by Scott (1976) but contested by Popkin (1979).

[22]With reference to experience in Latin America, the Middle East, and Southeast Asia, Goodell observes: "It simply cannot be assumed that all people, particularly the poor, know how to form committees to implement or lobby for their interests, how to pool resources to embark on an enterprise together or to support one another in doing so, how to manage common funds by holding leaders accountable, or how to sustain a local organization for common interests or ends—nor can it be assumed that they will be able to start doing so when the need arises" (1982:121–22).

most—the landless, tenants at will, marginal smallholders, craftsmen, day laborers, and especially the women in these groups. All of the inhibitions and disincentives mentioned above bear heavily on the disadvantaged, who are increasingly numerous and who have the most to gain from effective organization.

The Challenge

Our observations of trends in the "centers" and "peripheries" of developing countries persuade us that this survey and analysis may be especially timely because governments have begun to recognize that it is in their interest to tolerate and even assist rural local organizations.

It is true that some regimes do not tolerate rural local organizations of any kind, regarding all such associations as actually or potentially subversive. They consider atomized rural publics less threatening. Other regimes, including most Marxist governments, actually foster local organizations and make membership compulsory in order better to control and perhaps to serve their rural constituents. Still others neither proscribe nor require organizations among rural groups but patronize them, whether or not they were founded by government, in order to make them dependent and to curry favor with selected rural publics through subsidies and handouts. In such cases there are opportunities for local organizations to function autonomously, but there are heavy pressures to succumb to patronage and to be directed by the state rather than by their members. The majority of regimes, however, do not appear to be able or to desire to exercise such control over rural local organizations but seem prepared to tolerate or even to engage in active exchanges with them.

There are several reasons why supporting, or at least accepting, rural local organizations is in the self-interest of most governments.

(1) *Pragmatic politics.* Some government leaders know that attempting to repress or control all local organizations, some of which have deep and longstanding roots, would deplete their coercive resources and their legitimacy. Except where LOs threaten the regime, therefore, the prudent policy is to live and let live. If local people organize, government can deal with them to mutual advantage so long as they work within the rules of the system. Since they represent rural constituencies, LOs can even be helpful, providing services and support that the regime may find valuable in exchange for the right to operate and draw some benefits from the system.

(2) *Control of bureaucracy.* Some political leaders have more positive reasons for working with local organizations. Tension often results from the less than perfect trust between politicians and the specialized bureaus on which the former must rely for the implementation of policies and programs; politi-

cians often worry that bureaucrats are out of control and are not properly implementing public policy. These leaders may find local membership organizations a means for keeping bureaucracies responsive both to the priorities of government and to the needs of rural publics. Moreover, LOs can help to reduce the regime's problems of bureaucratic corruption or nonperformance of duties.

(3) *Information exchanges.* Resources and services provided by government are likely to be wasted unless they are appropriate to the intended recipients. This requires reliable sources of accurate information about the needs, priorities, and capabilities of specific rural publics. Such information is much more likely to flow, and to be a representative, from organized publics than from individuals. Thus a government agency may wish to have its rural clients organized, not necessarily to control them—though this may be either a motive or a consequence—but to enable the agency to perform its activities more effectively.

(4) *Resource mobilization.* Governments in most developing countries are short of resources, especially for rural areas, since their priorities often lie elsewhere. Organized publics can, if the incentives exist, mobilize considerable amounts of materials, money, labor, and even managerial skill, thus stretching the government's own limited resources and producing both economic benefit and political satisfaction for rural publics at relatively low cost. Actually, as we shall show, there are also valid reasons for governments to require counterpart contributions from local publics, so as to have more vigorous and responsible local organizations. Governments can multiply the resources they allocate to rural development by working through local organizations that evoke sufficient membership support to elicit matching resources and contributions.

(5) *Political support.* Most regimes are interested not only in maintaining but also in expanding their political support base. While this applies certainly to those with competitive political parties and periodic elections, authoritarian regimes too must take into account their standing with rural publics. To be recognized as helpful to local membership organizations, to move resources through these organizations, and to accord them legitimacy and respect may be a relatively low-cost way for politicians to protect and expand their support base among the majority of their rural constituents.

(6) *Social order.* Politicians and senior bureaucrats in power are concerned with avoiding trouble. Rural discontent frequently flares into violence, which can be economically costly and politically destabilizing. The price of a reasonably contented countryside may be the demonstration of interest in the welfare of disadvantaged publics, including the ability to communicate with them, to cater to their most pressing needs, to stimulate self-help among them, and to exchange services for support. Regimes that desire to maintain social order in the countryside may find it more profitable to work with and through organized rural publics than to rely primarily on the selective distribution of patronage or on coercion.

(7) *Donor agencies.* Many governments in less developed countries are interested in presenting a favorable image among development assistance agencies, on

whom they depend for important financial contributions and technical assistance. There has been a shift in the priorities of such agencies, including USAID and the World Bank, from urban infrastructure to agricultural and rural development, especially to activities that are likely to benefit the rural poor. One ingredient of a broad-based rural development strategy is local organization. Governments therefore may have an incentive to support local organizations in order to strengthen their claims for foreign assistance. Moreover, many of the younger development planners and political leaders have been exposed to currents of opinion that emphasize equity-based rural development and believe that such measures are not only morally right but politically and economically beneficial for their countries. They have identified their political and bureaucratic careers with this doctrine.

Our premise in undertaking this survey and analysis of local organizations is that a vigorous network of membership organizations is essential to any serious effort to overcome mass poverty under the conditions that are likely to prevail in most developing countries for the predictable future. Our previous research drew conclusions about local organization which we shall test and elaborate further in this study. While other components—infrastructure investments, supportive public policies, appropriate technologies, and bureaucratic and market institutions—are necessary, we cannot visualize any strategy of rural development combining growth in productivity with broad distribution of benefits in which participatory local organizations are not prominent. We concur fully with the conclusion of Johnston and Clark in this regard:

> One of the great challenges for policy analysis is the design of organizational structures which can mobilize local experience and integrate it with improved expertise. In seeking to promote this design function, we therefore emphasize the importance of local organizations for articulating needs and delivering services. We recognize, however, that organizational resources are at least as scarce and valuable as capital, land, and technical knowledge. [1982:34]

In this inquiry into experience with local organization, after reviewing what the existing literature offers, we shall be asking three sets of questions:

(1) What kinds of combinations of local organizations are likely to be most successful for what kinds of activities under what sets of environmental conditions? This is the focus of Chapters 3, 4, and 5.

(2) What practices of internal governance and processes of management are associated with successful local organization? In particular, how can the common problems affecting LOs—like elite domination, low participation, and financial irregularities—be offset? These issues we address in Chapters 6 and 7.

(3) What measures can be taken by governments, by nongovernmental bodies, and by international development agencies to promote, protect, and sustain local organization in the context of broad-based rural development? We look at responses to this question in Chapters 8 and 9.

Effective and enduring social structures are not built overnight. Therefore, there can be no quick fixes for local organization. But the time constraint can be reduced by drawing existing networks of trust and cooperation among rural publics into programs intended to enhance their productivity and well-being. Where such networks do not exist, efforts to introduce and assist new organizations are an important part of any government or private effort to promote broad-based rural development.

CHAPTER 2

Local Organizations in Development Theory and Literature

The considerations presented in Chapter 1, which attribute a key role to local organizations in rural development, are not entirely new. Yet they have not figured centrally in most of the literature, strategies, or policy formulations for rural development. Why this is so should be examined. We undertake here to trace the causes of the previous neglect of local organizations in most approaches to rural development and to account for the rise in the attention paid them over the past decade. Our concern with these issues of intellectual history is heightened by David Korten's observations in his review of the literature on agricultural and rural development:

> We might expect that the difficult problems of how to involve the rural poor in their own development, through local organizations or otherwise, would be receiving major attention in development journals and current policy documents. Yet this is not the case. . . .
>
> The recent 489-page Asian Development Bank review of Asian development gives the topic four pages. The 440-page presentation by Wortman and Cummings of a strategy for meeting the world food crisis based on small family farm production devotes two brief paragraphs to farmer associations and cooperatives with barely a mention of the impediments posed by village social structures to the implementation of their recommendations.
>
> The World Bank's 1975 *Rural Development Sector Policy Paper* gives five paragraphs to the importance of local participation and briefly acknowledges some impediments, but says little about how they might be overcome. [1980:480–81]

Our own review of the literature on agricultural development confirms Korten's observation that local organization has until recently been neglected in the mainstream literature; even when it is mentioned, impediments to its functioning and means for strengthening it are seldom explored. Eicher and Witt's *Agriculture in Economic Development* (1964) contains only three brief references to cooperatives (two referring to land reform and one to marketing) and nothing on local government or farmers' associations. Mellor (1966) devotes only a few paragraphs to rural local organizations.

The summary report of a Massachusetts Institute of Technology conference on agricultural development, by Millikan and Hapgood (1967), contains favorable comments on the contribution of local involvement to Japan's successful development and to the Comilla cooperatives in Bangladesh. But it assigns responsibility for the organization of supply, processing, and marketing to the state or the private sector, "leaving the grafting on of full participation by the people themselves to some later time when successful routines have been established" (p. 157). Otherwise, it notes that land reform often requires cooperatives (p. 93) and that cooperatives are often "imposed on the farmer without consulting him" (p. 87). The subject is not further discussed.[1]

A volume entitled *Agricultural Development and Economic Growth* (edited by Southworth and Johnston, 1967), which appeared about the same time as the MIT report, contains two short but serious discussions of local organization: in one, Raup stresses the importance of local peasant political organization for land reform but comments that "effective local participation in land reform is easy to prescribe but difficult to achieve" (p. 305); in the other, on marketing cooperatives, Abbott and Miracle note "disappointments encountered with government programs to implant cooperative systems," which they describe as "structures designed more for administrative and political convenience than for concrete local advantage" (p. 381). Otherwise, farmers' associations and cooperatives are mentioned only among possible institutional improvements (p. 567). No prescriptive or procedural suggestions are made. Brewster's discussion of peasant society views traditional social structures as presenting barriers to change. While the desirability of agricultural populations' organizing themselves into larger units for collective action is noted, the analysis evinces a pessimism similar to the "fatalism" that is attributed to peasants themselves.

The widely read book by Lester Brown, *Seeds of Change: The Green Revolution and Development in the 1970s* (1970), simply ignores the role of local organizations, even when it discusses land reform and small farms in Japan and Taiwan. Instead, it concentrates on how to improve the credit, marketing, and extension package and advocates consolidated farming with custom-hire tractor services (pp. 112–17). The lessons learned in Japanese

[1]The MIT conference, according to one of its participants, "had much more information about local action and organization than the Millikan-Hapgood popularization. The full report contained many chapters and appendices dealing with the problem, including an excellent statement by Akhter Hameed Khan; indeed the fact that Millikan was essentially uninterested in this kind of data only reinforces your argument . . . that the fads did not permit the mainstream writers to devote adequate attention to local phenomena" (John D. Montgomery, personal communication).

and Taiwanese experience with farmers' associations and cooperatives are omitted.

At the first Reading (U.K.) conference on "Change in Agriculture," there was little consideration of local organization. In an 80-page summary dealing with over 700 pages of papers contributed by leading authorities on agricultural development, Bunting (1970) devotes just two pages to cooperatives (pp. 769–70) and mentions but undertakes no discussion of "local capacity" as crucial for "change in agriculture."[2]

The review of agricultural development strategy by Weitz (1971a) has a brief, favorable discussion of farmer participation (pp. 70–71), but the kind of organization that is discussed specifically (p. 186) is not really local.

The most interesting theoretical treatment of agricultural development strategy in comparative international perspective is by Hayami and Ruttan (1971). But they mention cooperatives only tangentially, in relation to technological advances. For example, with regard to the centrifugal cream separator that facilitated the transformation of Danish agriculture, they write: "A remarkable aspect was that the new potential which emerged from this technical invention was exploited by an institutional innovation—the cooperative creamery. With this combination of technical and institutional innovations, 'the profitableness of milk production was raised on middle-sized farms and even on small-holdings, to the level of the big farms'" (p. 299). Similarly, their discussion of the development of the silk industry in Japan puts major emphasis on technical improvements and centrally sponsored institutions like silk inspection stations and research institutes, adding the brief comment that "in addition to these government institutions, the development of sericulture cooperatives was a crucial element. Their activities ranged from the transmission of technical information and the cooperative rearing of young worms, to the management of cooperative silk reeling mills and even the training centers" (p. 302). The authors' only other mention of local organizations concerns cooperative credit and marketing plans (pp. 294–95). Their theme of "induced institutional development" includes just one reference to what we call local organizations, even though in their two case studies they call the role of LOs "remarkable" and "crucial."[3]

[2]The second Reading conference made "organization" a major theme for analysis. See Hunter, Bunting, and Bottrall (1976).

[3]Goodell's critique (1982) of the "induced institutional development" theory as applied to local institutions was noted in Chapter 1. In their current revision of the book, Hayami and Ruttan take a more positive view of local organizations: "A clear inference from the literature on rural development is that efficient delivery of bureaucratic services to rural communities is very dependent on effective organization at the community level. Rural communities, operating through either the formal structure of local government or informal or voluntary institutions, must be able to interact effectively with the central institutions charged with responsibility for the delivery of services to local communities. They must be able to interact effectively in the establishment of priorities. They must be able to provide feedback to the agency management on program performance. And they must be able to mobilize sufficient political resources to provide incentives for effective bureaucratic performance."

The first extended discussions of local organizations we found in mainstream literature are by Wharton (1969) and Mosher (1969). In introducing the book he edited, *Subsistence Agriculture and Economic Development,* Wharton identifies as an "accelerator of agricultural development"

> voluntary farmer associations of various types: cooperative societies, 4-H clubs, farmer clubs and community construction projects. These have two advantages. One is that they can get tasks accomplished which individual farmers, operating alone, cannot achieve. And by involving group discussion of new ideas and ad hoc organizations around a specific interest, they can affect the local climate of public opinion with which the individual farmer must live and work. [1969:10]

The volume also contains positive discussions of the Bangladesh experience with cooperatives at Comilla. Otherwise, "local" matters are treated in the volume most explicitly as forms of "localiteness"—by Everett Rogers, who defines the term as the opposite of "cosmopoliteness" and thus as a negative feature of rural societies (Wharton, 1969:124–26).

In his analysis of the elements of "a progressive rural structure," Mosher deals mostly with administrative requirements, drawing on "central place theory." But he explicitly charts a major role for farmers' organizations in coordinating the various activities, citing as examples the Comilla cooperatives and the Farmers' Associations in Taiwan (1969:37–40). "Whatever the precise administrative arrangement," he writes, "it is important that the future of each headquarter be conceived as being to *serve* locality activities rather than administer all of them rigidly. It is *locality activities, and they alone, that comprise the 'cutting edge'* of any agri-support activities" (p. 35; emphasis Mosher's). This represents an important step toward a more positive role for local activity and local capability. It departs from the view articulated by Millikan and Hapgood that priority be given to agri-support activities from the center, with local participation "grafted on" later at the convenience of administrators.[4]

We should note here the importance in Mosher's thinking (and in that of Owens and Shaw, 1972) of central place theory as an aid to establishing a useful role for local institutions. This theory is associated particularly with the work of E. A. G. Johnson (1970), which also influenced a school of anthropologists who underscore the importance of "locality." Rather than treating villages in isolation, these anthropologists place them within sub-

[4]In an earlier work on the subject (1966), Mosher devoted eight pages to "group action by farmers," citing construction of community facilities, pest control, formal cooperative organizations, and local self-government as appropriate supportive activities for "getting agriculture moving." On the other hand, in a later book he does not include any explicit attention to local organizations in his strategy "to create a modern agriculture," even when discussing needed institutional changes (1971:120–21).

regional contexts of economic, social, and political relations, as discussed below.

Influential general treatments of peasant society have dealt mostly with traditional social structures and with political movements (e.g., Redfield, 1967; Wolf, 1966, 1969). Most of them have discussed relations between state and community in antagonistic zero-sum terms, at least until the more recent work on regional analysis (see the critique by Greenwood, 1973). There is not much concern with organizations oriented to self-help or to practical intermediation with governments. Studies of peasant ideologies such as those of Scott (1976) and Mintz (1979) emphasize (with considerable empirical justification) defensive strategies; they do not advance an understanding of active strategies that can promote peasant welfare, nor are they concerned with local organizations as vehicles. The literature stating the case for "peasant rationality," beginning with Schultz (1964), treats rural people as essentially individual decisionmakers and thus does not contribute much to theories of collective action. An exception is Popkin's recent book (1979), but this supports our general argument that the main themes in the literature on agricultural development and rural society have not focused on the role and dynamics of local organization.[5]

The main works that treat the value and problems of local organization address *rural* rather than agricultural development (Hunter, 1969; Barraclough, 1971; Owens and Shaw, 1972; Chambers, 1974; Lele, 1975; Jedlicka, 1977). That concern with local organization has not yet registered very clearly in the literature on *agricultural* development can be seen in the extensive survey by Eicher and Baker (1982). Though excellent in other respects, it discusses cooperatives only in passing (pp. 204–6) and omits the subject of local organization entirely in its summary discussion of research and development strategies in the 1980s (pp. 225–60).

It is worth noting that the literature contains little explicit opposition in principle to the establishment and operation of LOs, nothing comparable to the sophisticated intellectual case elaborated a century ago by neo-classical economists and lawyers against labor unions as combinations in restraint of trade or as violators of freedom of contract. Behind the prevailing neglect and lack of interest has been a concern that local organizations may contribute to economic inefficiency and market distortions, may be dominated by government or controlled by local elites, may complicate administration, or may be unable to perform because of the illiteracy and backwardness of rural people.[6] The vulnerability of service cooperatives to abuse, preemp-

[5]There have been, to be sure, treatments of local organization in the literature on specific communities. Good examples among those included in our review of case studies are Hamer (1976, 1980), Birkelbach (1973), and Edel (1969), dealing with local organizations in Ethiopia, Indonesia, and Colombia. But we are addressing here the general literature that may have influenced policymakers.

[6]Mellor, for example, saw cooperatives playing a positive role specifically in marketing "only if they render an efficient service." Because cooperatives are "largely controlled by the govern-

tion by the wealthy, and poor performance have been abundantly documented (e.g., UNRISD, 1975; Münkner, 1976; Lele, 1981). In Chapter 6 we examine the "pathologies" of local organization that have drawn justifiable attention, and in Chapter 7 we consider the means that may be used to counter them with varying degrees of success.

The romanticized view found in some of the community development literature, that village life is harmonious and equitable, has contributed to the disenchantment with local organization as an approach to rural development. As Blair (1982) shows, this conception was found in much of the writing and planning to "uplift" poor communities in the United States as well as southern Asia. When it appeared that these efforts were not bearing the anticipated fruit, or when they became too "successful" and stimulated demands and pressures from below—and were consequently throttled—the conclusion drawn was that the LO approach itself was faulty. As Blair further shows, however, despite the evident failings in many community development programs, there have been some longer-term successes that speak well for local mobilization and involvement. The gains are not as great as hoped for, and often not in predicted directions, but on balance the earlier verdicts that the attempts were failures seem exaggerated and even mistaken.

Development Theories of the 1950s and 1960s

A wide range of social science theorizing, which launched the developmental efforts of the 1950s and 1960s, emphasized the role of the central government and of nonlocal agents. Local communities were viewed by various writers as technologically backward; as traditional; as conservative or bourgeois; as controlled by parochial, reactionary elites; as disposed to consume rather than save and invest; as undisciplined; or as "peripheral," needing to be "penetrated" to become part of the modern nation-state.[7]

The approach to development symbolized by the Point Four program, stressing transfer of technology, saw local communities as constrained by their low level of technological development. Their production techniques were regarded as hopelessly behind the times, needing to be replaced by "advanced" methods, which peasants would be taught to "adopt." This technocentric view of development had little concern for local organization

ment . . . [their membership] tends to be quiescent, contributing little managerial talent, local know-how or even to guard against corruption. As a result, corruption and inefficiency are common, further reducing local interest in the cooperative" (1966:341–43).

[7]One significant exception was W. Arthur Lewis (1955). Despite his contributions, which led to a Nobel Prize, his concern with institutions and with farmer behavior was not pursued by the next generation of development economists.

except as a kind of transmission belt for technologies brought in from outside. Even then, unless and until they were educated, local people—it was thought—would be incompetent to do more than accept the new techniques and materials (Poats, 1972).

The most prominent sociological approach to development from the 1950s (e.g., Hoselitz, 1957) emphasized the difference between "traditional" and "modern" cultures, seeing the two in conflict so that the latter had to displace the former for progress to occur (Riggs, 1964; Apter, 1965). Indigenous local organizations were bound to be traditional and thus obstructive to the kind of change considered necessary by the agents of "modernization."[8] Thus, modernization theorists were no more sympathetic to local organizations than were those promoting Western technology as the key to development.

Marxian theorists rejected both of these approaches in favor of changing the class structure of society. Still, they were no more disposed to champion bottom-up local organization. Following Marx's view of the peasantry as basically conservative ("sacks of potatoes"—politically individualistic and inert as a class), mainstream Marxists looked to the urban sector and to the proletariat and intelligentsia for leadership in transforming the class structure by seizing the organs of the state. Those who followed Mao Zedong's analysis were more inclined to favor a progressive role for the peasantry, but it was to be guided and controlled by a vanguard revolutionary party.

The neoclassical economists who saw lack of capital formation as the main cause of underdevelopment stressed measures for increasing aggregate saving and investment. While they were highly market oriented and regarded the individual as the main unit of action, some of them advocated a strong role for central planning agencies, despite their support for a free market (e.g., Scitovsky, 1954, and Kindleberger, 1958; see the critique by Killick, 1978: ch. 2). Since reducing consumption was considered the key to more rapid GNP growth, it was up to the government to force such behavior on the public by taxes and other measures. Permitting local communities to have a greater voice in decisionmaking might result in increasing consumption, it was thought. If local organizations could marshall resources through self-help, that would be a positive contribution, but it was feared that they would more likely make claims through the political process, thus limiting the resources that could be squeezed out of agriculture for state-directed investment.

Equivalent theories in the political realm supported this position, including the view of Huntington (1968) that popular mobilization should be

[8]"Nothing is more desperate for progress-minded political leaders than to find that the public becomes not an asset, a pool of talent, and a reservoir of strength, but a weight to be shifted from one shoulder to the next, finally crushing those who are attempting to march forward with the burden" (Apter, 1968:82).

restrained; of Myrdal (1968) that the "soft state" needed to be hardened so as to enforce "discipline" on the unruly or impassive masses; and of Binder and associates (1971) that the state should "penetrate" the periphery to tie it into national development objectives. None of these theories welcomed a self-directed capacity for organization among rural constituencies. These writers represented much of the grand theorizing about development in the first two decades of purposive efforts by governments—in less developed countries (LDCs) and in donor roles—to accelerate economic growth and social change. It is thus no accident that little support for local organization was found in mainstream economic or political development theory during this period.

There were, to be sure, some less overarching schools of thought that did advocate local organization. The community development approach, regarding communities as natural units of organization, certainly supported strengthened capacity at the periphery. The more extreme views associated with Gandhianism in India or modern-day forms of *narodnikism* (populism) were also sympathetic but did not address seriously such obstacles as entrenched local elites, caste or class inequality, and the limits of self-help. There were practical arguments made for organizations like cooperatives, credit unions, and marketing societies, but these were not considered very interesting theoretically. In an era guided by grand strategic formulations—particularly those reflecting technological or economic determinism—little prestige or influence was accorded local institutions.

Emerging Theories of the 1970s

As we have previously suggested, the development thinking of the 1950s was largely influenced by technological concerns. A "technology gap" was identified between the advanced and backward nations, to be filled by the *transfer of technology* to the latter. This thinking underlay foreign assistance programs from Point Four onward until economists became ascendant at higher policy levels in the 1960s. Then various "resource gaps" were specified and measured: the budget gap between government expenditure and revenue, the foreign exchange gap between imports and exports, the capital formation gap between desired levels of investment and actual levels of national saving. These were to be filled by *transfer of resources* from richer to poorer nations—in sufficient amounts, it was hoped, for "take off" into self-sustained economic growth (Rostow, 1960; Chenery and Strout, 1966). Many extenuating circumstances could be pointed out to explain why neither of these theoretical formulations produced results: the technology was not appropriate; the social "preconditions" were lacking; there was not enough "political will."

In any case, as the 1970s began, a new development agenda was formulated, giving more thought to appropriate technology, labor-using strategies, self-reliance, and equitable growth and income distribution, as well as participation. As we have argued previously (Uphoff and Esman, 1974), it became clearer that there was an "organization gap" between central government agencies and the rural communities they were supposed to assist. One of the first statements of this view was by Owens and Shaw (1972), and it rapidly gained support. While D. Korten (1980) found the World Bank's sector policy statement on rural development (1975) inadequate in its treatment of local organization, decentralization, and participation, the mere fact that these were considered important represented a significant departure for the World Bank, which had previously been preoccupied with technological and, especially, economic resource transfers.[9] To be sure, organization is no more valid as a single-factor explanation than the ones preceding it. Technology, resources, and organization are like the economic factors of land, labor, and capital—complementary elements of larger processes. Any of them can constitute a bottleneck, but unless the other two are appropriately increased in amount and quality, increasing one by itself will produce diminishing returns.

This emerging view of the role of organization has not often been challenged, but difficulties in starting or sustaining effective local institutions have kept many agencies from making organization a central part of their development strategy. The direct *transfer of institutions* is even more dubious than that of technology or resources. The fact that establishing local organizations is a more organic than mechanical process—that it is not predictable, takes time, and does not obviously "move money" in large amounts—has kept government agencies and international donors from developing much enthusiasm for this approach. Technology and resource transfers, for all their demonstrated limitations, have remained more programmable and thus more popular with planners and budgeters.

Paralleling technocratic resistance to local organization has been the opposition from those who see any official development efforts as fated or intended to fail (e.g., Hayter, 1971; Paddock and Paddock, 1973). Vehement critiques of the whole "development" enterprise have come from left circles, which regard it as a palliative at best and a deception at worst, masking forces of concentration and exploitation that doom the Third World to underdevelopment unless and until radical revolutionary transformations are achieved (Frank, 1968; Cardoso, 1972; Rodney, 1972). Local organizations that do not mobilize popular sectors to overthrow the existing politi-

[9]When the World Bank and other donor agencies had supported the establishment of cooperatives or local governments in projects, for example, these had been regarded essentially as channels for technology transfer or distribution of credit rather than organizations of, by, and for their members.

cal-economic order have been thought to detract from the class struggle. From this perspective, only revolutionary movements can contribute to development. Actually, a few dependency theorists—Cardoso, for one—have seen a useful role for organizations in educating and giving more weight to the poor majority within the existing order; they may succeed in changing structural relations even if they do not overthrow the system (Kahl, 1976). Should revolution not be a realistic, imminent possibility, some amelioration of conditions and some increased competence for the lower classes would still be desirable, and organization would be a useful instrument.

Less extreme and more sympathetic critiques have come from liberal analysts (e.g., Blair, 1978; Holmquist, 1979) who see the possibility, even probability, that local organizations will be captured by more privileged local elements. Some of these critics placed more confidence in "targeted" programs of services and benefits, administered through a disciplined bureaucracy. Others recommended that standardized packages of innovations be distributed to the poor to raise their production and income. Since these could be used individually, local organizations would not be needed, though if LOs could facilitate adoption of the "techpacks" by aiding extensionists or by providing credit on a group basis, they would be judged to be useful.[10]

The importance of participatory local organizations as intermediaries between a government and its individual citizens—or clients—is not a new idea, and by the late 1970s it was gaining renewed support in a number of disciplines. Conservatives as well as liberals mustered arguments on behalf of "mediating structures" (Berger, 1977). But whereas emphasis had previously been placed on the lowest-level organizations—primary groups—it was now recognized that their effectiveness depends on the formation of federated or allied groups that reach more significant levels of membership and function. In anthropology, new methodologies of regional analysis,

[10]The "training and visit" (T & V) system of extension promoted by the World Bank during the 1970s presumed that the "contact farmers" through whom it worked were each in turn assisting a group and thus achieving economies of scale (Benor and Harrison, 1977). Skepticism about the efficacy of the system has mounted, however, precisely because no group basis for communication and innovation was established, and the contact farmers' linkage to other farmers has been weak. Simply choosing, or having the farmers elect, the contact did not assure cooperation when no functioning group existed to whom the contact farmer was accountable. Compton, reporting on T & V in Thailand, writes: "There was little or no evidence that farmers were speaking through the contact farmers, that farmers and contact farmers were being polled or, in most cases, even listened to by the extension agents, that extension agents were expected or encouraged to systematically study and report on farmer problems or innovative practices and knowledge, or that any kind of regular and systematic study, monitoring or evaluation was taking place. Instead all attention and energies seemed to be focused on promoting the adoption of established technologies" (1982:7–8). Similar experiences have been reported in Sri Lanka by Gunawardena and Chandrasiri (1981) and more generally by Howell (1982).

building on the central place theory of Johnson (1970) and the market structure analysis of Skinner (1964–65), showed the significance of socioeconomic units that operate beyond the village but are still subnational in function and benefit (C. Smith, 1976). Economists and planners began to argue for territorial units that draw strength from village units but integrate larger areas for decentralized projects (Friedmann and Weaver, 1979).

Previously, the focus of analysis had been on the state or the individual (or perhaps on the community as an aggregation of individuals). By the end of the 1970s, such a bifurcated view was giving way to a new appreciation of organizational structures that can not only mobilize local efforts and draw on local identities but also relate these to larger enterprises, on a district or regional level, responding to the impetus of constituent members. The state can—and does—interact with such multi-tiered units, which are composed of aggregated groupings of individuals. The idea of "local" organization takes on an expanded significance in such a theoretical and practical context, no longer being merely a matter of community representation but—through vertical and horizontal linkage—bridging household and regional activities. This is the point to which social science theory seems to have evolved, placing local organization in context. But the subject of organization has its own intellectual roots, which merit some explication.

Sources of Theory on Organization

The two most obvious, and opposed, theoretical positions concerning organization are those of the Marxists and the "pluralists." Marx saw classes as the primary forces working for or against technological and social change; if conscious of their own interests, classes would pursue those interests through organization. The ruling capitalist class, in Marx's conception, had little need for explicit organization; it was relatively small in number, very cognizant of its interests, and in control of the state apparatus. But the working class would have to organize for struggle against the bourgeoisie. A strategy of political agitation would lead to violent clashes; the proletariat would eventually win and seize state power through its vanguard political party, backed in particular by trade unions. In his later writing, Marx held out the hope that where there was universal franchise, the organized working class might gain power peacefully through the ballot box. In any case, the lesson Lenin drew from Marx's writing was that for accomplishing revolution, "organization is everything" (Selznick, 1952). It was to be created for the purpose of seizing and then consolidating state power, which would usher in true socialist development. Marxist theory, rejecting nonsocialist strategies and approaches as spurious, did not deal with the special tasks of economic development as generally pursued.

Even greater praise for organization can be found in the writings of Alexis de Tocqueville (1835) about America in the first part of the nineteenth century. "The art of associating together" was as crucial in his view for the vitality of democracy as was class-based organization in Marx's theory for the creation of socialism. The significance of organization for liberal democratic development was amplified in the work of Bentley (1908), an American political scientist contemporary with Lenin. Bentley was the founder of the influential "pluralist" school, which emphasized the importance of organization to political competition and political outcome: organizations were essential to democratic politics because they provided a balance between stability and evolutionary change (Truman, 1951).

A third school of social science, more neutral in its approach and more self-consciously theoretical, pursued the elaboration of what it called "organization theory." Representative works in this vast literature include those of March and Simon (1958) and Gross (1964). Many of this school's concepts and principles are relevant for us, particularly as they have evolved into theoretical statements about organizational structure, capability, and membership participation, such as those found in the work of Olson (1965) and Hirschman (1970), and extended to peasant collective behavior by Migdal (1974), Paige (1975), Popkin (1981), Bates (1981), and Russell and Nicholson (1981). However, the main intellectual roots of concern with local organization for rural development, while they have been enriched by these three schools of thought, are different.

The social theorist who has spoken with the most eloquence, insight, and anguish about "organization" is Robert Michels, whose book *Political Parties* (1915) is widely known for what he called the "iron law of oligarchy." In examining the trade unions and socialist parties in Europe at the turn of the century, Michels found, despite their ideological commitment to democracy, an inexorable tendency toward oligarchy, toward a leadership that would rise above the membership and become dominant and self-perpetuating. This meant not that organizations always failed to pursue or advance the interests of their members, but that leaders were likely to become unaccountable to their followers and self-aggrandizing in ways inconsistent with their democratic creed.

This was the bad news from Michels: "Who says organization says oligarchy." But there was some good news, too, the starting point of his concern. In the Europe of his day, there were stirrings of aspiration for a more equitable socioeconomic order, and the unions and parties he studied were means for supporting improvements in working conditions, wages, civil rights, and expanded education for industrial workers and their families. As classic a statement as Michels's pronouncement of the "iron law" is his dictum, "Organization is the weapon of the weak in their struggle with the strong." According to his analysis, reflecting the birth of the modern democratic era, organizations enabled individuals who were separately weak to pool their economic, social, and political resources and thus engage in self-help, improve their bargaining power, and acquire some weight within the system.

Organization could make numbers count as they could not when unaggregated, that is, unorganized. The cooperative movement launched in England and Germany in the nineteenth century (associated with the names of Rochdale and Raiffaisen), and later active in Scandinavia, showed that the "weak" could use their combined numbers not only to exert economic and political pressure but also to achieve a substantial measure of self-help. Its success, along with the growing influence of the labor unions, suggested that Gaetano Mosca's dictum—even more dismal than Michels's "iron law"—could be circumvented: "The dominion of an organized minority, obeying a single impulse, over the unorganized majority is inevitable. The power of any minority is irresistible as against each single individual in the majority, who stands alone before the totality of the organized minority" (quoted in Bottomore, 1964:9). If members of the majority could become organized, they could both acquire more resources to enhance their productivity and well-being, and deploy those resources more effectively.

Contemporary Perspectives on Rural Local Organizations

The state of the literature is in part a reflection of the different orientations of those who write on agricultural and rural development. To characterize the differences and to make our own position more explicit, we would identify four main schools of thought:

(1) *Marxists* and their intellectual associates regard the nonsocialist state as the instrument of the ruling class and hold that, despite differences that sometimes emerge within the political elite, the continuing function of such a state is to exploit the producing classes, including the rural periphery. Associated with the state are the dominant rural elites, who may still represent traditional landholding families but who increasingly are being supplanted by middle peasants or kulaks in Asia and by multinational agrobusiness firms in Latin America and parts of Africa. In this view, neither the state nor the rural elites are willing to tolerate local organizations of the marginalized rural poor that might act autonomously on behalf of their members. Those that emerge will either be co-opted and neutralized or suppressed, because the state will not deal with independent local organizations. The rural poor can achieve redress only by revolutionary peasants' and workers' movements (e.g., Feder, 1971; Petras and La Porte, 1971; Beckford, 1972; Griffin, 1976). The system is seen as closed to more moderate, gradualist activity because meaningful reforms are not acceptable to the

regime; thus it is not possible for organizations of the rural poor to achieve any significant progress within the prevailing system.

(2) A *liberationist* school has emerged more recently. Its members argue that while the contemporary state may tolerate some nonrevolutionary rural local organization, groups that genuinely represent and speak for the rural poor must avoid any association with the state. They regard the state as irremediably exploitative, incompetent, and corrupt in its relations with the poor. The only hope for the rural poor comes through greater self-awareness that leads to recognition of their inherent collective strength (Freire, 1970), then to confrontational—though not necessarily revolutionary—tactics, and eventually to a process of autoemancipation and liberation (Nerfin, 1977). Though outsiders can help local people achieve self-awareness and can lend their skills and commitment to the struggle, the motive force and the responsibility for strategy and tactics must come from the people themselves. At all costs they must maintain their distance from government and avoid any kind of dependency on its bureaucracies. In this view, the liberation and redemption of the poor can be achieved only outside the structures and rules of established political and economic institutions (Goulet, 1971).

(3) The *technocratic* label provides a catchall category for the large number of Western students and practitioners of agricultural and rural development who believe that rural progress depends primarily on economic growth and technological improvement (e.g., L. Brown, 1970; Wortmann and Cummings, 1978). They tend to regard the state as either benign or neutral, and they believe that such enlightened economic policies as keeping factor prices right, plus technological innovation and infrastructural assistance, can enable rural people—who, they imagine, are mainly owner-operators of small farms—to adopt measures that will improve their productivity and well-being. The state should assist small farmers with useful public services like extension information but avoid interfering in the operations of the market. Institutional and structural changes, such as land reform, tend to be disruptive or costly and should usually be avoided. Local organizations of a purely voluntary nature may be useful for pooling resources so long as they do not distort factor markets, and for articulating the needs and interests of farmers to the governments so long as they do not constrain "correct" technical decisions. These writers have not faced up to the reality of landlessness, the fragmentation and concentration of landholdings, and persistent rural poverty. Nor do they acknowledge the degree to which existing aggregations of power and unequal distribution of resources distort market mechanisms. They believe that eventually the rural poor will be absorbed into more productive employment in the cities, on modern farms, or in foreign countries as economic growth and technological modernization provide new opportunities. They see solutions to rural poverty primarily in terms of individual rather than collective action.

(4) Our approach, which we call a *structural-reformist* position, differs from each of the three above but accepts some of their premises. The state should not be seen as a neutral arbiter or broker in the tradition of American academic pluralism (Bentley, 1908; Herring, 1936; Truman, 1951); regimes tend to distribute benefits and costs unequally among component groups. At the same time, governments in most, if not all, developing countries are not monolithic. They do not necessarily, or in all matters, serve the interests of local elites at the expense of the poor, and their bureaucracies are not invariably corrupt, incompetent, exploitative, or self-serving. These tendencies may be present in certain agencies at particular times, but their presence is a matter of degree rather than an absolute. Regimes tend to protect the social and political status quo, but most of them are coalitions of varying and sometimes conflicting ideological, regional, bureaucratic, and factional interests. Their attitudes toward rural development and rural associations are not necessarily predetermined or fixed, so long as local organizations do not threaten to upset the existing system. Within the political framework of any regime, there may be opportunities for local organizations to function, to serve the social and economic interests of their members, to make claims on government, and even to be encouraged and patronized by government or by factions of the regime which for different reasons may advocate or be sympathetic to a bottom-up element in rural development. Where the opportunity to organize exists, as it does in most developing countries, rural publics can enhance their individual interests and collective well-being in various ways by taking advantage of these opportunities.

We characterize this fourth perspective as structural-reformist because it emphasizes the search for institutional and organizational changes that can cumulatively shift the balance of socioeconomic and political power. Modifications in the structure of landholding, in the bargaining power of those who have only their labor or their own produce to sell, in access to those in positions of authority and the ability to influence them, in the distribution of knowledge and technology through expanded literacy and education, in the social relations that allocate respect and opportunity between the sexes and among categorical groups—all these modifications require self-advancement as well as state action. Local organizations that strengthen rural people in large numbers through their own efforts are essential to the kind of incrementalism that this approach implies, building local power that can both limit and influence the actions of the state and of the private sector. We think that a variety of measures to alter the structure of assets and exchange through the gradual buildup of local institutions offers the most solid prospect for developmental progress.

The ongoing activities of local organizations in rural areas of many Third World countries are successful often enough to suggest that this perspective is a more tenable and realistic framework for rural development than are its alternatives. It avoids the naiveté and the excessive optimism of the technocratic position, the caricature of government and the romantic notions of solidarity in the liberationist perspective, and the pessimistic and dogmatic closure of the Marxian model. Those who are convinced that only revolutionary transformations can rescue the disadvantaged, and that half-measures undermine such an outcome, believe that any reformist approach betrays the interests of the rural poor. But those actually experiencing poverty find little virtue in failing to seize any feasible opportunity to struggle for direct improvements in productivity, welfare, security, and dignity and to achieve tangible benefits by their own collective efforts.

The perspective we have chosen does not expect governments to be benign or bureaucracies to be sympathetic. But it does not rule out the possibility that governments may find it useful for their own reasons to work with and through local organizations of disadvantaged rural publics. It recognizes some degrees of freedom, both for regimes and for local publics. Some tactics at some times in some systems may be unavailable or unavailing; but in most countries, opportunities for local organization are not foreclosed by the rules of the system, and pursuing them with some success can alter those rules in ways that are beneficial to the disadvantaged majority. We concur with Johnston and Clark that "the responsibility of the analyst—indeed his greatest challenge—is to discover and promote programs of incremental improvements which are feasible within the constraints of the social context that the poor actually face" (1982:168).

These are the premises that underlie our analysis of rural local organizations. The data and experience we present and the literature we review should demonstrate the utility, as well as the limitations, of this fourth approach. We would repeat that though local organizations are needed, they are not sufficient to produce rural development in the absence of sound economic policies, infrastructural investments, appropriate technologies, adequate public services. But these, without local organization, are themselves unlikely to achieve broadly based rural development through cumulative structural change.

In tracing the intellectual origins of concern with local organization, we have treated LOs as a relatively abstract and homogeneous category. Progress in understanding and implementation requires a more disaggregated view. In the next chapter we proceed to analyze local organizations as a universe of experience to be evaluated, looking at their types and tasks.

CHAPTER 3

Types and Tasks of Rural Local Organizations

Generalizations about local organizations flow freely in the literature and in conversations about what LOs can and cannot accomplish. Almost anything that one can say about LOs is true—or false—in at least some instance, somewhere. They present an immense array of efforts on the part of rural people, with or without outside assistance, to address a great variety of locally recognized problems. To deal with them in an analytical way, we need to delimit the universe of phenomena we will be considering and to structure this universe so that the diversity is rendered more comprehensible. This requires construction of appropriate typologies that bring some order to the subject.[1]

The kinds of institutions, formal and informal, that could be regarded as LOs are numerous. In our previous study for Asia, we considered a range of institutional structures that could facilitate communication and cooperation between the national center and local communities. These channels can be arranged along a continuum, from purely governmental or authoritative institutions (A) at one end to purely private or nonauthoritative ones (E) at the other.

(A)	(B)	(C)	(D)	(E)
Central government agencies	Local government units	Local organizations (cooperatives, clubs, interest groups, etc.)	Political organizations	Private enterprises
Governmental	....	....	....	.... Nongovernmental
(public sector)		(intermediate sector)		(private sector)

[1]The need for such an effort can be seen from the otherwise useful study by Development Alternatives, Inc. (DAI), which defines a local organization simply as "a grouping of small farmers, with some formal structure, directed towards increasing the agricultural production of members" (Gow et al., 1979, I:5). Contrary to the view that this is a "broad" definition, it is rather loose and too narrow.

Our earlier analysis focused on the contributions of the middle set of institutions—local government units (B), cooperatives and other associations (C), and political organizations (D)—which are generally more accountable to rural constituencies than (A) or (E). This important distinction can be a matter of degree, to be sure. Central government agencies (A) may be held responsive to local needs and demands by means of political processes, either through elected representatives or through lobbying. But the interests guiding the decisions of central authorities are bound to diverge often from local interests; thus one may speak of responsiveness more readily than accountability. Private enterprises (E), unless they are shielded by monopoly or monopsony power, will need to be reasonably responsive to consumer demands or to the needs of producers from whom they buy. But the accountability of such enterprises is to their owners, who desire to maximize profits, rather than to their clientele.

Our concern is with organizations that are accountable to their members and involved in development activities, which means essentially the complex middle range of local organizations (C). We have not focused on local government (B) for a number of reasons. There is already significant, if somewhat dated, comparative literature on local government as it relates to development tasks and accomplishments (e.g. Hicks, 1961; Maddick, 1963; Ashford, 1967). More important from our viewpoint is the fact that local governments are not really "membership" organizations unless one refers only to the local council, *panchayat,* commune council, or whatever the governing body is called. All persons living within the body's jurisdiction are subject to its authority, whether they like it or not.

It is often hard to distinguish how much accountability local government structures have to their constituents in developing countries. Sometimes what is called "local government" operates more as "local administration," enforcing laws, rules, and taxes decided on at higher levels. When that is so, local governments act more as part of the public sector than of the intermediate sector for rural development. We wanted to examine situations where special efforts could be made to mobilize resources, efforts, ideas, and support from rural publics by some combination of mutual interest and self-interest, rather than simply by required compliance. While we appreciate that local governments can play a useful role in rural development, indeed as intermediary institutions if they are active extensions of the community rather than of the central government, we decided not to include this category in our consideration. To have covered it adequately would have required all the time and resources we had and would have preempted analysis of other forms of local organization, which have received less systematic attention.

We also decided to leave out of our analysis the category of explicitly political organizations (D), even though they can be quite instrumental for rural development programs. In some countries, party cells or branches have been actively involved in rural development efforts: e.g., Egypt (Harik, 1974), Mexico (Hansen, 1971; Grindle, 1977), Tanzania (Bienen, 1967; Maeda, 1981), and China (Stavis, 1974b; Paul, 1982:79–90). As a rule, however, direct contributions to rural development are not the purpose of political organizations; their main purposes are to acquire influence and exercise governmental authority.[2] The desire for political power or advantage is a strong impetus and can be channeled toward promoting many developmental improvements, including self-help or other campaigns. But when this is the *primary* basis for forming and maintaining an organization, it is usually directed at the performance of *other* institutions, particularly government administration (A) or local government (B). Thus political organizations as a set are less often or less directly engaged in the work we associate with rural development.

Some organizations may be close to the boundary between categories (C) and (D). Some people might consider as political organizations, for example, the *harambee* self-help project committees in Kenya, which grew out of the movement for national independence and were politically instigated. They continue to be influenced by local political figures and to benefit from linkages to government. Nevertheless, the available studies of *harambee* suggest that these LOs are locally accountable and are sustained by motivations of local development more than by political partisanship (Bienen, 1974; Winans and Haugerud, 1977; Barkan et al., 1979; Holmquist, 1979; B. Thomas, 1980; Barkan, 1981). Similarly, the *Federacion Campesina de Venezuela* (FCV) was created through the efforts of party cadres. But the structures put in place—about 3,500 local unions, aggregated through regional organizations into a national federation representing over half a million peasants (Powell, 1971)—have performed needed services for members and given them some voice in national policy deliberations. Where the allegiance of a local FCV union's membership is divided between parties, the officers and representatives are chosen in proportion to the strength of each party. So while it has some important political aspects, the FCV operates to advance the interests of its members, not just of the parties as such (though the latter do benefit electorally from their improved connections with peasants). Using this criterion, we were satisfied to include some organizations having political initiative or linkage in our LO category, while leaving out of consideration those that are party branches or cells, subject to some political direction or control from above. As with local governments, we judged that including political organizations would

[2]H. Eckstein and Gurr (1975) distinguish between *structures of political authority* and *structures of political competition,* with party organization fitting the latter description. This corresponds to the definition of "political" in Ilchman and Uphoff (1969:50–51) as describing those activities and attitudes affecting the exercise, acquisition, or influence of authority.

stretch our analysis beyond the coherence we could achieve with a narrower spectrum of cases, a category between those institutions to be classified under (B) or (D).

This middle category (C) is large and heterogeneous, quite a sufficient challenge to deal with cross-nationally. The organizations it subsumes are not the only ones important for development, but we focus on them because they are, first, *membership* organizations, unlike (B), and second, primarily *development-oriented,* as distinguished from (D). As such, they are potentially useful vehicles for augmenting rural development efforts from above and from below.[3]

To make some sense of the variety of associations in this broad middle range, we need to distinguish various types of local organization. This can be done descriptively or analytically. The latter kind of typology is preferable, but we found that the features or aspects of LOs most salient in analytical terms were better treated as variables than as defining characteristics. So we arrived at a typology that highlights different purposes and bases of local organization, treating their structural variations as matters for subsequent empirical analysis.

A Typology of Local Organizations

The range of local organizations itself can be divided into three categories: (1) local development associations (LDAs), (2) cooperatives, and (3) interest associations (IAs).[4] A common distinction between formal (or "modern") and informal (or "traditional") LOs was not made typologically, as the degree of formalization is better treated as a variable cutting across all other types. The cases we covered in our review did not include many of the most informal groups, since direct involvement with development activities was a criterion for consideration. But all three of these categories include relatively informal as well as formal associations.

The first type, *local development associations,* shares some characteristics with local government (B). First, LDAs are *area-based,* bringing together all or most of the people within a community or region to promote its development by direct self-help or other means, such as lobbying for needed ser-

[3]Limiting our consideration to LOs with developmental involvement does not deny the value of LOs with purely social or cultural orientation; as far as we know, most of the principles formulated here would apply. But for our purposes of analysis and prescription, we have defined LOs as those with some developmental orientation. A number in our sample, such as *Ayni Ruway* in Bolivia, have both developmental and cultural activities. On the value of cultural LO programs for the rural poor, see Breslin (1982).

[4]This designation is not to be confused with "interest groups" in the "pluralist" tradition of American political science. The interest associations included here undertake self-help activities as well as seeking to promote, by whatever means are feasible, the interests of their members.

vices or raising funds to pay for new construction. The *servicios* of the National Community Development Service and the *Ayni Ruway* in Bolivia are representative of this type, as are the Village Development Committees in Botswana, Tanzania, and Zambia, the Local Development Associations in Yemen, and the *Sarvodaya Shramadana* organizations in Sri Lanka.[5]

Membership in LDAs is as heterogeneous as the communities involved, since the only common characteristic that members share is their place of residence. There can be considerable ethnic, religious, or economic homogeneity, depending on the locality. But this is a variable, not a defining characteristic. The organizations are generally *multifunctional* in that they can undertake a wide variety of tasks—from supporting education to building roads, assisting agriculture, maintaining churches or mosques, or possibly even regulating social conduct (for example, temperance activities). They are not comprehensive in their responsibilities, as are local governments, nor do they have the same legal or taxing powers. They are extensions of the community more than of the government, though they may be government instigated and assisted. The authority they exercise is likely to be extra-legal as well as legal in origin, since they draw legitimacy less from formal charters than from expressions of community need. Certain LDAs may be so much the creatures of government that it would be better to regard them as fitting in category (B). But judging by the criteria indicated above, we found a substantial number of LOs—19 percent of the cases in our sample—which are best regarded as LDAs, clearly differentiable from local government.

The second type, *cooperatives,* is extremely varied and has many subtypes. At one level it is a purely nominal category, as thousands of LOs around the world are called "cooperatives."[6] One can usefully distinguish this set of LOs from the rest, however, by focusing on their economic functions for their members. The defining characteristic of cooperatives, as Galjart (1981b) has suggested, is the *pooling of resources* by members. Using this criterion, we classified 35 percent of the LOs in our sample as cooperatives. The resources involved may be capital (savings societies or rotating credit associations; e.g., C. Geertz, 1962), labor (rotating work groups; e.g., Seibel and Massing, 1974), land (production cooperatives; e.g., Wong, 1979), purchasing power (consumer co-ops), or products (marketing co-ops). In addition, there can be pooling of financial resources and labor to secure production inputs like fertilizer or power tillage through service

[5]Studies that have been instructive concerning these are Savino (1984), Healy (1980), Chambers (1974), NIPA (1976), Cohen et al. (1981), Ratnapala (1980) and C. Moore (1981).

[6]For example, the local organizations in North Yemen referred to as *ta'awun* in Arabic are labeled in English both Local Development Associations (their national organization is called the Confederation of Yemeni Development Associations) and cooperatives. In fact, *ta'awun* have all the characteristics of LDAs and few features of cooperatives as we define these categories (Cohen et al., 1981).

cooperatives. There is a democratic principle operating in most cooperatives which presumes that even if resource contributions are not equal, all members should have an equal voice in decisions—with purchase or input of even one share entitling the member to a vote like other members.

Various distinctions can be made between co-ops and LDAs. Co-ops are usually more limited in the scope of their activities, though the multipurpose cooperative societies found in numerous countries of Asia and Africa generally include credit, agricultural inputs, and marketing services along with consumer goods, as do the SAIS organizations set up in Peru after the land reform (McClintock, 1981). Having single or few functions is thus not a defining characteristic of co-ops; rather the number of functions is a relevant variable. Membership in cooperatives is usually more limited or selective than that in LDAs, though co-op membership can be quite encompassing within a community and can thus be rather heterogeneous. While homogeneity of membership is often thought to be an important feature of cooperatives, we need to examine this as a variable rather than take it for granted. The most crucial difference between co-ops and LDAs is that the latter contribute mostly to "public goods," accessible to all, while the benefits from co-ops are usually of a more private nature, most accruing directly to members. To be sure, co-ops can provide gains to non-members as well, for example, those in Bolivia studied by Tendler (1983).

One reason for considering cooperatives separately as a group is to examine the arguments of some observers that co-ops are more likely to disadvantage poorer sectors of a community than to help them, whereas by working for public goods, LDAs provide benefits that should be accessible to all.[7] Our findings are not much more encouraging; still, there are enough exceptions to indicate that this form of LO can be useful for disadvantaged groups. What needs to be better understood is how to work within different kinds of environments. The success of a leather workers' co-op for untouchables in Barpali, India, of the *bams* (co-ops) among Tiv small farmers in central Nigeria, and of Andean Indians' cooperatives in highland Ecuador suggest that this category should not be written off as unable to help the poor (F. Thomas, 1968; Morss et al., 1976, II:231–41; Meehan, 1978:134–37). We need to know more about what structures, incentives, and outside support will be most conducive to the desired developmental results.

The third type of LO, *interest associations,* is the most diverse of the three. What we call interest associations are defined not by geographic boundaries as are LDAs, or by pooling of economic resources as are co-ops, but by certain *common features of their membership.* In some IAs, persons come

[7]See Fals Borda (1976), Münkner (1976), ICA (1978), and Lele (1981). Fals Borda summarizes the negative findings against cooperatives in this regard from a major comparative study by the U.N. Research Institute for Social Development covering Africa, Asia and Latin America (see UNRISD, 1975).

together for the sake of performing some particular function better, perhaps water management or public health or primary education. Water users' associations, health committees, and parent-teacher groups are examples of *functional* interest associations. In other IAs, people join together on the basis of some personal characteristic—such as their sex, ethnicity, religion, or economic status—to promote common interests. Women's associations, mothers' clubs, tribal unions, mosque committees, church groups, and landless laborers' organizations are examples of what might be considered *categorical* interest associations. As a rule, IAs will be less encompassing than LDAs, which are inclusive and multifunctional by definition, but more so than co-ops, since they are concerned with social as well as economic interests and with public as well as private goods. Nearly half—46 percent—of the LOs in our sample fell into this category.

A specific interest association can verge on either of the other two categories. A water users' association operating in an irrigation scheme may be fairly similar to a cooperative. But to the extent that the association is occupied with deciding on water rotation schedules, resolving disputes, and carrying out channel maintenance, it is *servicing* an economic activity more than performing it. Traditional water users' associations like the *subak* in Bali (Birkelbach, 1973) and the *zanjera* in the Philippines (Coward, 1979b; Siy, 1982) involve some pooling of labor to keep the channel systems in good order, and they operate in a highly equitable manner. But contributions of resources and voting are likely to be proportional to landholding, rather than equal as in a co-op, since the organizations' purpose is to provide goods or services to members more than to share inputs, risks, and benefits. Similarly, the members of an ethnic association might in effect constitute an LDA, if the whole community is ethnically homogeneous and the group engages in many activities, as did the Ibo State Union in Nigeria (Sklar, 1963; Smock, 1971). The difference remaining, in this example, is that non-Ibos in a community would not become members as they would in an area-based LO like the Onitsha Community League.[8]

The continuum of membership differences between homogeneity and heterogeneity is particularly complex. At the extreme of heterogeneity, both sexes, all ages, all social classes, and any diversity of races and religions would be represented in a group—a rare circumstance. Homogeneity in all these regards, on the other hand, is not much more likely. A group which is homogeneous with regard to ethnicity, for example, is likely to be heterogeneous in terms of age, sex, and probably some other characteristics. As a rule, a cooperative would not be extremely heterogeneous, as members are usually a subset of the general population motivated to come together by

[8]Consider, for example, the differences arising between the Onitsha Ibo Union and the Non-Onitsha Ibo Association, both formed according to ethnic criteria but the latter made up of "strangers"—Ibos from outside this town. A third organization also formed in the town of Onitsha, the Community League, was open to anybody living in Onitsha (Sklar, 1963:151–57).

similar economic or social interests. Of course, some cooperatives, like the multipurpose cooperative societies found in India and Sri Lanka, have quite heterogeneous memberships. By definition, an IA would be homogeneous in some important respect but not necessarily in others. Only LDAs would tend always toward the heterogeneous end of the spectrum (except with regard to members' place of residence).

The variable of membership homogeneity, even if multidimensional, is important because of its relationship to an organization's resource-mobilizing capacity. Other things being equal, we would expect members to identify more closely with a more homogeneous group and to be willing to make more contributions to its operation. Unfortunately, this relationship is confounded by the fact that as a rule homogeneous groups are usually smaller—or put the other way, larger groups are usually more heterogeneous. Homogeneous groups are thus likely to have a smaller resource base on which to draw. Even if they can expect greater contributions from each member, the total will probably be smaller. Larger groups, despite their heterogeneity, have more potential for mobilizing and deploying resources, either their own or those they can gain from others, including government. Whether they will accomplish this or not is an open question. What counts in actual cases is whether larger, more heterogeneous groups capitalize on the potential that their size gives them or whether smaller, more homogeneous ones benefit from greater solidarity. In all three kinds of organizations, members' perceptions of how tightly their own interests are bound with the LDA, the co-op, or the interest association are crucial for organizational maintenance and performance. Prediction from any single variable is difficult, because many factors interact and because some can cut either way, as we shall see in Chapters 4 and 5.

Table 3-1 summarizes the main lines of variation within the universe of local organizations and shows local governments (B) and political organizations (D) as they bracket the continuum of local organization. The defining characteristic of each group is boxed. IAs, for example, may have some pooling of economic resources and may have single or multiple functions, but they are defined in terms of a specific common characteristic of members; they all perform or relate to a common function, or they all have some feature such as sex, occupation, or status in common.

Classifying specific cases can present some ambiguity, which is why we have not subdivided the IA category into two groups. For example, a fishermen's organization might be classified as a functional interest association, but if the fishermen constitute a caste or an ascriptive occupation, it would correspond more closely to our definition of categorical interest associations. If some shared characteristic is the basis for the organization and it seeks to benefit its members, we are satisfied to call it an interest association. If the LO in this example engages in economic tasks with members pooling resources, the group may qualify as a cooperative—a

Table 3-1. Variables and characteristics distinguishing types of local organization

		LOCAL ORGANIZATIONS (C)			
	LOCAL GOVERNMENT (B)	LOCAL DEVELOPMENT ASSOCIATIONS (C1)	COOPERATIVES (C2)	INTEREST ASSOCIATIONS (C3)	LOCAL POLITICAL ORGANIZATIONS (D)
RATIONALE	Perform government functions at local level	Improve income, services, etc., for area residents	Increase benefits from economic production or consumption	Advance members' common interest and/or perform specific activities for members	Influence or acquire authority
AUTHORITY	LG is structure of political authority*	Quasi-authoritative; may have government authorization	None, but may be registered and regulated	Only de facto authority which members may concede to it	LPO is structure of political competition*
FUNCTIONS	Comprehensive	Multiple functions on an area basis	Single or multiple	Single or multiple, depending on agreement of members	Primarily political; may perform others
MEMBERSHIP COMMONALITY	Residence (degree of heterogeneity depends on area)	Residence (degree of heterogeneity depends on area)	Contribution of economic resources (land, labor, or capital)	Personal characteristics (e.g., sex) or activity (e.g., irrigation management)	Political allegiance or ideology (degree of heterogeneity varies)
RESOURCES	Taxation and grants from central government	Assessments and contributions (often in-kind); may have government subventions	Pooling of member resources as shares; may have government subsidies	Fees, dues, levies, etc., set by members for group activity or interest; may get outside resources	Dues and donations; likely to have some outside financial support

*As defined by Eckstein and Gurr (1975).

classification to be made sometimes according to degree rather than kind. If the organization includes women as well as men and deals with problems of health, schooling, and so forth, as well as economic livelihood, it could qualify as an LDA.

For purposes of analysis, we were concerned with whether the group in question was not a form of local government (B) or political organization (D) and met the general criteria of "local organization"; that is, was it a membership organization, and did it undertake development tasks? Even this can admit of some ambiguity, but we found in practice very few organizations that were borderline between (B) and (C) and (D). Where there was doubt about classification, we left the case out of consideration in our statistical analysis.[9]

This typology brings some ordering to the otherwise amorphous category of LOs. But such a nominal classification based on observable characteristics does not provide much power for predicting or explaining performance. We find successful and unsuccessful LDAs, efficient and inefficient cooperatives, effective and ineffective IAs. We tried to formulate typologies based on variables that were more analytical than descriptive, but were not able to arrive at satisfactory formulations.[10] Our analysis, it turned out, was paralleled by that of a group working at Berkeley on problems of "managing

[9]Our construction of these categories represents a clustering of cases along dimensions in much the same way that factor analysis is done as a statistical exercise, though we could not operationalize such analysis for our cases. LOs can vary along dozens of variables, many of them overlapping. Our typology in effect identifies "vectors" of variation that are in principle orthogonal to one another as in factor analysis. LOs that score high on both multifunctionality and heterogeneity tend toward the LDA category; LOs that score high on pooling economic resources would go into (or toward) the cooperative category; and LOs having some particular kind of homogeneity of membership, based on assisting a specific group or a given activity, would be regarded as interest associations. Cases "loading" high on one of the three vectors would tend not to do so on the other two, though some cases could be relatively equidistant from two or even all three vectors. From observation, we can say that no LOs load high on all three, and all will correspond to at least one of these characteristics.

[10]We considered formulating a typology based on the different kinds of functions that LOs performed. A typology in terms of the number of functions broke down because two types—single- and multifunctional—would be too gross, and the number of functions performed by LOs varied from one to many (half a dozen or more). In the literature, the *degree* of single- versus multifunctionality is sometimes proposed as a significant variable to analyze, so we wanted to avoid making it a typological determinant.

The nature of activities was often too indistinct to be used as a criterion. One could, for example, propose that LOs making one-time efforts like building a road or an irrigation system would be different from those having continuous responsibility, such as maintaining a road or operating an irrigation system. But few LOs fit clearly into one category or the other.

A distinction between LOs involved with physical facilities and those involved with social services might contrast an irrigation LO with one providing immunizations. But if the health committee were also maintaining a clinic and the irrigation LO had any operation and maintenance tasks, such a typology would not hold up.

Some effort to derive typologies based on function should be continued, since it is theoretically challenging and could have some significant practical payoffs if successful. But we think our analysis here will be more fruitful in terms of variables, taken singly or in combination, than in terms of idealized "types."

decentralization." They identified similar structural criteria for differentiating kinds of local organizations that might interact with government or higher-level participatory organizations.[11] Our simpler classification scheme should suffice to delimit and map the universe of local organizations to be studied. Assessing the elements of successful LO performance in any case requires analysis that moves beyond classification.

In the remainder of this chapter and the following two chapters, we undertake a systematic analysis of the characteristics of local organizations and of what accounts for their varying contributions to rural development objectives. We have tried to employ quantitative techniques to a degree that is illuminating and defensible, while recognizing that—given the complexity of LOs and the limited established theory about their operation—our discussion is more often exploratory than definitive. We have classified LOs in our sample according to a five-point scale of "overall performance," reflecting the extent to which a number of concrete objectives were accomplished. We also compare subsets of LOs according to "summary scores," which reflect such performance. For example, the summary scores of overall performance for the three nominal types of LOs were as follow:

Local Development Associations ($N = 29$)	125
Cooperatives ($N = 52$)	13
Interest Associations ($N = 69$)	106

This indicated that, on the whole, cooperatives were the least and LDAs the most successful kind of LO. But we observed the whole range of performance, from outstanding to very poor, for all three types, including the excellent examples of very successful co-ops mentioned above, as well as the Bolivian cooperatives analyzed by Tendler (1983). From the outset of our consideration of cases, it was clear that we needed to go beyond any nominal classification and delve into the variations in structure, function, performance, and environment. How we approached this task is discussed next.

An Analysis of LO Variables

There are many possible explanations for the fact that some LOs make a greater contribution to rural development than do others. These explana-

[11]The first characteristic specified by David Leonard and his colleagues (working under a cooperative agreement with USAID similar to ours) was whether membership is inclusive or exclusive, the first corresponding to LDAs, and the second to interest associations (they used the term "interest organization"); cooperatives would fall in between, depending on the breadth of their membership. Their second characteristic was whether the organization has a generalist or a specialist orientation, which clearly distinguishes LDAs from cooperatives in their scheme. (See Leonard, 1982a:32–33.) Considering also whether there is support from a higher-level agency and whether there is interaction with the private sector (the latter either "philanthropy" or "marketization"), they classify 24 types of "local participatory organizations." We find their analysis rigorous and rewarding, but considerably more complex and no more explanatory than ours.

tions can generally be categorized in terms of (1) the performance of certain organizational *functions,* (2) the *environments* within which they operate, (3) the *structures* of the LOs, and (4) the kind and extent of *participation.* (The fourth set of variables is difficult to measure and assess, since participation is affected by an organization's goals, its task environment, and its internal structuring.) These different sets of variables bear on (5) *performance,* which itself presents a real challenge for measurement and assessment. These different sets of variables, plus exogenous factors, are represented collectively in Table 3-2. The sequence for considering them in this and the following chapters is not a simple left-to-right one, however, because the relationships are interactive, not linear.

We will begin with consideration in the latter part of this chapter of operational variables: that is, the functions that LOs commonly carry out. To use simple terminology, we usually refer to them as "tasks." The examples given in Table 3-2 are planning and goal-setting, resource mobilization, service integration, and claim-making. In social science it has been common to distinguish between structures and functions. One can use these designations as descriptive terms without accepting the theoretical baggage of "structural-functionalism." In trying to assess the contribution of LOs to rural development, one needs to understand the functions (operations or tasks) they undertake. To the extent that these are accomplished, they represent the principal *outputs* of LOs as organizations, according to a distinction made in organization theory between outputs and outcomes.

Outcomes are those benefits desired from LO activity, which we will deal

Table 3-2. Variables affecting the contributions of local organization to rural development

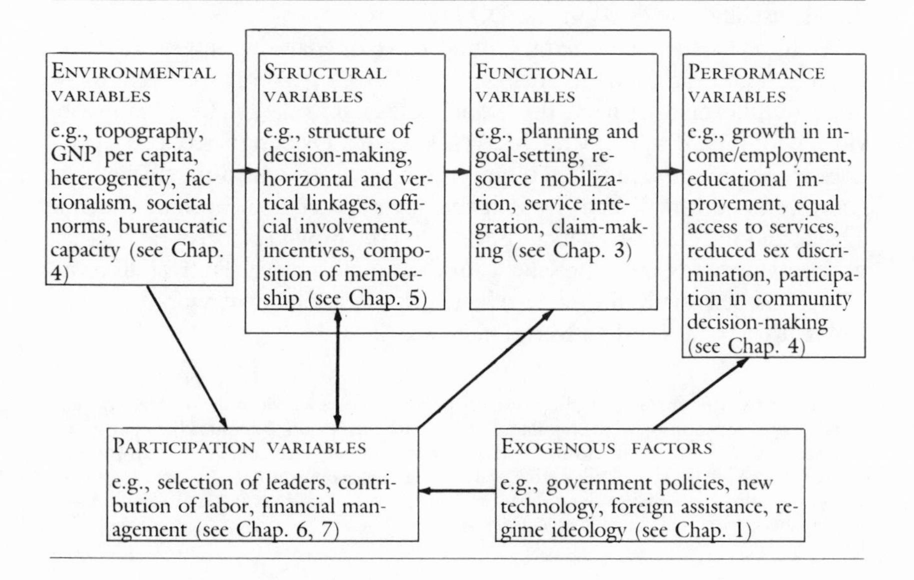

with in terms of *performance*. Beneficial changes in the economic, social, or political environment of LOs are not attributable solely to the workings of local organizations. As emphasized in Chapter 1, factors like government policy, public investment, and improved technologies all contribute to the achievement of rural development objectives. What we are interested in here are the outcomes that can be associated with LOs, and these we consider both in terms of disaggregated, specific indicators and in an aggregated assessment called "overall performance." These will be discussed at the beginning of Chapter 4.

LOs not only may have an impact upon their multifaceted environment but are themselves affected by it. As will also be addressed in Chapter 4, there are many ways in which variations in the environment of LOs are thought to condition or constrain their performance. The category of variables considered here is rather encompassing. The examples given in Table 3-2 are topography, which is a *physical* feature; income level, which reflects *economic* resource availability; heterogeneity, which makes *social* relations more complex and possibly difficult; factionalism or partisanship, which has *political* aspects; societal norms, which represent *ideological* influences; and bureaucratic capacity, which grows out of *institutional* development.

This concept of the "environment" is broader than that commonly associated with ecology and nature. It comes from the literature on organization theory, though it is compatible with the more common definition of environment if one thinks of the "ecology" of local organizations. One of the major conclusions of our study, which will be amplified and supported in Chapter 4, is that local organizations, while not independent of their environment, can function effectively within a wider variety of "ecologies" than is usually suggested in the LO literature.

We have already suggested the importance of analyzing internal features of LOs, since variations in their structure and operation are likely to be more significant than what the organizations are called. This is most obvious with regard to cooperatives. While most LOs designated as cooperatives will have something in common, the extent to which they have been conceived and established by their members rather than by some outside agency and the extent to which they rely on their own resources rather than government funds are quite salient for their performance. To treat all cooperatives as essentially the same when such variations exist would obscure factors that significantly affect performance.[12]

[12] "In a cross-cultural context, the word 'cooperative' covers a wide range of local organizations. In one organization in the highlands of Peru, members work communally to pay for the land received under agrarian reform. In another organization in Yemen, every man works for himself, leadership is in the hands of the local sheik, and the cooperative has its own accountant and shareholders, pays dividends, and has established a thriving export market in Saudi Arabia for members' produce." Gow et al. (1979, I:5)

The ways in which LOs are structured internally—their decision-making process, their relations with other LOs at the same level (horizontal linkage) or with higher-level organizations (vertical linkage), their connections with government agencies, the incentives or sanctions that tie members to the organization, the composition of membership (homogeneous or heterogeneous)—could all be expected to have some bearing on the performance of such organizational tasks as planning and goal setting, resource mobilization, and so on. Beyond these organizational outputs, we would expect some association with the achievement of developmental outcomes, analyzed as manifestations of LO performance. Because structural variables are much more amenable to policy initiatives than are environmental ones, we are most concerned with the analysis of structural alternatives, discussed in Chapter 5. There we try to assess what *forms* of local organization seem most likely to support desired development goals.

Our concern with participation variables has given impetus to this study from the outset. Our analysis grows not only out of our 1974 comparative analysis but also out of the work of the Cornell Rural Development Committee in half a dozen countries—Botswana, Costa Rica, Jamaica, Sri Lanka, Tunisia, and Yemen (Y.A.R.)—between 1977 and 1982.[13] We have thus observed experience or commissioned studies of the dynamics of local organization with regard to participation factors on a rather broad basis. Unfortunately, measurement of these factors is made difficult by their inordinately contextual nature (Cohen and Uphoff, 1977). Even a factor as concrete as meeting attendance is exceedingly difficult to interpret and compare across cases.

We recognize that participation variables are likely to be affected by environmental factors, and themselves to influence both structural and functional elements of local organizations. We have addressed participation variables in our analysis of LO structures and functions, trying to minimize any tautological relationships, recognizing that these can arise wherever effects are interactive rather than simply one-directional. Any analysis and assessment of participation is made more complex by the fact that it represents an objective as well as a means of development in many definitions (e.g., Seers, 1969; Goulet, 1971; Grant, 1973; ILO, 1975, 1977). To regard participation only as a means or only as an end in itself is a point of view based on ideology more than on reality, since most understanding of development acknowledges both instrumental and intrinsic values in the broader

[13]This activity was part of the Rural Development Participation Project (RDPP), funded under a cooperative agreement with USAID. In addition to the interdisciplinary work covering two to four years in each of these countries, there was shorter-term involvement concerning participatory rural development in Bolivia, Cameroon, Dominica, the Dominican Republic, Egypt, Guatemala, Honduras, Indonesia, Liberia, Nepal, the Philippines, Senegal, South Korea, Tanzania, Upper Volta, and Zimbabwe.

participation by people in decision-making, implementation, evaluation, and—certainly—benefits.

We have tried to keep the ends-means distinction clear in our analysis, but doing so also limits the extent to which, quite apart from difficulties of measurement and interpretation, we are able to deal quantitatively with participation variables. In addition to our review and analysis of 150 cases of local organization, discussed later in this chapter, and our observation and reading of the LO literature, we have drawn on a wide range of experience with participation—its problems as well as payoffs. These factors enter into our discussion particularly in Chapters 6 and 7, since the "pathologies" of local organizations as well as the "cures" revolve so often around the extent and quality of participation: contributing resources, holding leaders accountable, sharing responsibilities, minimizing or exacerbating factional divisions, and so forth.

Obviously, understanding and assessing local organizations in relation to development is an unusually complicated task. It requires some strategic simplification and, sometimes, assumptions that press the limits of established methodology so that their complexity can be taken into account—not simply dismissed—without overwhelming the subject. We present later the sources of our data and information and the methodologies employed. This discussion of sets of variables and of their relationships provides a conceptual overview of the analytical exercise we have undertaken. With this background, we will turn to consider the functions that local organizations can perform in support of multiple development objectives.

Local Organization Tasks

When dealing with the whole range of LOs, one can identify a number of operations that represent the generic outputs of these organizations. In our 1974 review, we specified six tasks that were evident in the operation of LOs: planning and goal-setting, resource mobilization, provision of services, integration of services, control bureaucracy, and claim-making on government. In this analysis, to focus on the working of specific LOs more than on whole LO systems, we have seen a need to add two more: conflict management and resource management. These eight tasks can be viewed as four pairs, which constitute a continuum from initiating and maintaining internal organizational activity (A) to influencing the external political-administrative environment (D). Activities pertaining to resources and services, the inputs and the outputs of organization, come between.

(A) Intra-organizational tasks
  Planning and goal-setting
  Conflict management

(B) Resource tasks
  Resource mobilization
  Resource management
(C) Service tasks
  Provision of services
  Integration of services
(D) Extra-organizational tasks
  Control of bureaucracy
  Claim-making on government

We will discuss each of these in turn as basic LO functions. Then, after describing briefly the methodology we used for making quantitative assessments, we will indicate observed relationships among them.

Planning and goal-setting is logically the first task, though any strictly sequential view distorts the reality that this task must be undertaken continually to determine the relevance and precision of other tasks. We were impressed with the way some of the most effective LOs made explicit and thorough surveys as part of their planning and goal-setting process—in particular, house-to-house interviews supplemented by group discussions to ascertain the most urgent needs of individuals and groups, what resources they controlled, and what they would be willing to contribute toward collective efforts. This function, well performed, can have the side effect of educating the community, as in the case of setting up the Banki water supply committees in India (Misra, 1975), or of drawing forth effective new local leaders, as in the organizational efforts in Tambon Yokkrabat, Thailand (Rabibhadana, 1983). Few LOs that we reviewed have had formal plans such as would please the eye of a planning commission member, though some do have fairly detailed plan documents.[14] When LOs took this task seriously, the most important result was a shared knowledge among members of their needs and capabilities and a grounded consensus that could buoy their performance of other tasks. In this way, the *process* of planning and goal setting may be more important to the success of LOs than the specific outputs of that process.

The means for eliciting planning and goal-setting are varied. The Small Farmer Development Programs (SFDP) established with FAO support in

[14]The difference between planning and accomplishment was demonstrated to Uphoff in a study (1979) of 16 rural communities in Sri Lanka. Half were selected for their record of development activity through LOs and the other half were matched neighboring communities with no such record. One of the latter had a rather detailed plan for local development, with a handsome hand-illustrated cover. The Rural Development Society was prepared to provide all the labor needed for ensuring cultivation of some 300 acres if it could get technical advice and some material inputs (mostly cement) to link seven small reservoirs by canals to a large one that overflowed each rainy season. But no work followed when the community became divided over alleged misappropriation of "food for work" allocations and when the government failed even to acknowledge receipt of the request for technical assistance.

Nepal, Bangladesh, and the Philippines (which we will refer to often in this book because of their organizational innovation and frequent success) developed methodologies to be introduced by group organizers for group problem solving (FAO, 1978–79). Some outside "catalysts" for LOs focus their efforts on "consciousness-raising" among members or potential members as part of, or even as a preliminary requirement for, planning and goal-setting. Apart from having some outside agent prompting the effort, it appears that the need to cope with problems can precipitate planning and goal-setting within the community. Hyden (1981b) has even suggested that some "obstructive power" can be an advantage for starting viable rural organizations, a theme we will return to several times. Both the Chipko and the Bhoomi Sena movements in India initially mobilized marginal populations to defend their access to land and natural resources, and then proceeded to plan and carry out productive activities as well (De, 1979; de Silva et al., 1979). The same dynamic was observed in the Kagawasam case in the Philippines, as confrontation over land rights prompted collective planning and self-help (Hollnsteiner, 1979). On the other hand, systematic planning was developed under more "normal" conditions with the Sukuma cotton cooperative in Tanzania (Lang et al., 1969), the Amul milk cooperatives in India (Paul, 1982), and the Baglung district bridge construction program in Nepal (P. Pradhan, 1980). In these cases, energetic leadership initiated a process that grew well beyond the initial LO effort.

It is tempting for outside agencies to do the planning for LOs, intending to leave ongoing decision-making to the organization after initial decisions about priorities and program support have been made. Indeed, one of the few quantitative studies of local organization concluded that project success was more positively affected by the resource contributions of small farmers than by their initial participation in project design (Morss et al., 1976). This conclusion, however, may have been influenced by the large proportion of agricultural credit and production projects in their sample.[15] In our quantitative analysis we found that planning and goal-setting had the same correlation with overall performance as did resource mobilization, suggesting that participation in initial decisions be treated at least as seriously as resource contributions.[16]

It may be thought that planning and goal-setting requires a certain level of sophistication on the part of LO leaders and members. However, we

[15]The DAI sample included a large number of projects where subsidized credit or other inputs were involved, so "planning" decisions might be less important if the activity was fairly standard and an attractive opportunity was being offered. Then the "matching" resources contributed by participants would be more significant. For a reanalysis of the DAI conclusions, see Young (1980).

[16]In the more recent DAI study, a statistical analysis of the overall impact of LOs found participation of the poor in decision-making to be significant, but not resource commitments (contributions) of the poor to the program (Gow et al., 1979, I:232).

found no correlation at all between effectiveness in this particular task and levels of literacy or per capita income. This suggests that at the local level planning need not require many technical skills or resources. Rather, in information gathering and consultation, local knowledge is needed more than scientific training. Technical knowledge can be added to local planning efforts to enlarge the range of alternatives, to achieve some internal consistency in plans, or to reconcile them throughout a larger area. Assistance from a higher-level organization (an LO federation or a government agency) will be more effective once the planning process has been started on a firm foundation of local knowledge and commitment to collective action.[17]

Conflict management is probably the most "internal" of tasks in that its purpose is to maintain group solidarity for achieving common purposes.[18] Accounting for degrees of success in this task is difficult because effective conflict management is not as evident an activity to outside observers as is planning. Indeed, when explicit efforts at conflict management come into operation, it may mean that the more important informal measures have failed. It may be that conflict management is performed best where it is least visible.

We presume that in most settings there is potential for conflict within the group, though how incipient it is will vary. To some unavoidable extent, judgments about success in conflict management will reflect the environment: low-conflict situations may offer no opportunity to develop or show such capacity, and high-conflict situations may overwhelm it. We have tried to minimize this problem in our analysis by looking for indirect as well as direct indications of conflict management. If the environment or the development activity under consideration is potentially conflict-producing (for example, because there are caste or class differences or because loan repayments are being enforced) but there are no evident conflicts, one may assume some success in conflict management. Such an inference might be incorrect, but it would be even more misleading to assume that the amount and success of conflict management is proportional to what is observed.

Some conflict within LOs should be regarded as normal and, within limits, as useful. Social scientists who study conflict have shown how, if successfully resolved and limited, it can mobilize resources and build larger, broader, and deeper commitment to common purposes (Coser, 1956). This view is more than a Toynbean tautology, which suggests that conflict—like challenge—is good in moderate quantities, defined by whether or not one

[17]One might regard a subject like rural road construction as a fairly technical and sophisticated activity with little scope for LO assistance in the planning process, but Tendler's study of rural road projects (1979) highlights the contributions that LOs can make at the planning stage to get an "optimal fit" between local uses and road layouts.

[18]We are not using the term to refer to handling conflict between the LO and others in its environment, except insofar as this creates problems of group cohesion and cooperation.

"survives" it. Rather it is a recognition that within any organization interests are likely to be in some respects divergent. At the same time, there are likely to be some common interests which, if activated and emphasized, can compensate constructively for differences. Acceptable rules and legitimate procedures for dealing with conflict can serve to protect and accommodate divergent interests, channeling them into joint efforts for agreed objectives.

We find instructive the analysis of conflict and cooperation in a dozen Peruvian rural communities between 1964 and 1969 by our colleague William F. Whyte. It has usually been presumed that these are inversely related—the more conflict, the less cooperation. But the two forces operate in the real world as distinct variables. One can find not only high conflict–low cooperation and low conflict–high cooperation situations, but also low conflict with low cooperation, and high conflict with high cooperation. A low conflict–low cooperation situation represents stagnation, whereas conditions of high conflict and high cooperation can energize relations and lead participants toward economic progress and increasing satisfaction (Whyte, 1975).[19] The absence of conflict is not necessarily productive, and its presence not always destructive.

The significant circumstance is how the conflict is handled, whether energies are channeled through reconciliation of interest or dissipated in mutual sabotage. The two factors that appear to be most associated with successful conflict management are the quality of leadership (an explanation always verging on tautology but a reality nevertheless) and the existence of "informal" modes of organization. Unfortunately, neither of these factors is easily amenable to outside instigation or improvement. Persons who work with LOs have often pointed to the lack of ability to manage internal conflicts as a key cause of LO ineffectiveness, and have often felt themselves unable to improve this capability. We will consider this issue in Chapters 6 and 7, noting here that one of the encouraging findings of our quantitative analysis was that the degree of economic or social heterogeneity within LOs was not significantly associated with successful conflict management. Those LOs with substantial heterogeneity were as often as not able to handle the conflicts that they confronted.

Resource mobilization is probably the task that government agencies most value in LO performance. Our assessment of this task took a community

[19]Two of the 12 communities studied fit the last category. Whyte's distinction between conflict and cooperation, using operationalized scores for each variable, arose from his observation that whereas in 1964 only one of the 12 had low conflict with low cooperation, five were below the mean on both indexes five years later. All three communities that remained low in conflict and high in cooperation were small, relatively "traditional" ones. The factor of socioeconomic differentiation (which increased with size of community) correlated with conflict as would be expected (though not when the smallest villages were omitted); it did *not* correlate one way or another with cooperation, which is quite interesting.

perspective. Estimates of success included not only the raising of local resources but also the acquisition of outside resources—from government or external agencies—if these were given to the LO as a result of activity on its part. We included as resources not only money and labor but also various material or in-kind contributions and the mobilization of political resources for voting or lobbying to promote LO goals.

In the literature the most specific analysis of this function has been done by Morss et al. (1976) in terms of "resource commitment" by participants in development projects, in conjunction with some form of local organization.[20] By "resource mobilization," we refer to all resources channeled as a result of LO activities into development efforts, not simply *to* or *through* the organization, so the two terms are comparable. Resource mobilization is always a relative matter, since its value depends on how much or how well it meets the needs of the community. Fairly small amounts of resources put to use in alleviating urgent problems can represent great success from the perspective of members and also of assisting agencies. Assessments should reflect such judgments rather than simply absolute amounts.[21]

Among the issues to be resolved with regard to resource mobilization, is the balance to be struck between outside and local resources. Communities, especially poor ones, can benefit from external assistance, but to rely very much on it creates a dependency that may prove to be counterproductive. The concomitant paternalism is likely to inhibit self-help and even undermine long-standing patterns of community initiative.[22] The total volume of resources (and ideas) for local problem solving would thereby become less than would otherwise be attainable, and the use of whatever resources are made available would not be as likely, for lack of local involvement, to meet priority needs efficiently or to be providently managed. A good example is the finding by the World Bank in its study of village water supply (1976) that the facilities were maintained better over time when villagers not only

[20]The findings of that study are summarized by its successor study as follows: projects intended to benefit small farmers are more likely to succeed when small farmers participate in project decisions and make resource commitments to project activities; organizations can facilitate small farmers' participation in decisions and resource commitments (Gow et al., 1979, I:7).

[21]Resources mobilized could be compared on a per capita basis or per community or per area, as a percentage of per capita income or of LO budget. Each standard would be valid for some kinds of comparisons and not for others. Since resource mobilization by any one criterion is seldom measured or reported for more than a few cases, systematic assessments must rely on grosser, composite comparisons. Any single quantitative measure will limit the range of comparison and reflect only one aspect of resource mobilization, not representing its utility relative to need.

[22]Gow et al. (1979, I:58–59) discuss the case of a successful cattle producers' cooperative in Peru that was able to pay substantial dividends to its members as well as sell meat, cheese, and wool to the community at reduced rates; it was disbanded after government agencies approached it "with an open checkbook" and made large loans. The resulting farmers' organization was judged no more than average in its success.

participated in decision-making about projects but also contributed resources toward construction and operation. This supports the finding of Morss et al. (1976) about "resource commitment."

Our own finding, discussed later in this chapter, that resource mobilization correlates more highly with overall LO performance than does claim-making, suggests that self-help activities are a more important part of LO functioning than outside resources. The paradoxical twist on this relationship is that there is a positive association overall between self-help efforts and resource mobilization from outside the community. We found that if *only* outside resources were involved, performance was relatively poor except in the area of services. Some combination of local and outside resources is generally to be preferred, with enough outside resources to encourage and extend local resource contributions, and with enough of the latter to justify and multiply the former. This is a very important subject with regard to assessing and assisting LOs and will be taken up in later chapters.

Resource management is perhaps the least glamorous of LO activities, but the fact that in our statistical analysis it turned out to have the highest correlation with overall LO performance suggests that it may be the most influential. (It also had the highest average correlation with success in other tasks.) This function involves keeping track of funds, collecting loans, maintaining buildings and equipment, operating irrigation structures, repairing roads, and the like. In assessing the performance of this task, we focused on the extent to which LOs are able to manage their resources, whether mobilized locally or acquired from outside the community, in ways that increase the volume of resources subsequently available to the LO and its members. Where we are considering the management of natural resources, particularly forests, soil, and water (an increasingly important kind of resource management in the rural sector), the criterion is preserving the resource base while it supports productive activities.

A prime example of good resource management is the building up of a revolving loan fund by enforcing repayment and by making productive loans to members, so that more persons could be assisted over time. Examples of poor resource management include corruption by officers who deplete LO treasuries, squandering resources on unproductive ventures, and failure to harvest crops planted on communal fields. The tasks of resource management and conflict management are related in that LO conflicts are more likely to arise if there is poor management of local resources.[23] Indeed, in our statistical analysis, the highest correlation was between these two tasks.

[23]A dramatic example of this interaction was seen in a pair of SFDP projects in Bangladesh, where landless persons were given loans to buy rickshaws with which to earn nonfarm income. One LO had a leader who was honest and dedicated, but it also had a member able and willing

Since most of the reported experiences of poor resource management involve ineffective or dishonest handling of *financial* resources, this is one area toward which government and other outside training and support efforts could usefully be directed. Training, of course, is likely to be more effective with regard to skills like bookkeeping or maintenance of equipment than to attitudes like honesty. Still, constructive skills and attitudes can be reinforcing. In Chapters 6 and 7 we address some of the problems of resource management and suggest, on the basis of LO experience, some things that may be done to reduce the problems. Since resource management is central to LO performance, we need to draw on whatever means can increase effectiveness in this function. Remedies can be considered with relatively few environmental constraints, as indicated by the fact that *none* of the environmental variables analyzed has a significant correlation with LO success in resource management.

Provision and integration of services are tasks easy to conceive and describe though not so easy to assess, because seldom are LOs solely responsible for agricultural or social services. If LOs are involved in service delivery or coordination, it is usually in conjunction with some government or private agency (though we do find some cases where LOs are wholly responsible for domestic water supply or make loans from their own funds mobilized from savings deposits). We do not count as LO provision of services any coming entirely from outside agencies. But one can give LOs credit for involvement in the delivery and monitoring of health services, for example, where they schedule visits with doctors at a clinic or follow up on patients to see that medicines are taken, so that outside services become more accessible or effective. One should also credit LOs that help to coordinate such services as approval of credit applications and timely delivery of fertilizer.

Service provision is the "bread-and-butter" task for most LOs; it occurred in about 90 percent of the cases we studied. The exceptions were organizations concerned mostly with lobbying or legal redress—though access to the legal system itself is a service usefully performed by the Sidamo associations in Ethiopia, for example (Hamer, 1980)—and organizations involved in planning and even coordinating but not delivering services.

---

to do maintenance and repair work on the group's rickshaws. It prospered, while the other languished. Conflicts arose in the latter LO over who was responsible for the deteriorating condition of their rickshaws. Nobody in that group was able or willing to repair them; eventually they became unusable, income ceased, and the loans could not be repaid. If the group had been better able to manage its internal conflict, the resource management lapse might not have been so devastating. But also if members of the group had taken better care of their equipment, some of the controversy could have been avoided. Lack of attention to equipment maintenance was a deficiency in project design (Islam, 1979). These cases are discussed again in Chapter 6.

Service integration, on the other hand, was attempted in only about half the cases reviewed, and only one-tenth of these were judged to be "quite effective" in such integration, compared with one-quarter judged "quite effective" in provision. This may reflect the resistance of bureaucratic agencies to any horizontal coordination of their activities at the local level.

We found at least some LOs that were able to increase the relevance, timeliness, and efficiency of services by being involved in their coordination. Engaging LOs in the task of service integration appears to offer government and private agencies opportunities for improving the benefits to be derived from their services. This would involve LOs not only in decisions as to the timing and level of services but also in the evaluation and modification of services in relation to local needs. LOs are an underused channel for service delivery, according to a study of rural organizations in the Philippines done for the Asian and Pacific Centre for Development Administration (Montiel, 1980:184–85). An excellent example of the possibilities for service delivery and integration is the Kottar Social Service Society in India (Field, 1980).

Control of bureaucracy and *claim-making* are perhaps the most difficult tasks for LOs, since strengthened capabilities along these lines are likely to constrain government to some extent. There can be differences of interest, of course, between a government (the political leadership) and its agencies. Insofar as priorities and resource allocations are established, the government has an interest in efficient administrative implementation. Are extension agents actually visiting the farmers, and as often as they are supposed to? Do medical assistants show up for their clinic work as expected? Are loan officers certifying credit-worthy farmers without insisting on bribes?

Local people and leaders can know more precisely than any central government officials just what lapses or misdeeds are occurring in program operations, and local organizations are in a much better position to insist on improved performance of local staff than are individuals. Perhaps the most instructive case of LO control of technical and administrative staff has been that of the Farmers' Associations and Irrigation Associations in Taiwan, where the LOs have actually hired and supervised their own field-level staff.[24] Control over bureaucracy can be indirect rather than direct if a

[24]The most thorough study of these associations is by Stavis (1974a). Changes in their organization are discussed by Gilbert Levine in an appendix to the Stavis chapter on Taiwan in Uphoff (1982–83, 2:248–252), and also by Moulik (1981). During the period of greatest agricultural dynamism in Taiwan—through the early 1970s—the FAs and IAs although closely supervised from the center, had a good deal of discretion in undertaking local agricultural development activities. Their employment of staff evidently contributed to both the level of effort and the responsiveness of trained personnel. Increased central control over association activities was introduced in 1975 in the name of greater efficiency and equity in resource use (farmers' groups in richer areas could afford to pay staff more and could therefore attract and keep better-qualified personnel). Much of the staff and budget control was returned to the associations in 1982. Additional sources on these LOs are Abel (1975), Bottrall (1977), Kwoh (1964), and M. Moore (1983).

government solicits the opinions of farmers, irrigators, or mothers on a systematic basis. Through organizations, idiosyncratic views can be sifted out to get representative assessments, which deserve to be treated seriously by higher political and administrative echelons. The limits of control and coordination of bureaucratic behavior from above are increasingly apparent.[25] The orderly logic of hierarchical direction according to Weberian theory is better for getting staff to follow superiors' orders (or at least to appear to do so) than for getting staff to work together across administrative boundaries and to be responsive to clients' requirements. With functioning local organizations, greater control and coordination from below should be available to supplement if not replace the oversight of political and administrative superiors.[26]

The relevance of such a strategy for improving governments' development performance depends, as indicated in Chapter 1, on the goals and values of the top political leadership. They have it in their power to frustrate, if not always to promote, effective LO activity. One of the main findings of theoretical significance in our earlier study was the importance of congruence between the development objectives of national leaders and those of local communities. To the extent that their goals are compatible—that is, that leaders want for communities what communities want for themselves—the distribution of power between them is positive-sum, not zero-sum. In this situation, more power at the center contributes to furtherance of local goals, and power at the local level increases the center's ability to achieve its objectives. If there is a divergence of objectives, however, strengthening LO capabilities will detract from central power (Uphoff and Esman, 1974). Thus, one cannot look at claim-making as an LO activity without reference to the orientation of the regime's leaders. Enhanced claim-making ability should be positively regarded by political leaders insofar as they desire greater satisfaction of rural people's needs, and approve of greater capacity on the part of rural people to articulate and meet those needs by self-help and in cooperation with the government. The LO outcome of empowerment, discussed in Chapter 1, must be evaluated in this context.[27]

[25]See Leonard's excellent empirical study of the agricultural extension service in Kenya (1977), which accounts for the observed relationships by drawing on organization theory.

[26]This is discussed in Uphoff (1983a) in terms of creating LO capacity to make demands to balance the "supply-side" orientation of prevailing administrative doctrine and practice in most developing countries.

[27]It should be borne in mind that regimes are seldom monolithic; some top leaders may not be sympathetic to advancement for the rural majority, while others are. Or some elements of the bureaucracy, on which the regime depends, could take a more positive-sum view than political leaders toward central vis-à-vis local capacity. To be sure, a leadership sympathetic to enhancing local capacity could be thwarted by less supportive elements of the bureaucracy. Our discussion here of respective gains and losses of power refers to the purposes of national leaders generally, vis-à-vis rural communities, without elaborating on the implications arising from pluralism within regime structures.

A government must expect certain costs from claim-making. Even if the demands coming from an LO are regarded as reasonable by its members, a regime having limited resources or different priorities may consider them excessive, though LOs are seldom powerful enough to impose significant costs on a government. It should be understood on both sides that the government cannot always respond favorably. The consequence should be a continuing dialogue that helps each to understand the priorities of the other and to make feasible adjustments. Shortfalls in what government can provide may then occasion more local resource mobilization on a cost-sharing basis.

This set of functions of local organization we think encompasses the main tasks of concern analytically and for policy purposes. As suggested previously, these functions do not represent the *outcomes* of LO activity, which are desired for their own sake; those we will address in terms of LO performance. Rather they constitute *outputs* of organization. Our analysis here reflects the concepts and experience reported in the literature plus our own observations. An effort has been to provide a more systematic and quantitative basis for making generalizations and for inferring appropriate strategies of support. Before we discuss our statistical findings concerning LO functions, we need to describe in brief the methodological basis for these findings.

Quantitative Analysis of Local Organizations

To assess systematically the various factors bearing on LO performance, we undertook to analyze a broad cross section of LOs from all parts of the developing world. The literature on local organizations is full of hypotheses and speculations about the relationships between variations in LO structure and task performance, or between environment and effectiveness; generalizations about "participation" abound. Unfortunately, few studies address these relationships with regard to more than one type of LO or in more than one country. As we will discuss with regard to our own work, venturing beyond a single variable or a single case presents serious problems of comparison and evaluation. The studies done by Development Alternatives, Inc. (DAI), which have paralleled our own, were pioneering but not without methodological difficulties.[28] We have tried to learn from their efforts,

[28]Two analyses by Sobhan (1976) and Young (1980) of the first study (Morss et al., 1976) have raised some objections but they have not negated the general line of conclusions. Unfortunately, Gow et al. (1979) combine both "output" and "outcome" variables into their measure of performance. Both studies are limited by a "small farmer bias" that leads them to exclude a substantial proportion of relevant rural LOs. Charlick's treatment (1984) of some 40 LOs in five countries, in connection with *animation rurale* programs, parallels Gow's and our own analysis here in its substantive concerns and methodology; it is noted often in the following discussion.

as they did from ours. Working with small samples or with only a few variables is easier to defend according to standard canons of investigation, but the results can support only very tentative conclusions based on simple bivariate relations. With small samples, one cannot control for the effects of the multiplicity of other variables suspected to have some influence.

The possibilities of mistaken inferences from limited samples or casual observation—with unfortunate implications for policy—can be seen in two studies that Uphoff consulted in preparation for fieldwork on rural LOs in Sri Lanka. One study of four multipurpose cooperative societies done for the U.N. Research Institute for Social Development (Fernando, 1977) found that the two more successful co-ops were in relatively rich communities, the two less successful in resource-poor environments. The conclusion drawn was that co-ops will perform better where resource endowments are better and, further, that a strategy for development relying on cooperatives would widen the gap between richer and poorer communities. If true, this is an important finding for poverty-oriented development, suggesting that co-ops—and perhaps LOs generally—should not be encouraged. Uphoff's subsequent study (1979) covering 16 communities (eight pairs of "active" and "typical" communities classified in terms of their development efforts through LOs) disclosed no such general relationship; indeed, it indicated that LOs were often used by more disadvantaged groups to narrow the gap. Still, one would want to know on a wider basis how important a factor affecting LO performance is the natural resource endowment or the level of economic development already achieved.

One of the first major studies of poverty in any less developed country was that by the Kandyan Peasantry Commission in 1951 (discussed in Uphoff and Wanigaratne, 1982). It observed in passing, without any systematic investigation, that the recently-established Rural Development Societies (LDAs in our typology) were most effective in communities where the level of education was relatively high and not very effective where literacy was low. If this is correct, it would suggest that resources to support the work of LDAs should be channeled to more educationally advanced communities, and that resources intended for the other communities might pay off better if spent directly on schooling. Uphoff's analysis of LO performance in Sri Lankan communities (1979) did not confirm this relationship, but it deserves further consideration because it would be very significant if true. Each of the variables in our analysis represents some hypothesis about LO performance. We have discussed the "functional" variables already. The others—of environment, structure, and performance—are operationally defined in the coding protocol in Appendix B and are treated in the next two chapters.

There is no way one could undertake a systematic sampling of the universe of Third World LOs, because no complete inventory has ever been done. Given the number and changing status of LOs, some of them ephemeral, this would be an impossible task; one would need some standard definition and typology, such as that introduced at the beginning of this chapter, just to begin such a cataloguing. So we went through bibliographies, journals, books, and theses on the subject of local organization and rural development to pick out case studies with enough data to permit systematic comparisons of LO experience. This was the bibliographic equivalent of a "random walk" (Mitchell, 1965), taking case studies without knowing in advance how successful or unsuccessful the LOs in question were, and discarding them only if they did not meet the definition and typology introduced above or if they did not have enough data for purposes of comparison. The case studies were sometimes for a single LO and sometimes for a number of LOs of the same type (e.g., mothers' clubs). Since we were not trying to describe statistically the universe of LOs but rather to ascertain whether, for example, illiteracy in the population was associated with poor LO performance, this did not present a debilitating problem. What we have is a very large set of case studies from which we can identify relationships. In a proper statistical sense, it is not a random sample of LOs, though it comes as close to approximating one as we could make it, since there was no "bias" in its construction apart from whatever exists in the literature.

There is some bias in the literature toward reporting relatively successful LO experience; however, a good number of cases that were unsuccessful by almost any standard have also been documented. So the set of cases we are considering fairly certainly encompasses the range of experience, if not its exact distribution. The latter limitation is not of major significance for our purposes, since we want to ascertain what factors appear to be associated with effective or ineffective LO performance. For this purpose, subsets or subsamples of the whole, broken down by degrees of good or poor performance, are more important. We think the sets of cases represent reasonably well, if not perfectly randomly, the different levels of LO performance. We want to have some idea of what are the most probable relationships, but we also know that the most likely ones are seldom the ones contributing most to rural development. In this sense, it is as useful to consider the range of LO characteristics as their distribution, to know not only what is probable but also what is possible in desired directions.

Since the size of the universe of LOs is not known, there was no scientific way to determine the size of a set of cases that would constitute a reasonable sample. Our main requirement was to have a large enough set to provide subsamples large enough for analysis and inference in their own right. This seemed to indicate a need for at least 100 cases, and possibly for 200 or more. When we had done 150 cases, we analyzed them according to the 55 variables being coded, splitting them into three subsets of 50 cases each: (a) the first 50 cases coded, (b) the next 50, and (c) the next 50. This was conceived as a "split-half" test of reliability, but done with three groups to see whether there were any *trends* in the sample as we expanded it, having three rather than two points of comparison. A chi-square test for significance of differences among the three subsets was done.[29] For only one of the 55 variables did a possible trend from the first to the third subset appear, and this could be purely a chance occurrence. So we inferred that expanding the sample by, say, another 50 cases would not affect the analysis significantly. The cases that were included in the analysis are listed in Appendix A.

Five research assistants in our working group read the cases and entered all relevant quantitative and qualitative information concerning the variables on standard forms, including statistics where reported and whatever description was given in the case study. Each case was then scored by a principal investigator on a scale of 1 to 5, using the criteria stated in the protocol reproduced in Appendix B, for all variables on which data were recorded. Putting all the information on a standard form was intended to minimize bias and any value judgments in the language used by the case studies' authors. The research assistant who had read the original case material reviewed the scoring, and any differences in assessment were reconciled by discussion or reference to other possible sources of information, such as the World Bank Atlas.

In such a procedure, translating mostly qualitative data into quantitative form for analysis, there are bound to be some judgmental decisions; but where the situation was relatively ambiguous, no scoring was attempted. All data on performance were reviewed and scored only *after* the scoring of task, environment, and structural variables had been completed, to minimize "halo effects" (or the opposite). Assessing variables in terms of as few as five categories (values) made relevant and reasonably reliable distinctions possible. As seen from the protocol in Appendix B, most of the variables are continuous, and clearly ordinal if not always cardinal in their distinctions.[30] The ordinal categories were defined to approximate cardinality as much as possible. This enabled us to employ, with appropriate qualifications, analytical techniques that went beyond simple cross-tabulation of variables, though that was our first step and our most basic technique, accepting Tufte's advice on data analysis (1969).

Since cross-tabulations are complicated to present and explain, a more

[29]For this analysis, since many variables were nominal rather than cardinal (though almost all were ordinal), we used a nonparametric statistic. The statistical confidence level used was .05, as it was throughout this study unless otherwise indicated.

[30]Categories representing degrees of social heterogeneity vs. homogeneity, for example, were ordinal, though for some variables the categories could be cardinal: e.g., levels of literacy by quintile. This latter is an example, however, of what may be misleading cardinality, since percentage increases in literacy may not represent equal increments in terms of social relations or factors promoting/inhibiting participation. Going from 10 to 30 percent literacy is a more profound change than going from 50 to 70 percent. So when it comes to representing the variables, cardinal data are not necessarily better than ordinal rankings.

summary statistic was needed to indicate both the direction and magnitude of association. After consulting statistical references, we decided to use product-moment correlation (*r*), which is a common statistic and more widely understood than many of the nonparametric measures that could have been used.[31] Because stricter statistical assumptions are involved, we are cautious about inferences from multiple regression analysis of variables, though we found stepwise regressions a useful way of seeing which variables accounted best in statistical terms for observed variance in the cases. Some of these multivariate results are reported in appendixes to Chapters 4 and 5, with appropriate caveats. We also present in appendixes some disaggregated analyses of relationships bearing on LO performance, which are more detailed than some readers may want to tackle.[32]

The most illuminating analyses were usually the simplest—cross-tabulations of subsamples controlling for a particular variable like the level of literacy, when subsample sizes were large enough to permit some confidence in the observed distribution. When interpreting correlational measures, we saw the ***direction*** of association (positive or negative) as the most important thing. The ***relative*** size of coefficients is more important than absolute levels of correlation, since the latter might be affected by the method of scoring, but not the former. We also devised a simple method, discussed below, for calculating a summary score, which reflects in a single number the distribution of cases for any particular variable.

In dealing with correlations or any other statistical measure, it is desirable to have some way of distinguishing between those results that appear worth taking seriously and those that might have occurred by chance. We are not dealing here with a strictly random sample, although it is as close as we could come to one. Nor are all the data parametric, though given the way the categories were constructed and scored, most distributions approximate

[31]With scoring limited to five levels, using *r* with our ordinal data is reasonably defensible. Concerning *r* compared with Kendall's *tau,* for example, Nambodiri, Carter, and Blalock say: "In general it was found that if one uses as few as 5 levels for each variable, one obtains very good approximations to zero partials even where the original associations are quite high. This is also true for non-linear but monotonic relationships" (1975:477). This is supported further by O'Brien (1979) who used extensive simulation techniques to test the proposition of Labovitz (1970).

Comparing our results of *r* with nonparametric correlations indicated that differences were minimal. Comparing *r* with Spearman's *rho,* a nonparametric correlation coefficient, for the matrix of correlations among environmental variables showed that only 14 of the 153 coefficients were different by as much as .10. Nor did correlation coefficients for *gamma* and Kendall's *tau,* other nonparametric statistics, produce different patterns or results. Indeed, the *tau* values for correlations between task variables and overall performance averaged slightly higher (.10) than with *r,* though their rank-ordering of correlations was the same (*r* is usually thought to give somewhat higher correlations with nonparametric data). The average difference between *r* and *gamma* coefficients for structural variables was only .03, for example.

[32]Exploratory factor analysis did not produce results having much apparent theoretical validity, perhaps because of the diversity of cases.

"normality." Under such circumstances, we felt it reasonable to use a standard test of statistical significance (at the .05 level of confidence) for identifying the associations we would consider seriously and those we would discount. While such a use of significance tests does not follow strict methodological canons, we think it is defensible as a method for screening statistical results.[33]

Our analysis of LO experience and our conclusions do not rely entirely upon the statistical work. The other two "legs" that our analysis stands on are the growing body of literature on LOs, which we have been following for more than a decade, and the field observations connected with the Rural Development Participation Project (RDPP), which we directed between 1977 and 1982. Nevertheless, we do not want to understate the significance of the 150 case studies analyzed. They contributed not only to the quantitative treatment of LO variables in this and the next two chapters but also to the qualitative discussion of LO problems and remedies in Chapters 6 and 7. We believe they represent a reasonable cross section of LO experience.

Presenting the descriptions and data on which the analysis was based would take as much space as this entire book. To give readers a better idea of what the organizations and coding were like, we have included brief write-ups of 20 cases in Appendix C. We have reproduced in Appendix D an article describing the groups of the Small Farmer Development Program in Nepal to illustrate the kind of original source materials we drew from, as

[33]Two colleagues more methodologically advanced than we reviewed our methodology. The first encouraged us to go forward, "despite the fact that the materials probably will not meet strict methodological standards. It is a Catch-22 situation: Most researchers would not attempt to study these problems because it is impossible to meet high standards, yet we will never learn how to study them unless somebody wades in and tries." The other said, "What I like about the methodology is that you do not strive for too many 'significant digits' in your coding scheme, but reduce the data to a reasonable five-point scale. It seems to me that approach is entirely reasonable. Moreover, your defense of the use of the product-moment coefficient is persuasive; I would have done the same thing."

One of the readers for the publisher made the following comment: "Some purists would argue that the entire effort to undertake and present the sort of statistical analysis contained in these chapters was misguided. My view is that the authors are to be congratulated for their painstaking effort to carry out a systematic and disciplined analysis of the 'evidence' provided by 150 LO case studies. They recognize that there are very serious problems involved in applying correlation analysis to this sort of 'data set.' I agree with them that it was nevertheless worth the effort."

Another reader wrote: "No methodological purist will be entirely happy with what Esman and Uphoff have done. The 150 cases are not a random sample drawn from the population of LOs. Instead we have a non-random sample of the population of published and ephemeral literature on LOs, produced by taking the first cases encountered. Thus, the probability theory which underlies most statistical analysis does not apply. There also is no quantitative way to draw inferences to the population of LOs, rather than to the literature, from this sample. Although the study is not statistically pure, it still is of interest. We would welcome a qualitative analysis of 150 case studies; I see no reason why the additional insights generated by quantitative analysis shouldn't be welcomed as well. Nevertheless it is imperative that [the authors] remain acutely aware of the limitations of what they are doing and that they continue to remind the reader of these limits."

well as to convey to readers in some detail what one of the more successful LOs is like. The 20 cases, which are listed in Table 3-3, were selected to give readers a representative view of the kinds of LOs we found in the literature and of the sources from which the cases were drawn.[34]

Assessment of Task Performance

The associations among performance of various tasks by the LOs in our sample are shown in Table 3-4. The scoring of task performance was done according to the definitions given in Appendix B. After considering the reported evidence from the case materials, each LO's operation in regard to these eight functions was judged and coded on a five-point scale.[35] After all the material had been considered and all specific performance measures coded, the cases were also categorized in terms of overall performance. The correlations of task performance with this summary indicator are also shown in Table 3-4.

It is not surprising that all the correlations are in a positive direction and relatively high. As noted above, one cannot interpret the absolute magnitudes of correlation unambiguously, because the data were themselves relative. Still, the differences in coefficient size warrant attention. Internal tasks (planning and goal-setting, conflict management) are generally more highly correlated with other tasks than are the external ones (control of bureaucracy and claim-making). Resource management is the task most highly correlated with other task performances on the average; conflict management comes next.

The correlations *between* the pairs of tasks described above—planning and goal-setting with conflict management, resource mobilization and management, service provision and integration, and control of bureaucracy and claim-making—are somewhat higher than other correlations, which suggests that we may be dealing basically with four general functions.[36] Even so, it is instructive to consider the eight tasks separately, since there appear to be some substantive bases for divergence of correlations observed within a pair when these occur; for example, conflict management is more closely associated with service provision than with service integration, and control

[34]In this set of cases, we have included a disproportionate number of LOs which scored highest (5) on overall perfrmance, so as to communicate to readers some of the substance of successful LO operation at the same time that we illustrate our methodology. The set of cases was constructed to give some balance among regions of the Third World and maximum spread of countries. The range of performance found for each type is also represented within the set.

[35]The coding was as follows: 5, quite effective; 4, effective; 3, average; 2, ineffective; 1, quite ineffective; 0, not relevant or no information.

[36]The correlations between these pairs as seen in Table 3-4 (0.71, 0.68, 0.63, and 0.62) are consistently higher than the average correlation of tasks in these pairs with other task variables (0.62, 0.61, 0.56, and 0.50 respectively).

Table 3-3. LO cases presented summarily in Appendix C

Country	Name	Type/Score	Source
Latin America/Caribbean			
Bolivia	Centers for Social and Economic Development	LDA/5	Gow/Development Alternatives, Inc.
Ecuador	Small Farmer Cooperatives	CO/5	Meehan/Inter-American Foundation
El Salvador	San Luis Reconstruction Committees	LDA/2	Huizer/American Friends Service Committee
Jamaica	Morningside-Delightful Buying Club	CO/1	Gow/Development Alternatives, Inc.
Mexico	Puebla Farmer Committees	IA/4	CIMMYT (international research center)
Peru	Anta Cooperative	CO/3	Roca/International Labour Organization
Venezuela	Federacion Campesina de Venezuela	IA/5	Powell/Harvard Center for International Affairs study
Africa/Near East			
Ethiopia	Sidamo Mahabar Associations	LDA/4	Hamer/anthropological field studies
Malawi	Self-Help Water Supply Committees	IA/5	Robertson/Ministry of Community Development; Liebenow/American Universities Field Staff
Niger	Village Health Care Teams	IA/1	Charlick/Cornell RDC study
Nigeria	Tiv Farmers' Associations	CO/5	Morss/Development Alternatives, Inc.
Tanzania	Sukuma Cotton Cooperative	CO/5	Lang et al./anthropological studies
Yemen	Bani Awaam Local Development Association	LDA/3	Swanson/Cornell RDC study
Asia			
India	Banki Water Supply Project Committees	IA/4	Misra/Planning, Research & Action Institute
Indonesia	Subak Irrigation Associations	IA/5	Birkelback/Cornell Southeast Asia Program
Nepal	Baglung Bridge Construction Committees	IA/5	Pradhan/East Consulting Engineers, Inc. and Cornell RDC study
Philippines	Banes Irrigation Association	IA/2	de los Reyes/Institute of Philippine Culture
Philippines	San Jose Credit Cooperative	CO/4	Hollnsteiner/Institute of Philippine Culture
South Korea	Mothers' Club of Oryu Li	IA/5	Kincaid et al./East-West Center
Taiwan	Farmers' Associations	IA/5	Stavis/Cornell RDC study; also others

Table 3-4. Relationships among performance of LO task variables

	Planning/ goal-setting	Conflict management	Resource mobilization	Resource management	Service provision	Service integration	Control of bureaucracy	Claim-making	Overall performance
Planning/goal setting	—	.71	.67	.72	.59	.54	.55	.53	.71
Conflict management		—	.64	.78	.70	.54	.61	.57	.75
Resource mobilization			—	.68	.61	.53	.47	.40	.71
Resource management				—	.71	.71	.57	.47	.80
Provision of services					—	.63	.57	.39	.69
Integration of services						—	.55	.34	.65
Control of bureaucracy							.52	.62	.68
Claim-making on government							—	—	.58

All coefficients significant at .05 level.

of bureaucracy is more associated than is claim-making with resource management. Such distinctions among tasks shed light on the way they relate to one another and, taken together, constitute a network of organizational outputs.

When considering how these different tasks relate to overall LO performance (see the last column of Table 3-4), we find internal and resource tasks somewhat more important than service and external tasks. The differences are only relative and not very great, but they are consistent and thus worth considering. This observation is supported by stepwise regression analysis, which we felt could be used at least in an exploratory way, since the task variables and overall performance had been scored around a central value, and the distributions were close to normal.[37] We wanted to see which tasks, when performed well as independent variables, would best account for overall LO performance as a dependent variable. The high intercorrelation (collinearity) of task variables shown in Table 3-4, meant that they could not be treated as making truly independent contributions. But we wanted to know which were most associated with overall performance when the influence of other task variables was statistically controlled for. Regression analysis permits inference of *ceteris paribus* influence that simple correlation cannot.

Resource management and mobilization proved to have the strongest relationship with LO performance in this mode of analysis. Planning and goal-setting, and control of bureaucracy were also significantly associated with overall performance.[38] That the other four task variables were not included in this equation does not mean that they had no influence but rather that their statistical associations with overall performance were not as strong in a *ceteris paribus* analysis, despite their high simple correlations.[39] Certainly all the tasks identified and analyzed here are interrelated, and little independent causation can be inferred for any task variable. Still, the picture that emerges from both simple correlation and multiple regression

[37]The average score for tasks was 3.4, with a standard deviation of 1.0.

[38]These four task variables together in a multiple regression equation produced an R^2 of 0.71. The beta weights and levels of signficance for this "best" equation, where all independent variables passed the 0.05 test of significance, were as follows:

Resource management	Resource mobilization	Planning/ goal-setting	Control of bureaucracy
.41 (.00)	.28 (.00)	.18 (.01)	.14 (.01)

[39]The effects of claim-making, for example, could be reflected in the coefficient for resource mobilization, as could the results of conflict management and service provision. The benefits of service integration could be represented statistically at least in part by the variable of resource management.

analysis is that *internal operations are somewhat more strongly associated with LO performance than are external tasks.* This same relationship holds for disaggregated LO performance (see the appendix to this chapter).

One should not make too much of any single number in this analysis. But taken together, the coefficients and scores reported here illuminate what might be described as the topography of relations among LO task variables and with the promotion of rural development objectives. Although our analysis has focused, for the sake of simplicity and relative brevity, mostly on aggregated performance, we have examined—as the reader may also want to do—the various dimensions of rural development performance. In the following appendix, we observe how the different LO tasks related to disaggregated outcomes. We begin the next chapter with a more detailed and theoretical consideration of performance indicators that represent improvements generally sought by rural people and the agencies seeking to assist them.

APPENDIX TO CHAPTER 3

Disaggregated Analysis of LO Tasks and Performance

Since we assessed discrete kinds of LO performance before classifying LOs in terms of overall effectiveness, we have a basis for looking at relationships between degrees of success in carrying out the various LO tasks addressed here and specific kinds of LO performance—economic, social, and political outcomes. This more detailed analysis will be of interest to readers with special concerns for certain aspects of rural development or those who wish to consider relationships beyond the summary, aggregated ones.

The performance measures analyzed encompass various developmental objectives and can be grouped into five sets of outcomes: economic gains, social benefits, equity effects, reduction in discrimination, and participation. The specific indicators and the criteria for scoring them are described in Appendix B and will be discussed more in the next chapter. The correlations between the different tasks and specific performance outputs are shown in Table 3-5.[40]

[40]Services are represented only by education and health outcomes; other outcomes in terms of nutrition, water supply, transportation, and public facilities did not have any coefficients significant at the .05 level, as there were a smaller number of cases where these outcomes could be observed and assessed. (If a case could not be scored on *either* variable, it was eliminated from the correlation.) The average number of cases for which the different outcomes could be correlated with tasks is shown in Table 3-5. The average number of cases for these other services was 12, 22, 19, and 13 respectively.

Table 3-5. Correlations of LO task performance with development outcomes

	(*N*)	Planning/ goal-setting	Conflict management	Resource mobilization	Resource management	Service provision	Service integration	Control of bureaucracy	Claim-making
ECONOMIC GAINS									
Agricultural production/income	(82)	.55	.54	.48	.66	.59	.54	.48	.39
Nonagricultural production/income	(30)	.35	.42	.50	.54	.47	.46	.14*	−.13*
SOCIAL BENEFITS									
Education	(39)	.38	.26*	.39	.24*	.44	.22*	.36*	.34
Health	(32)	.34	.30*	.18*	.32*	.55	.50	.16*	.25*
EQUITY EFFECTS									
Income for poor	(83)	.36	.50	.38	.36	.43	.33	.52	.37
Assets for poor	(29)	.29*	.50	.36	.11*	.37	.28*	.34*	.37
Access to services for poor	(91)	.47	.58	.48	.40	.53	.30	.50	.39
REDUCTION IN									
Sex discrimination	(30)	.43	.33*	.24*	.33*	.34	.34*	.31*	.37*
Social discrimination	(28)	.48	.59	.30*	.60	.43	.74	.42*	.53
PARTICIPATION IN									
Government decision-making	(112)	.37	.43	.36	.35	.26	.25	.47	.53
Community decision-making	(114)	.58	.60	.53	.50	.49	.41	.55	.45

*Coefficient not significant at .05 level; level varies according to number of cases for which scoring could be done.

For increasing agricultural or nonagricultural production and income, resource management appeared to be the most important, and claim-making the least. The task of service provision is understandably most highly correlated with educational and health outcomes, but service integration appears important for health and not for education. On the average, the other six tasks are less important for social improvements than they are for other kinds of gains. Achievement of equity effects is most associated with successful conflict management, though service provision and control of bureaucracy come close behind. We find resource mobilization (reflecting self-help efforts), planning and goal-setting, and claim-making well ahead of service integration or resource management with respect to the several equity improvements.

Achieving reduced discrimination against women is not correlated with any one task much more than another, but reduction in discrimination against ethnic or other socially disadvantaged groups is associated with service integration, resource management, conflict management, claim-making, and planning and goal-setting. Resource mobilization is not much associated with reductions in discrimination. It is interesting that both resource mobilization and resource management appear to figure more prominently in gains in participation at the government level than at the community level, a relationship we have observed qualitatively in the literature and in our own RDPP experience. This would be consistent with the strategy of assisted self-help discussed in Chapter 8. A community that mobilizes and manages its own resources well can thereby increase its bargaining power with government authorities.

These are the results of simple correlational analysis, which amplify and in some instances somewhat qualify the general pattern reported in this chapter. Because there are relatively high correlations among the task variables (average correlation, .59), apart from the fact that the data are not parametric, multiple regression analysis needs to be approached with some caution. On the other hand, precisely because of the high association among variables, one would like to determine what their relative contributions to performance might be if one could control for the other variables. This is precisely what multiple regression analysis is intended to do. Understanding that the analysis was exploratory—rather than being a test of hypotheses formulated from preexisting theory—we undertook it, using a statistical package that entered independent variables sequentially (stepwise) to explain variance in the dependent variable (the different performance indicators). We took as the "best" regression equation the one that offered the largest explanation of variance (highest R^2) with independent variables that were all significant at the .05 level. The beta weights for these variables represent the relative contributions to explanation of variance and, perhaps most important, indicate whether the association with the dependent variable (performance) is positive or negative for each independent variable (in this case, task).

In Table 3-6, we show the beta weights and significance levels for task variables vis-à-vis performance outcomes, and also the R^2 for the "best" equation, emphasizing in bold face the coefficients for those task variables significant at the .05 level or better. (Because one wants to look for *patterns* among variables, we have included also the beta weights for variables—not boldface—that had significance levels up to .33, which means they have a one-in-three chance of being wrong. Including them helps to fill in a picture of relationships; we leave out of consideration any that do not have at least a two-in-three chance of being correct.) As already noted, to the extent that one independent variable is associated closely with another, including it in the equation to explain the dependent variable reduces the amount of "explanation" it can add, and it is thus "demoted" in the implied hierarchy of causation that regression analysis constructs statistically. If a task here has a lower coefficient and significance level than in correlation analysis, it means that part of its implied causal effect has been assigned statistically to another variable (with which it is correlated and which has a stronger association with the variable to be explained).

In regression analysis, we find resource management the most important task for agricultural and nonagricultural production gains, similar to the results of correlation. One difference is that conflict management is negatively associated with agricultural production in regression analysis when other variables are controlled for, and it is nonsignificant even at the .33 level for nonagricultural production. Both internal and service tasks also lose significance for nonagricultural production in regression analysis, as does resource mobilization for agricultural production. It is worth noting that making agricultural gains is better explained (attains the highest R^2) than any other individual performance measure and by a broad range of almost all the tasks.

For social service gains, control of bureaucracy and claim-making show up somewhat more strongly in regression analysis (the smaller Ns for these reduced the confidence levels), and understandably, provision of services is very important. Resource mobilization is more evidently important for education than for health improvements, also an understandable difference (the correlation coefficients suggested the same difference, but regression analysis reversed the sign for health). Education was more influenced by planning and goal-setting, and by conflict management, *ceteris paribus,* than was health. That the highest beta weights should be provision of services for health and claim-making for education is reasonable. Integration of services was important for health outcomes in correlational terms but did not show up in regression.

Equity improvements for the poor were most strongly associated with

Table 3-6. Regression analysis of task variables on performance indicators

	R^2	Planning/ goal-setting	Conflict management	Resource mobilization	Resource management	Provision of services	Integration of services	Control of bureaucracy	Claim-making
Overall performance	.71	**.18** (.00)		**.28** (.00)	**.41** (.00)	.12 (.07)	.06 (.25)	**.14** (.01)	.08 (.13)
Agricultural production	.47	.13 (.19)	−.11 (.30)		**.50** (.00)	**.26** (.01)	.08 (.33)	.13 (.09)	.08 (.33)
Nonagricultural production	.26			.28 (.14)	**.51** (.02)				
Education improvements	.20	.17 (.22)	.18 (.19)	.20 (.13)		.25 (.07)		.22 (.12)	**.45** (.00)
Health improvements	.20			−.22 (.16)		**.45** (.00)		.16 (.26)	.23 (.10)
Income for poor	.23		.16 (.20)	.16 (.14)		**.30** (.00)		**.28** (.00)	
Assets for poor	.17	−.21 (.30)	**.42** (.01)		−.26 (.12)	.16 (.28)			.25 (.13)
Access to services for poor	.31	.14 (.20)	.20 (.07)	**.25** (.02)		**.37** (.00)		.14 (.10)	.13 (.11)
Reduced sex discrimination	.15	.23 (.15)	.22 (.14)	.14 (.33)	.18 (.22)	.27 (.06)	.16 (.29)	.21 (.17)	**.39** (.01)
Reduced social discrimination	.30		**.55** (.00)	.30 (.10)			.27 (.07)	.19 (.22)	.27 (.13)
Participation in government decision-making	.29		.09 (.32)	**.18** (.02)				**.17** (.04)	**.33** (.00)
Participation in community decision-making	.40	**.25** (.01)	**.21** (.01)	**.19** (.04)				**.15** (.04)	

R^2 shown for "best" regression equation, having largest number of independent variables significant at .05 level; the beta weights for these variables are in boldface; beta weights for variables having significance level up to .33 are also shown; significance levels shown in parentheses.

conflict management, as was indicated by correlation, with service provision also showing up strongly. Control of bureaucracy and claim-making were both important for improving access to services for the poor; but control of bureaucracy improved income for the poor, and claim-making improved asset distribution in an interesting distinction. Resource mobilization, reflecting self-help efforts, was important for both income and access to services for the poor, but less so for asset distribution. This outcome was associated negatively in regression analysis with planning and goal-setting and with resource management, and its R^2 was lower than most of the others, suggesting that factors apart from task performance are relatively more important for this outcome. The equations for income and for access to services are generally quite reasonable, with the main surprise being that resource management does not show up significantly, as it does for some other outcomes and for overall performance.

For reducing discrimination, conflict management is quite important, as are control of bureaucracy and claim-making. So is integration of services, though this may be more a matter of concomitance than of causation. That resource mobilization is not positive for reduced social discrimination is the only other surprise in this set of results. Reduced sex discrimination, interestingly enough, has significant association with all the task variables at the .33 level, though claim-making is the only one significant at the .05 level.

For participation outcomes, both resource mobilization (representing a degree of self-help) and control of bureaucracy (representing empowerment) are associated with gains both at community level and in government decisionmaking. Claim-making, as it did in correlation analysis, understandably showed up for enhancing participation in government affairs, whereas the internal tasks of planning and goal-setting and conflict management showed up more strongly for participation in community affairs. The relative importance of task variables appears somewhat different for certain disaggregated performance measures in regression compared with correlation analysis. But the patterns are generally quite similar, as they are also with the explanation of overall performance.

Another way of looking at the relative importance of internal tasks became apparent when we analyzed differences in task performance by region of the Third World. In general, we found LOs in Asia performing tasks better than those in Latin America and Africa. (The number of cases from the Middle East was not large enough to warrant disaggregated analysis; our "random walk" produced only four. This may reflect fewer LOs there or less coverage of such LOs in the English literature; in any event, our main concern was with structural, not geographic, variation.)

As explained in the next chapter, we developed a simple method for summarizing in a single score the distribution of degrees of performance for any set of LOs. This permitted comparing in a weighted way the distribu-

Table 3-7. Net scores of performance on task variables, by region

Task	Asia	Latin America	Africa
Planning/goal-setting	164	116	93
Conflict management	144	150	73
Resource mobilization	173	155	148
Resource management	152	128	91
Provision of services	182	195	139
Integration of services	139	165	110
Control of bureaucracy	106	38	3
Claim-making	140	104	80
Internal tasks (average)	154	133	83
Resource tasks (average)	163	142	120
Service tasks (average)	161	180	125
External tasks (average)	123	71	42
All tasks (average)	150	132	93

tions of LO task performance in each region. These scores are shown in Table 3-7, both by specific task and averaged according to the four sets of tasks. When the average of all tasks is taken, it corresponds quite closely to the average score for the internal tasks—planning and goal-setting, and conflict management. Interestingly enough, Latin American LOs come out ahead of their Asian and African counterparts in the area of service tasks. The performance of African LOs is relatively poor in both internal and external tasks. These numbers are descriptive artifacts of much larger sets of numbers, but they present in synthesized form some regional variations that are evident impressionistically from reading the cases studied and also the general literature.

CHAPTER 4

Performance and Environments of Rural Local Organizations

Just as rural development encompasses many different aspects of improvement in the productivity and well-being of rural people, so do local organizations—at least potentially—have multiple objectives according to which their developmental performance may be assessed. No single criterion of development can be persuasively advocated now that increased economic income per capita has been displaced as the proxy for development. An appropriate conception of development entails more than economic considerations, though as Meehan (1978) reminds us, it would be as much a mistake to neglect the economic aspect as to overemphasize it.[1]

Recognizing this, we developed a set of performance indicators that included economic, social, and political gains for rural communities. Some expressions of psychological or cultural benefit are included under these headings. The indicators were defined as shown in Appendix B and can be grouped into five major categories:

(A) *economic gains* through increased agricultural and/or nonagricultural production and/or income;

(B) *social benefits* in terms of education, health, nutrition, water supply, transportation, and/or other public facilities;

(C) *equity effects,* resulting from increased income, assets, and/or access to services for the poor;

(D) *reduced discrimination* on the basis of sex or other ascriptive categories, so as

[1]Drawing on his study of experience with local organization and people's participation in projects supported by the Inter-American Foundation (IAF), Meehan says: "Any program that neglects the economic dimension will almost certainly fail. Moreover, the populations receiving assistance are acutely concerned with and very strongly influenced by economic considerations . . . [however,] the influence of economic considerations is almost always limited, often in quite surprising ways. The organized poor will forgo economic benefits willingly and even enthusiastically for psychic, cultural and other considerations. The discernable benefits to the individual from such collective goods may be minimal" (1978:89).

to increase social opportunities for women and/or other disadvantaged social groups where discrimination has existed;

(E) *participation in decision-making* at the community and/or governmental level.

These variables represent the same range of outcomes—from efficiency improvements to equity to empowerment—that we discussed in Chapter 1. The specific measures are tangible expressions of three overarching objectives according to which LO performance may be judged. The LO "impact" measures used in the DAI study (Gow et al., 1979) paralleled these but were not as complete, nor did they distinguish between LO outputs and outcomes.[2] In his analysis of different kinds of *animation rurale* projects and rural development outcomes, Charlick (1984) deals with the same kinds of measures for evaluation of performance as did Gow and associates and as did we.[3]

Like the DAI researchers, we saw a need for some summary index of performance. Although it is instructive to look at disaggregated relationships, as we did in the appendix to the previous chapter, these can be overwhelming in their complexity and can also obscure overall relationships. Having scored each case on as many of the 15 disaggregated performance measures as there was documented evidence for, we classified each in terms of *overall performance,* an aggregated assessment.

This was not done by simply adding up the discrete performance scores, because such scoring would bias the judgment necessarily in favor of those LOs that provided the greatest number of benefits. Any reasonable judgment would take qualitative as well as quantitative effects into account. On the other hand, simply taking the *average* score for performance measures would give LOs no credit for producing a broader range of benefits. Some consideration of both the number and the value of economic, social, and political changes was in order, taking into account factors like economic

[2]Their measures of income changes (x_1), equity (x_4), and influence (x_5) correspond fairly closely to our categories (A), (C), and (E), while their measure of services (x_3) is really representative of service provision, not necessarily of benefits that result from these. Their measures of financial viability (x_2) and leadership (x_6) refer specifically to LO operations and not really to LO impacts. When these measures are included in a summary measure of "overall impact," much analytical clarity is lost. See Gow et al. (1979, I:219–29).

[3]Charlick's variable of "technical outcomes" covers the objective of efficiency as we defined it, including increases in income and improvements in health or literacy. His variable of "equity outcomes" corresponds to our second objective; it also includes access to benefits for women, which is the same as reduction in sex discrimination in our analysis. He has three participation or empowerment variables, referring to participation in implementation (mobilization of local resources and management talents), to empowered local participation, and to empowered extralocal participation. He adds two variables for institutional development that match fairly closely the DAI variables of "financial viability" and "leadership" (see footnote 2). Fortunately, Charlick did not sum all of these into one overall outcome variable, because the latter two are organizational outputs rather than outcomes.

returns, improved quality of life, enhanced self-reliance, special assistance to the poorer sectors, and expanded opportunities for those previously held back.

It might be thought that some of the specific performance measures would not be correlated and might even be inversely related. There has been, after all, a mainstream school of economic theory holding that productivity and equity gains are likely to be incompatible (though its credence has been declining; see, for example, Adelman et al., 1976, and Cline, 1975). And participation is regarded by some as endangering the stability necessary for economic growth (S. Huntington, 1968). Our data on disaggregated measures of development performance associated with LOs cannot be taken as proving the contrary, especially since they deal with outcomes at the micro level. But they are consistent with the emerging view in the literature that growth, equity, and participation are compatible. The simple correlations among performance indicators, and their respective correlations with overall performance, are shown in Table 4-2 (in the appendix to this chapter) for those who wish to inspect the relationships indicated. Of the 55 correlations, only three are negative (none significant at the .05 level); the average correlation is .35, indicating generally positive relationships among the different aspects of LO performance.

The five categories among which the cases were distributed are as follows:

5 = *outstanding,* multiple successes, no significant failures;
4 = *very good,* a number of successes or a few solid successes, no significant failures;
3 = *good,* some successes, few if any failures;
2 = *poor,* few successes, some obvious failures;
1 = *very poor,* no real successes, multiple failures.

These classifications were made with a broad-based conception of rural development in mind. The classification of cases was ultimately judgmental, but we think there will be general agreement among persons who read through all the documentation. The summary judgments did not rely on any particular weighting of the performance indicators, because such weightings could differ from person to person. Rather it reflected the breadth of efficiency, equity, and/or empowerment associated with a particular LO. The disaggregated indicators were not so much components of the overall performance measure as they were objective bases for comparing the scope and extent of development contributions associated with particular LOs.

Readers can get a better idea of these differences in specific cases by reviewing the summary descriptions offered in Appendix C. The distribution of cases turned out to be somewhat bell-shaped: category 5 had 23 percent; 4, 36 percent; 3, 43 percent; 2, 26 percent; and 1, 17 percent. As noted in Chapter 3, the universe of LO cases in the literature probably differs from the distribution of successful and unsuccessful LOs in the real world. We have no way of knowing what the "real" average score for LOs is, but it is surely less than 3.0.[4] This sample distribution is adequate for our purposes of analysis because we are interested in accounting for the differences in performance *between* the higher and lower categories. Even if the whole sample is not representative of the real world, the subsamples are reasonably representative of the LOs that would fall in each category. Analyzing the relationships they exhibit should provide insights as to what does or does not buttress the contributions of more successful LOs to rural development.

The Relationship of Environment to Performance

One of the most common concerns in the literature on the place of local organizations in rural development is the extent to which their environment, broadly defined, affects or even determines their ability to perform effectively. Very often, the issues are framed in terms of the hindrances or barriers that environmental factors may present. Hyden's analysis of cooperatives in East Africa (1973), for example, maintained that they were extremely weak vis-à-vis their sociopolitical environment and that their ineffectiveness was caused by their vulnerability to external manipulation. Other treatments have found certain environments conducive to greater success. Hildred Geertz's comparison (1963) of Balinese versus Javanese communities within Indonesia attributes the greater success of irrigation associations in the former to the surrounding social institutions.[5] One of the most graphic discussions of the subject suggests that the influence of the social structure and the values of a society "will show through, like a stain that cannot be painted out" (Hunter, 1976:202).

Few discussions treat the environment as either posing absolute constraints or guaranteeing good performance. For example, Johnston and Clark speak of the chances for effective LOs being "greatly enhanced wherever social conditions provide a relatively unstratified population, a rela-

[4]The average for our sample was 3.11 (S.D. = 1.26). In a paper presented to a workshop on water resources held at Los Banos in the Philippines, December 1976, Thomas Wickham and Kunio Takase suggested: "There are some extremely successful village water systems, but these are usually older ones with assured water supplies. There is a tendency, especially among those interested in social organization, to overlook the many unproductive village systems and to overestimate their actual operating performance" (cited in Robinson, 1982:47).

[5]The indigenous irrigation associations in Bali known as *subaks,* to which Geertz was referring, are among the more successful LOs we came across in our study (see Appendix C).

tively equitable distribution of assets, or a means of holding elites at least partially accountable in their leadership" (1982:168), while Hunter suggests that the kind of activity around which some form of grouping may be founded will be affected by the general conditions in the environment—high or low population density, settled or migrant population, irrigated or rainfed or semiarid agriculture, good or very poor physical infrastructure (1980a:3).[6] But beyond such qualified generalizations, we would like to have a more systematic and empirical understanding of the linkages, if any, between environmental factors and LO performance.

Some generalizations (noted in the previous chapter) based on cursory examinations of LO experience in Sri Lanka contended that good performance depended on, or was at least strongly influenced by, favorable endowment of natural resources and higher levels of literacy. But in the one case the sample on which the argument was based was very small (only four LOs), and in the other the observations had been made only impressionistically. Neither generalization was supported by Uphoff's larger and more systematic study in Sri Lanka, though the opposite relationships did not necessarily appear to be more assertable. One of the most rigorous analyses of relationships between environment and LO performance was done by Adelman and Dalton (1971). They found, with regard to community development programs in India involving *panchayat* local governments or LDAs, that higher population density was associated with greater success. However, when Charlick (1984) examined this environmental variable with regard to *animation rurale* programs in West Africa, no such association was evident. Did the divergence in findings result from studying different kinds of LOs, or from looking at LOs in different parts of the world? One cannot say.

To get a better grasp on such questions, it is useful to examine environmental variations and LO performance for a full range of LOs across different parts of the Third World. The studies by Gow et al. (1979) and Charlick (1984) each had a sample of 40 LOs, in seven and five countries respectively. They reached the general conclusion that environmental factors are neither often nor strongly associated with performance, measured in terms similar if not identical to ours. This is essentially our conclusion also, based on a much larger and wider-ranging sample. (We will explore this conclusion and the reasons undergirding it later in the chapter.) Such a finding, if correct—and we think it is—has substantial implications for development strategy. It suggests that environmental constraints are not as determinative as may have been thought by academics and policy-makers alike. Because this is such an important issue, and any generalizations about the environ-

[6]This is argued also by Young, Hebert, and Swanson (1981) on the basis of investment patterns of local development associations in North Yemen; they found road construction much more common in some environments than in others, for example.

ment need to be examined carefully, we will first consider how to test them, which factors are the most prominent in the literature, and what our data indicate.

Analysis of Environmental Variables

Obviously, one cannot deal with all the possible aspects of the environments of LOs. Sets of factors that are prime candidates as influences on LO performance need to be identified, defined, and then assessed in a variety of circumstances to see what relationships hold, if any. Unfortunately, a number of potentially important factors cannot be studied because of limitations on research resources or methodology. The psychological environment of local organizations, including "expectations for the future," could very well affect the willingness of members to contribute resources. Survey questions inquiring whether people think their lives will be better, worse, or the same in five or ten years can tap this variable, have been widely used, and are by now quite standardized. But such data were reported in only three of the case studies and then not in identical form. The distinction elaborated by Shadid, Prins, and Nas (1982:44–46) between "high-trust" and "low-trust" environments, with evident implications for LO operation, is easier to state analytically than to measure, and in any case would require field studies at great expense. The same is true for what Dore (1971) calls "propensity to organize."

Some environmental factors identified in the literature are too complex to be operational—such as "high solidarity," which Inayatullah has characterized in his UNRISD study as involving homogeneity of belief, uniformity of status, flexibility in social structure, predisposition to collective action, extensive external exposure, and an activist orientation to nature (1972:268). If all of these occur together, one may indeed have an environment where cooperatives, for example,will function well. But these characteristics may diverge in specific situations, in which case one would have to deal with them separately instead of as a highly aggregated factor like "solidarity."

Other factors may be limited in extent and thus not amenable to comparative analysis. Seibel and Massing (1974) make a good argument, supported by data on cooperatives in Liberia, that co-ops perform better where the indigenous social structure has been organized along segmentary rather than hierarchical lines. But this generalization would be difficult to test in situations where principles of "tribal" organization had been attenuated in people's consciousness. Applying it in a country like India would be difficult because the social system of caste relations has elements of both segmentation and hierarchy.

Obviously, there are limits on how broadly and how systematically one can examine environmental influences on LOs. What we have done is to identify a set of factors that have been asserted or speculated on in the literature, each one representing a hypothesis about environmental influence on local organizations. Not all possible factors could be studied, but we selected a representative set of 18 environmental variables that we thought could be made operational, that were believed to be important in the literature, and that could be scored on the basis of information from the case materials or other knowledge about the area.

The general categories of environmental variables studied were three:

Physical-Economic

(1) Topography: how little or how much difficulty it posed for travel and movement of goods;

(2) Resource Endowment: how ample or poor were the natural resources available, particularly for agriculture;

(3) Infrastructure: how adequate or inadequate were facilities for transportation, communication, and so forth;

(4) Economic Diversification: how small or large a proportion of the labor force was engaged in agriculture;

(5) Income Level: how high or low was per capita income;

(6) Income Distribution: how unequal or equal was the distribution of income and assets, particularly land.

Social-Societal

(7) Settlement Pattern: how compact or spread out were village/household units;

(8) Social Heterogeneity: how diverse or uniform was the society in terms of ethnic, religious, language, or other characteristics;

(9) Social Stratification: how hierarchical or egalitarian were social relations between groups;

(10) Sex Discrimination: how much or little restriction there was on economic, social, and political opportunities for women;

(11) Social Discrimination: how much or little restriction there was on opportunities for disadvantaged groups;

(12) Literacy: how large or small a proportion of the adult population was literate.

Political-Administrative

(13) Partisanship: how factionalized or not were group relations with regard to decision-making;

(14) Group Patterns: whether the bases for group action were voluntaristic (associational), paternalistic (patron-client), immediate kinship (familistic), broader kinship (clan or tribe), or broadly ascriptive (caste);

(15) Community Norms: how supportive of participatory/egalitarian outcomes were the prevailing norms within the LO's community;

(16) Societal Norms: how supportive of participatory/egalitarian outcomes were the prevailing norms within the society at large;

(17) Political Support: how positive or negative was the government's orientation toward the LO;
(18) Administrative Support: how willing and able or not was the bureaucracy to assist the LO.

These variables were scored for each case on a five-interval scale (a score of 5 represented the end of the continuum stated first in the descriptions given above), according to the criteria listed in the protocol in Appendix B. Where the case materials did not give specific numbers or descriptions, data were sought from other sources to fill in the information. Because the scoring was by fairly gross categorizations, precise information on factors like per capita income or community norms was not needed. While more precision would have been desirable, to have required it for all variables would have greatly reduced the number of cases we could analyze. The trade-off was between precision of measurement and comparison on one hand, and the scope and extent of experience considered on the other. We opted for the latter.

The simple correlations among these environmental variables and between the variables and overall performance are given in Table 4-1. Panels (a), (b), and (c) present the correlations within each of the three sets; panels (d), (e), and (f) show correlations across sets. The direction and magnitude of correlations that would have been predicted, such as the high associations of social heterogeneity, stratification, and sexual and social discrimination in panel (b), or between these and unequal income distribution in panel (d), give face validity to the operationalization of these variables, as do the low correlations for group patterns in panel (f).[7] With this background, we will consider in turn each of the environmental factors treated in our study.

(1) *Topography*. We began with this variable partly because it was the most unambiguous, and we will discuss the methodology of analysis briefly to indicate how the other variables were also dealt with. Even if they were not reported for each case, the terrain and the extent of mountains, rivers, and other natural features that might impede physical movement and thereby affect the formation and functioning of LOs could be readily determined. Almost three-quarters (73 percent) of the cases were scored as 3, a neutral value, where nothing in the case materials or in our knowledge of the country or region indicated anything unusual about the topography. In

[7]Group patterns (14) were scored as if they represented ordinal variations, but the categories are distinct and do not necessarily represent a continuum. The regional association of voluntaristic and particularly patron-client patterns with Latin America, which has higher per capita income and greater economic development generally, affected the correlations for this variable with economic diversification and income level in a spurious manner.

22 percent of the cases, there was indication of somewhat unfavorable conditions for travel or communication; there was just 1 percent at either extreme; and 3 percent were reported as somewhat favorable. The hypothesized relationship would associate unfavorable topographical features with poor LO performance.

The distribution of cases (in percentages) according to overall performance, divided between those with unfavorable and those with neutral or favorable topography, is shown below.

	5 (outstanding)	4 (very good)	3 (good)	2 (poor)	1 (very poor)
Unfavorable topography (1 + 2) (N = 34)	29%	24%	26%	15%	6%
Neutral or favorable topography (3 + 4 + 5) (N = 116)	13%	22%	29%	21%	15%

The relationship between topography and overall LO performance is in the opposite direction of what was expected. This is expressed as a correlation coefficient of −.15 (see Table 4-1).[8] Since we are using statistics here descriptively, rather than to test a hypothesis formally, the correlation is interpreted as simply indicating both the direction of association (positive or negative) and the relative magnitude of association. Another way of comparing the two distributions is to calculate a *summary score* for each; it comes out to 134 for the LOs with less favorable topographical surroundings, compared with 61 for those with neutral or favorable topography (the average net score for the sample as a whole is 77).[9]

According to our test of statistical significance, set at the .05 level, we would say that there is no association between favorable topography and LO performance; a correlation of this magnitude, given the sample size,

[8]This is significant at the .07 level of confidence, not the .05 level used here according to common convention. The significance level for this distribution using chi-square, a nonparametric statistic, is .14 (x^2 = 6.85, df = 4), not much different from that for r.

[9]The method for calculating summary scores was to add together the highest category triple-weighted, the next highest category double-weighted, and the middle category unweighted (since it is the modal value), and to subtract from this sum the next category (which is negative) and double the final category (which is more negative). The calculations for these two distributions are as follow:

Unfavorable topography: 87 (3 × 29) + 48 (2 × 24) + 26 − 15 − 12 (2 × 6) = 134
Neutral/favorable topography: 39 (3 × 13) + 44 (2 × 22) + 29 − 21 − 30 (2 × 15) = 61

Such scores represent how positively or how negatively skewed a distribution is. *An even distribution with 20 percent in each category would produce a summary score of 60, close to the latter score here, which reflects an almost symmetrical distribution of 13, 22, 29, 21 and 15 percent.* We experimented with several other methods for weighting and summing the distributions, but this one proved to be both sensitive to and representative of differences in distribution, capturing such differences in a single score.

Table 4-1. Correlations of environmental variables and overall LO performance

(*a*) PHYSICAL-ECONOMIC VARIABLES

	Overall performance	Topography	Resource endowment	Infrastructure	Economic diversification	Income level	Income distribution
Topography	−.15	—	51*	.60*	.26*	.13	−.04
Resource endowment	−.03		—	.68*	.39*	.31*	.14
Infrastructure	−.03			—	.42*	.42*	.00
Economic diversification	.03				—	.38*	−.03
Income level (GNP)	.11					—	−.05
Income distribution (unequal)	.03						—

(*b*) SOCIAL-SOCIETAL VARIABLES

	Overall performance	Settlement pattern	Social heterogeneity	Social stratification	Sex discrimination	Social discrimination	Literacy
Settlement patterns	.01	—	−.04	−.04	−.11	.01	.23*
Social heterogeneity	.18*		—	.71*	.50*	.65*	−.16
Social stratification	.20*			—	.64*	.68*	−.09
Sex discrimination	.06				—	.73*	−.34*
Social discrimination	.14					—	−.31*
Literacy	.08						—

(*c*) POLITICAL-ADMINISTRATIVE VARIABLES

	Overall performance	Partisanship	Group patterns	Community norms	Society norms	Political support	Administrative support
Partisanship	−.03	—	.12	−.07	.14	−.03	−.02
Group patterns	−.01		—	.30*	−.04	−.03	.03
Community norms	.20*			—	.20*	−.09	−.07
Societal norms	.12				—	.21*	.11
Political support	−.02					—	.68*
Administrative support	.14						—

Table 4-1. Continued

(*d*) Physical-economic and social-societal variables

	Settlement patterns	Social heterogeneity	Social stratification	Sex discrimination	Social discrimination	Literacy
Topography	.36*	−.17*	−.24*	−.20*	−.23*	.29*
Resource endowment	.25*	.05	.03	−.12	−.08	.36*
Infrastructure	.23*	−.01	.00	−.14	−.18*	.52*
Economic diversification	.16	.01	−.07	−.19*	−.17*	.45*
Income level (GNP)	.16	−.20*	.00	−.34*	−.37*	.69*
Income distribution (unequal)	.14	.57*	.70*	.50*	.48*	−.21*

(*e*) Social-societal and political-administrative variables

	Partisanship	Group patterns	Community norms	Society norms	Political support	Administrative support
Settlement patterns	.04	.10	.12	.01	.02	.02
Social heterogeneity	.25*	−.27*	−.38*	.11	.03	.05
Social stratification	.31*	−.10	−.34*	−.17*	.00	.06
Sex discrimination	.02	−.45*	−.41*	−.18*	.15	.03
Social discrimination	.06	−.37*	−.35*	−.09	−.07	−.07
Literacy	.16	.48*	.21*	.25*	.11	.17*

(*f*) Political-administrative and physical-economic variables

	Topography	Resource endowment	Infrastructure	Economic diversification	Income level	Income distribution (unequal)
Partisanship	.15	.20*	.16	.13	.24*	.35*
Group patterns	−.05	.00	.09	.17*	.68*	−.10
Community norms	−.11	−.20*	−.15	.00	.20*	−.42*
Societal norms	.21*	.08	.21	.15	.02	−.23*
Political support	.20*	.14	.17*	.10	.08	.02
Administrative support	.23*	.19*	.25*	.18*	.13	.03

*Significant at .05 level.

could occur by chance one time in 14, not the one in 20 required with the .05 level. These odds do suggest that a negative correlation is more likely than the hypothesized positive one. Examination of this environmental variable in a number of other connections indicates that it tends to be negatively associated with LO performance measures, a surprising pattern. In any such statistical analysis, *patterns* of association are more important than any individual statistic. Of course, correlation does not demonstrate causation; thus one could not argue on the basis of the data that adverse terrain *leads* to more effective LOs. We will see in our subsequent analysis that a number of environmental conditions generally thought to be undesirable correlate with better LO performance, presenting a curious but also encouraging situation to be explored further.

The relationship reported here may be more representative of this sample than of the rest of all possible cases, since it includes a number of LOs in mountainous areas (the Andes and Himalayas in particular) that were quite successful. We cannot extrapolate directly from the relations in these 150 cases to the real world because we cannot establish the randomness of the sample. There would be no point in arguing that adverse topography is conducive to LO success. We would be satisfied with evidence that it poses no insuperable barriers to success, since we are not just interested here in what is probable but also in what is possible. The correlation suggests, as do the net scores, that LOs facing difficult physical circumstances *can* (not necessarily will) perform as well on the average as local organizations more benignly located.

(2) *Resource Endowment.* A more interesting and significant environmental factor, one debated more vigorously in the literature, is the extent to which a beneficent natural resource endowment contributes to successful LOs. Hyden (1970) in his study of cooperatives in Kenya found the least successful ones in poor, remote areas with low agricultural potential. H. Lewis (1971), on the other hand, attributed the success of the indigenous *zanjera* irrigation associations in the northern Philippines (similar to the *subaks* in Indonesia) to the comparative difficulty of life and of obtaining irrigation water in their area. Apparently, LOs can be effective in spite of sparse endowment; perhaps they are spurred on by unfavorable conditions. Indeed, Adelman and Dalton (1971), as a result of their study of community development in India, suggested that the response to self-help opportunities may be greater in poorer communities. This might mean that more poorly endowed communities would develop more effective LOs, reversing the proposition advanced by Fernando (1977) in his study of Sri Lanka cooperatives.

Our analysis of resource endowment, assessed in terms of soil and climate advantages that make agricultural production easier and more profitable, did not find any association with overall LO performance ($r = -.03$). Neither did the DAI analysis (Gow et al., 1979, I:108). Their measure of potential for increased agricultural productivity "did not correlate significantly with any of the impact measures."[10] Charlick (1984) did not study this environmental factor directly, but his variable of economic diversification reflected exploitation of more favorable natural environments for agricultural cash cropping. It did not correlate significantly with his LO efficiency or equity measures, and only weakly with two of the empowerment measures—ostensibly for the reasons that Gow and associates suggest: "A favorable environment tends to attract interventions in the form of development projects. In an unfavorable physical environment, a local organization may be able to attract and provide scarce resources which would otherwise be unavailable. In addition, it is precisely in such an environment that the rural poor are most likely to be found" (1979, I:108). As this last observation points out, there is more scope for LO success where conditions are less favorable. But equally important, local residents in such disadvantaged circumstances may conclude that if they do not help themselves, nobody else will. This is a hypothesis that we will have reason to return to again. Certainly some specific connections between resource endowment and LO performance can be identified, as King (1981) has done for half a dozen cooperatives operating in northern Nigeria. Resource endowment can be very important in microcontexts, but these vary greatly, as King documents, reducing one's ability to predict LO performance from a variable like resource endowment; how people respond to favorable and unfavorable conditions will differ.

(3) *Infrastructure.* A number of the environmental factors are themselves correlated; infrastructural development is rather closely associated with both topography and resource endowment, as shown in panel (a) of Table 4-1. Our concern with infrastructure was first focused by the Philippine case studies done for our comparative analysis in 1974. Simpas et al. (1974:61–109) looked at LOs in four rural towns in two different provinces. Panitan, which had been the more developed of the first two in terms of infrastructure and amenities when Luykx studied it 15 years earlier, did not have more active or effective local government than Jamindan, the "backward" com-

[10]Of the 15 LOs that demonstrated low resource potential (out of the 41 in their sample), the rankings in terms of overall impact ranged from first to fortieth. Illustrative of how natural circumstances can be overcome is the Portland–Blue Mountain Coffee Cooperative Society in Jamaica, which Gow and associates studied and which we included in our analysis, scoring it as a 5. They describe it as "an example of successful organization in a generally unsuccessful agricultural environment" (1979, I:67–68). It succeeded through good leadership and extremely capable management, which helped members get premium prices for the especially flavorful coffee they were able to grow and instituted a rigorous grading system and efficient processing plants.

munity with which it was compared. The local organizations in San Quintin, located near a large city and enjoying easy access to institutions there (as well as having a more literate and relatively "modern" population), were languishing compared with those in Sallapadan, located in the mountains with little access to the outside world (and having lower income and literacy levels). Remoteness and lack of infrastructure in Sallapadan did not appear to be a barrier but rather a stimulus to effective LO activity, spurred by a community-minded priest. San Quintin, despite its better facilities, appeared to suffer the backwash effect of being so close to an urban area. The transportation and communication opportunities seemed to draw away local leadership and to orient most LO members to the city rather than the community where they lived. In this case, good infrastructure was not even a mixed blessing, but rather a detriment, when it came to mobilizing local resources for self-development through LOs.

Perhaps because good infrastructure can have supporting or detracting effects for local organization, the correlation we observed in our sample was −.03, not at all significant.[11] The possibility that this and some other environmental factors show essentially no correlation, because they can have *both* positive and negative implications, will be taken up later in the chapter.

(4) *Economic Diversification*. One can argue, on the basis of different strands of sociological theory, that stronger or weaker LOs can be expected with increased diversification of the economy. With diversification—some say—comes differentiation of occupational and social roles, a more "modern" outlook, and greater interdependence according to the division-of-labor paradigm, contributing to the establishment and maintenance of voluntaristic organizations to solve emergent problems. On the other hand, those who stress the value of community solidarity for local organization see diversification as breaking down the preexisting bonds and values that can make for vital, if relatively more traditional, associations. Interestingly, Whyte and Alberti (1976:230) found no significant correlation between the level of structural differentiation and the degree of cooperation in the dozen Peruvian communities they studied over a five-year period.

Environments were scored on this variable according to the proportion of the labor force engaged in agriculture, ranging from (1) over 80 percent to (5) under 20 percent. Since the rural environments being studied were largely agricultural, the distribution was rather skewed, with 80 percent falling in the least diversified category. Still, no significant correlation (.03) was found, and the summary score for LOs in areas with more than 80 percent in agriculture (N = 113) was 75, compared to 81 for those with less than 80 percent (N = 37). The study by Gow et al. (1979) found a similar

[11]The summary score of overall performance for LOs scored as 1 or 2 on infrastructure (N = 59) was 82, compared with 72 for LOs rated as 3, 4, or 5 (N = 91)—essentially no difference.

lack of correspondence (.05) between subsistence vs. commercially oriented agricultural economies on one hand, and overall LO impact on the other. Charlick (1984) found no significant correlation between his measure of economic diversification and either his technical (efficiency) or equity outcome variables, and only a weak correlation with two of his empowerment variables: he found somewhat better performance where there was *less* diversification.[12] This environmental variable, like the previous one, may cut both ways. We find no evidence that, on the average, it has an effect on LO performance one way or the other; neither does it show significant correlation with specific LO performance measures.

(5) *Income Level.* As with economic diversification, one can imagine ways in which higher or lower income levels might be associated with better LO performance. Where financial resources are more abundant, members will have more available to contribute to LO operations, and their opportunity cost of doing so will be less. On the other hand, with lower income levels, members may have greater need for LO performance and may value more the outcomes associated with LO activity. So the effects could work in either direction, or both.

One of the most interesting empirical analyses bearing on this question was done in Thailand by the American Institutes for Research (1973) from the mid-1960s to the early 1970s.[13] The analysis distinguished between individual investments which benefited only the investor (private goods) and community investments which benefited all (public goods): for example, investments in improving rice production compared with those for improving roads. The latter involve—even require—LOs, while the former do not. The AIR study found that the ratio of individual vis-à-vis community investment went up in villages as the level of income (and of income disparity) rose. The salience of LOs and their activity would correspondingly decline. To the extent that LO activity is producing collective goods on a self-help basis, it would be most relevant in poor communities with low levels of public facilities and services that could be increased by group action. This kind of relationship has been observed and reported with regard to the *harambee* self-help campaigns in Kenya by B. Thomas (1980). Such efforts are more needed at a lower stage of economic development than later on, when the resource and technology requirements of additional improvements are greater. One agreeable implication of this, if true, is that encouraging local self-help organization would benefit poorer communities relatively more—other things being equal. Of course, other things that are not

[12]The correlation with locally empowered participation was significant at the .05 level, and with extralocally-empowered participation, only at the .10 level.

[13]The specific study referred to here covered 49 villages in the north, where interviews were done during 1972–73.

equal are the economic resource base and the extent of outside support they can draw on. Thomas found in Kenya that richer communities gained more from self-help campaigns, in large part because they were better able to tap outside resources.

We would like to have had as detailed data to work with cross-nationally as AIR and Thomas had for Thailand and Kenya. Unfortunately, to get greater cross-national breadth of coverage, our measure had to be fairly gross. Still, the fact that the observed correlation is only .11 suggests a very ambiguous general relationship. In the context of specific countries, and with regard to certain tasks or kinds of organization, a significant relationship (one way or the other) between income level and LO performance may appear when measures are refined. But overall, we found no such relationship, and none appeared when we compared income level with specific LO performance measures, using either correlation or regression analysis (see the appendix to this chapter).

(6) *Income Distribution.* Even though the level of per capita income does not show up in our analysis as significant, one might expect income distribution to do so. In our earlier comparison of *systems* of local organization (1974), we found a definite association between more equitable distribution of income nationally and the effectiveness of LO systems. One would expect this to hold true for the performance of specific LOs. Reflecting others' work as well as our own, Johnston and Clark (1982:168) state that relatively equal distribution of assets—particularly land, which translates into corresponding distribution of income—will increase the chances for effective local organization.

A countervailing influence can be proposed, inasmuch as the leadership (and problem-solving) function in local organization, may be enhanced by unequal distribution of assets and income (Gow et al., 1979, I:232). It can be objected that much of the leadership which comes forward from inegalitarian conditions will not be supportive of developmental changes that include the poor in new streams of benefits. This may or may not be true. It is evident that unless there is leadership of some sort, the benefits flowing from LO activity are not likely to be very great, for the poor or anyone else. In communities where all are relatively equal, it may be difficult for members to vest leadership status, and the advantages that usually go with it, in a peer. The acceptability of egalitarian leadership may itself be something that varies among different sociocultural environments. At least we should entertain the possibility that some degree of inequality can contribute to LO performance. Such circumstances may also provide more impetus to interest associations that seek to promote the well-being of less advantaged groups, in conflict or at least competition with those benefiting from an unequal status quo.

We could get specific income distribution data from only a few case studies, but we could classify environments into one of five ordinal categories ranging from relative equality to high inequality (see Appendix B). If there were a real relationship between income distribution and LO performance—even with some errors in scoring—we would not expect a correlation as low as .03, which was our result. The DAI study measured distribution of land rather than income (in rural societies these are closely related), hypothesizing that the more equitable the distribution of land, the greater the overall impact of the LO intervention. Nevertheless, Gow et al. (1979, I:107) report that "this factor did not correlate significantly with any of the impact measures." The half of their 41 cases classified as having an equitable distribution of land at the time of intervention ranged in their overall success ranking from second to forty-first: "This would suggest that an equitable distribution of land, while highly desirable for many reasons, is not a necessary condition for a successful intervention."

We believe that while our earlier observation about an association between relative equality and LO performance at the macro level probably remains correct, the relationship does not necessarily hold at the micro level. Many others have supported our earlier observation, but our project-level findings are consistent with those of Gow et al. and Charlick. At the local level, the fact of inequality might give impetus to the poor to make compensatory efforts through organization to advance their productivity and well-being. This might (we cannot say will) offset the negative influences on local organization of an inegalitarian environment. It is worth noting that the most successful LO that Gow and associates studied—a cooperative at San Martin Jilotépeque in Guatemala—grew up in one of the poorest areas of that country's highlands, where economic inequality was pronounced and 60 percent of the cooperative's members were landless. Yet it grew from 32 members in 1972, with $177 in capital, to 732 members six years later, controlling $38,000 in share capital. The group received assistance from Oxfam and World Neighbors, but its leadership came from within the community. Not only did the co-op increase corn production, but it also introduced soil erosion control, a future-oriented task often thought beyond the capacity or interest of poor people (1979, I:52–53). There are enough cases like this in our sample that despite possible negative linkage between inequality and LO performance, the success stories in the midst of inequality produce a nil correlation.

(7) *Settlement Patterns.* It has been an article of conventional wisdom that introducing and sustaining organizations is more difficult where people live in scattered homesteads or hamlets and, conversely, easier where they have clustered settlement patterns. Linear patterns, as along roads or rivers, would be intermediate between dispersed and concentrated living areas.

The idea behind this generalization is that people will have less sense of common identity if they do not dwell together, and that meeting and working together will involve greater costs in time and effort. As Silberfein (1982:10) correctly points out, however, the form of settlement—clustered or dispersed—does not necessarily predetermine the amount or kind of social interaction. Contacts between dispersed relatives or members of a community can be regularly maintained, especially during the dry season when cultivation is least demanding. Moreover, a clustered village setting is more likely than a solitary homestead to be the site of interpersonal conflicts, witchcraft allegations, and other antisocial behavior.

Scoring this variable was difficult because reference was seldom made in case studies to settlement patterns, and these are not as ascertainable from secondary sources as is topography or infrastructure. Still, the observed correlation with overall performance is so negligible (.01) that scoring errors are unlikely to have obscured much of whatever relationship exists. It may well be that in specific countries and with regard to certain kinds of LO activity, settlement patterns have some influence. But we did not find any apparent effects across the full set of cases studied.[14]

(8) *Social Heterogeneity.* This factor is thought to have a negative impact on LO functioning, on the grounds that diversity of ethnic, religious, or other groups within a community makes for more division and conflict. Conversely, social *homogeneity* is regarded as promoting unity and cooperation. Social homogeneity is a major aspect of the variable of "high solidarity" proposed by Inayatullah (1972). It is hard to construct a counterargument, that heterogeneity might be supportive of better LO performance, but that is what our statistical analysis suggests. The correlation of .18 is not very great, but it is positive and passes our test of significance, as few others did.

The result is puzzling unless viewed in the context of other relationships that emerge from within our sample, particularly the positive association that social stratification also shows with overall LO performance.[15] To be sure, the coefficient would account in statistical terms for less than 4 percent

[14]If there is a positive association between more concentrated settlement and LO performance, it may be offset by "exceptions" such as reported by King (1981:270) for communities in northern Nigeria. The most dynamic cooperative of the six he studied was in a community described as "a lengthy straggle of compounds," unlike others which had "a tight concentration of buildings within walled boundaries . . . symbolic of a community of very different character."

[15]Social heterogeneity and stratification are highly correlated (.71), as seen in Table 4-1, panel (b). When environmental variables were put into a stepwise regression, we found that social heterogeneity was statistically less important than social stratification as a factor to explain variance in overall performance. These variables are linked, inasmuch as there cannot be social stratification without some social heterogeneity. But with heterogeneity there can be varying degrees of stratification, from low to high.

of the variation in LO performance. We do not ignore the observed relationship, however, because it is higher than most others and because the result is consistent with the larger pattern of relationships emerging from analysis of the sample. What can be conservatively inferred from the data is that social heterogeneity per se does not appear to be a critical barrier to LO success. This can be seen in the set of six cases in northern Nigeria studied by King (1981), where the most heterogeneous community had the most successful cooperative.

(9) *Social Stratification.* This environmental variable is much discussed in the LO literature, with a strong consensus that stratification makes effective local organization less likely, certainly with regard to the kinds of development that will benefit the poor rural majority.[16] An opposite view is occasionally voiced with regard to LO promotion of technical objectives. For example, Fresson (1979) argues that associations in a Senegalese irrigation scheme had more success where stratification was greater because leaders had more control over members. But more often, because such stratification usually means that "traditional" leaders will be in control, the common suspicion is that the results will not be very desirable for the poor (Dore, 1971).[17]

Given the general view in the literature, we were not expecting the positive and statistically significant correlation (.20) observed between the degree of social stratification and LO performance. Gow and associates (1979) did not analyze this variable specifically, and Charlick found that none of his impact measures was significantly correlated with his measure of stratification. His interpretation of the data was as follows: "Although animation projects seem to produce better overall outcomes where the level of local social stratification is low, in some instances the best projects can be found where there is a high degree of political and social inequality at the local level, as is the case, for example, in rural Haiti" (1981:24). In our sample, we also found a number of LOs where success was attained in spite of a high degree of stratification in the local environment. There were enough such exceptions to the apparent general rule that the correlation was in the opposite direction from what was expected. The coefficient is not very

[16]Johnston and Clark (1982:168) identify "a relatively unstratified population" as one of the main factors conducive to effective local organization. Leonard (1982a), in analyzing how LOs can serve the rural poor, treats stratification as one of the major hindrances. The same view is offered by Meister (1970), Inayatullah (1972), and Hunter (1976). Our own work (1974), which analyzed relationships in terms of income distribution rather than social stratification (the two are closely related) reached the same general conclusion.

[17]Charlick (1984) found that the weakening of traditional political roles and institutions did not correlate significantly with his efficiency or equity measures, but did with his participation and LO development measures. This finding does not disprove but neither does it support the conventional view of traditional leadership.

great, even if statistically significant. What is substantively significant is that the data in neither Charlick's study nor ours supported the usual generalization.

We would not argue that greater social stratification as such contributes to LO success or that it is a favorable condition. Our sample may have contained a higher proportion of cases like the ones Charlick reports in Haiti, where LO leaders and members were able to prevail over social rigidities and resistance, than would be found in the "real world." There may well be instances where social stratification is a barrier to LO progress. But on the average it does not appear to have such an effect. We are prompted to treat the observed correlation between stratification and overall LO performance seriously because more often than any other environmental factor, stratification was significantly related to specific LO outcomes.

Two analyses coming from the working group on "participation and access" at the University of Leiden may help to account for the observed relationship. Stratification is a very complex variable, with effects resulting more from the way it is perceived and acted upon than from the fact of stratification itself. Grijpstra (1982a) describes four possible situations with regard to stratification: (a) no large social differences exist; (b) they exist, but the poor wish to follow the example of the rich; (c) they exist, but the poor do not feel exploited; (d) they exist, and the poor do feel exploited. In situation (b) the possibility of mobility mitigates the effects of stratification (the extent of immobility was one of the criteria we used for judging the degree of stratification, the other being social distance between groups). In (c) sociocultural norms legitimize stratification and make the situation different from (d), the stereotyped image of stratification. Boer (1982:155) suggests that conflict with other groups in the LO's environment may produce greater solidarity among its members and encourage them to more energetic and sustained effort.[18] "Polarization" within the environment, setting the group apart from and probably at odds with other groups, does not guarantee success, to be sure. But we can see why LOs in a nonconflictual situation might fail to attract and hold strong member commitment.

(10 and 11) *Sex and Social Discrimination.* We have not found much discussion of these environmental factors in the literature—the extent to which social norms and structures derogate the status of women or other social groups—but they are recognized as possible impediments. These factors were difficult to score because there was little reference to them in the case studies. Still, norms and structures of this sort do not originate in specific

[18]He also cites Hyden (1981a) to the effect that the absence of an "obstructing power" may be a disadvantage rather than an advantage when setting up viable rural LOs.

communities, but in the society at large; thus it was possible to score cases according to five ordinal categories based on knowledge of the larger societies involved. As with most of the other environmental variables, no significant relationships with performance, overall or specific, were found in either correlation or regression analysis, contrary to our expectations. We cannot say that there is no relationship, but we found no indication of it.

(12) *Literacy*. Our interest in this factor was stimulated by a statement in the Report of the Commission on the Kandyan Peasantry for the colonial Government of Ceylon, Sessional Paper, 1951 (quoted in Uphoff, 1979): "In backward areas . . . it was evident that although some [Rural Development] Societies existed, they were not sufficiently active or progressive enough. In the areas where educated men and women were available, the Societies once organized are capable of managing their own affairs." A study of experience with cooperatives for UNRISD (Fals Borda, 1976) indicated, on the other hand, that education and literacy levels per se were not relevant. Since literacy was one of the environmental variables that could be scored most accurately, our finding of no significant correlation (.08) between literacy and LO performance (or any specific LO outcome) appears reasonably solid. The same kind of compensatory dynamic appears to be possible here as with some other factors. Other things being equal, literacy is a more favorable condition for LOs than illiteracy, but enough LOs in largely illiterate communities are made into vehicles for improvement that the average relationship is not significant either way.[19]

(13) *Partisanship*. If literacy was relatively easy to assess, partisanship was the opposite. Certainly the literature has many references to this environmental factor, practically all regarding it as obstructing LO success. After studying LOs assisted by the Inter-American Foundation, Meehan says: "Factionalism, for example, appears at times as a very serious problem" (1978:104). Lowdermilk, Early, and Freeman (1978), studying irrigation as-

[19]About half the environments of LOs in the sample had estimated adult literacy rates of 20 percent or lower. Their summary score of performance was 69, compared with 87 for those with higher literacy rates, which would account for the low but positive correlation. The respective distributions were as follow:

	Outstanding	Very good	Good	Poor	Very poor
Low literacy (under 20%) $N = 79$	13%	23%	32%	16%	16%
Higher literacy (over 20%) $N = 71$	21	21	25	23	10

A chi-square test shows that the difference in distributions would be significant only at the .39 level ($x^2 = 4.14$, df = 4).

sociations in Pakistan where kinship groups known as *biradiris* are prevailing patterns of collective identity and action, found "polarization" in villages to be a major predictor of organizational ineffectiveness.

Environments can certainly be conflict-prone, as Williamson (1982) documents with several community studies in Nepal. In an RDPP study of Development Committees in Jamaica, Blustain (1982b) found partisan factionalism to be an accepted principle of group action, with a winner-take-all division of benefits. It was possible to get farmers' associations started under these circumstances but seldom with full or equal participation by all concerned, and the mentality of "clientelism" kept them from embarking on any sustained, self-reliant development efforts.

On the other hand, we have found a few examples where cooperation was fostered and effective in the midst of a factionalized environment. Wade (1982a), for example, tells of two villages in India, one of which was fiercely divided—yet they came together to form a water users' association that was unusually successful in getting water and distributing it among all who needed it.[20] In a pair of village-level studies done for the RDPP in North Yemen, it turned out that the village with the higher level of factionalism (Bani 'Awwam) had a more effective local development association than the other (Maghlaf), which was dominated by its traditional leader (sheik) and was more unified politically (Swanson and Hebert, 1982).

Getting a quantitative "fix" on the variable of partisanship was very difficult and not very satisfactory. Our scoring from the case studies proved to be more subjective than we were comfortable with. The correlation that emerged (−.03) is based on relatively "soft" data.[21] When we correlated partisanship with specific LO performance outcomes, still no significant relationships were found. However, as discussed in the appendix to this chapter, some significant negative coefficients emerged in multiple regres-

[20]The villages, cultivating an area of 600 acres together, needed to cooperate to ensure that enough water would reach them, given their downstream location. They formed a council and charged all farmer-members 100 rupees per acre per season (with a promise to contribute additional money if needed, such as for bribes). With a fund of 60,000 rupees they hired their own watermen (34) at a salary almost twice the average daily agricultural wage to acquire and distribute water to all the fields. The council met on alternate fortnights in each village. The president was the leader of one faction, but if disputes developed, they were taken to the head of the other village for settlement. The need and demand for water, a valued resource, encouraged organizational expedients to minimize the long-standing effects of partisanship.

[21]Where a case study explicitly reported group conflict or factionalism, whether or not this affected performance, we scored it 5, and where there were suggestions of factionalism, we scored it 4; otherwise the case was scored 3, unless indications of little intragroup conflict warranted a 2 or a 1. The distribution of scores was as follows: 5 = 11%, 4 = 43%, 3 = 34%, 2 = 11%, 1 = 1%. Where an environment was known to be high in partisanship—like Jamaica, where we were studying local organizations in the field—or where the literature on a country itself indicated high or low factionalism, we used such information to augment the case study. This was a more approximate way of scoring than for other variables.

sion analysis, which would support the general proposition about factionalism's deleterious effect on local organizations, other things being equal.

(14) *Group Patterns.* It might be thought that LOs would be more successful in environments where tribal principles of group formation prevail, or less able to achieve developmental goals in settings where caste relations structure interpersonal bonds. We wanted to consider such differences as they might bear on LO performance. As noted already, our five coding categories—caste, tribe, family, patron-client, associational/voluntaristic—constituted at best a very approximate continuum, ranging from highly ascriptive to highly individualistic ties, from highly prescribed relations to those more freely chosen. Actually, coding turned out to be more difficult than expected, because there are some elements of patron-client organization in most rural societies (Scott, 1972), and because when a whole rural area is tribally homogeneous, family or kinship groups become more important. Principles of organization apply only within a population that is heterogeneous in some dimension. In a community where all are of one caste, there are often familistic or patron-client patterns, though under some circumstances the caste group may remain united vis-à-vis a differentiated outside social world.

Despite some ambiguous cases, most of the classifications were quite straightforward. We knew that this variable could not be regarded as continuous. When a correlation of −.01 emerged between group patterns and overall performance, it seemed to vindicate the use of parametric statistics, because no association should have emerged. When we did a nonparametric test for association between LO performance and group pattern, the relationship was in no way significant.[22] In other analyses, a few low but significant correlations did emerge. But these could be attributed to the scoring of patron-client and voluntaristic patterns (which are somewhat more common in Latin America with its higher income and literacy) as 4 and 5, while caste and tribe (respectively found more often in Asia and Africa, with lower socioeconomic levels) were scored as 1 and 2. When group patterns per se were considered cross-nationally, we found no association.

(15 and 16) *Community and Societal Norms.* One might expect that the normative environment of LOs—the extent to which participatory practices and egalitarian outcomes are valued—would affect the extent to which they can operate effectively and achieve development objectives; Belloncle (1979:22–23), for example, lays some emphasis on this idea. When we tried

[22]The size of sample for caste was $N = 34$; tribe, $N = 46$; and patron-client, $N = 34$. Chi-square with 8 degrees of freedom was 5.26; $p = .73$.

to assess the normative environment for specific LOs, it quickly became apparent that we had to distinguish between the prevailing norms at the community level—an LO's immediate surroundings—and the societal level. It was possible to have supportive local norms and unsupportive national ones, or vice versa. One would expect the two sets of norms to be similar as a rule. The correlation we found between them was positive, but not very large (.20, significant at the .05 level).

Different sets of factors operate at each level. Community norms are usually traditional, legitimating unequal rank and power *or* validating the equal involvement of all in decision-making and benefits, while national norms are likely to be more politicized, reflecting the ideological orientation of the regime. Community norms correlated significantly, .20, with LO performance in our sample, while societal norms correlated positively but not significantly, .12. That the relationship is not stronger appears to reflect the fact that a number of LOs among the 150 were achieving good results despite unsupportive conditions in their environment. There is a generally positive but not very strong association between norms for participation and egalitarian outcomes on one hand, and specific LO outcomes on the other. Unfortunately, since this relationship has not been looked at systematically by others, we have little to compare our findings with. Gow et al. (1979) looked at the extent to which decision-making for the community was done by outsiders or by the community as a whole. They found no association with overall impact (−.03). We are surprised that normative orientations outside the LO did not have more influence on their performance.

(17) *Political Support.* One of the factors often mentioned as affecting the operation of LOs is the extent to which the government favors and assists them. It may be presumed even more strongly that if the government opposes LOs, their prospects of success are greatly diminished. But there is still another way of looking at this relationship: Meister (1969) suggests that LOs will do best in areas that have been neglected by the government, where people will be spurred on to self-help because they *have* been bypassed. Charlick (1984) did not find that LOs were more successful in neglected areas. A positive association was found by Gow et al. (1979) between the government's giving LOs freedom to operate and their overall impact (.17) but it was not great enough to be significant.

One of the principal criteria we used for gauging political support was the resource commitment that government made to the LOs in question. We looked also at favorable policies for LO operation. The correlation we observed between political support and LO performance was −.02—really none at all.[23] In multiple regression analysis (see the appendix to this chap-

[23]Belatedly we drew some encouragement from the fact that Charlick (1984) in his study found no significant correlation between any of his LO outcome measures and the variable of government commitment to rural development—contrary to his expectations.

ter) the coefficients for political support with various performance indicators was negative. This surprising result prompted considerable reflection and reconsideration of the cases. As discussed in Chapters 8 and 9, it became apparent that the *extent* of government support per se was not particularly helpful to LOs if the support was given with many strings attached or with preconceptions that distorted the shape and content of the LOs. Rather, one needed to consider *how* the support was provided, since a small amount given appropriately could yield more stimulus and strength than larger amounts given heavy-handedly. To have tried to introduce such qualitative judgments of support into our coding, however, would have built in the kind of circularity we wanted to avoid. One could hardly determine how "good" the support was without considering its results. So we held to our scoring criteria, which reflected mostly the extent of support. The results suggest that what is called "commitment," "support," or "political will" is more multidimensional and more conditional with respect to effects at the local level than many writers have presumed.

(18) *Administrative Support.* From his assessment of IAF experience in Latin America, Meehan (1978:104) concluded that rural LOs are much more vulnerable than others to the detrimental effects of "bureaucratic malaise." On the other hand, bureaucratic paternalism has been found to inhibit the development of effective LOs (Hunter, 1976). We are faced with alternative views of whether strong or weak bureaucratic capacity will benefit LOs. Recognizing that the effects could run in either direction, Gow et al. (1979, I:103) hypothesized that the potential for LOs would be greater where bureaucratic capacity for delivering services was ineffective or nonexistent, assuming that this would give impetus to more self-help efforts. Contrary to their expectation, there was a .22 association between *effectiveness* of administrative delivery of services and overall LO impact. Perhaps because of the way their "impact" variable was constructed, this was higher than the relationship we observed, a correlation of .14 between administrative support (willingness and ability of the relevant bureaucracies to assist the LO in question) and overall LO performance. Still, the analytical results are in the same direction and range.

When disaggregated, few LO outcomes were associated with administrative support, though increased agricultural production and income was significantly correlated (.20). The DAI study, contrary to its hypothesis, found administrative service capacity correlated with increased income, with equity impacts, and with "influence" (participation or empowerment). Charlick (1984) did not find a significant association between "technical competence of local bureaucracy" and efficiency or equity outcomes, but did with two participation measures and with enhanced LO institutional capacity. Taking all three studies together, there seems to be a clearer message with regard to this environmental variable than most of the others.

Competent and supportive administration is likely to have a positive influence on the performance of local organizations, though the margin of difference does not appear to be very great, unless perhaps one takes qualitative factors more into account.[24]

The relationships discussed here have been mostly simple, bivariate ones. One might like to know what associations result when other variables are controlled. In the appendix to this chapter, we review the conclusions of multivariate regression analysis. The basic finding can be reported as follows. In statistical terms, no more than one-fifth of the variance in overall LO performance could be accounted for by all environmental variables taken together, and no more than five (less than one-third) of the variables were found significant at the .10 level in a stepwise regression equation. Most interesting was the fact that *none* of the physical-economic factors proved significant in this analysis, whereas four of the six political-administrative factors did. A number of the social-societal factors were correlated quite highly with social stratification, which could explain why the others did not show up statistically. Literacy, for example, one of the best measured variables in our study, was not at all significant. This analysis, exploratory though it is, reflects the limited effect of environmental factors that was observable in the sample of 150 cases.

Environmental Influence on Local Organizations

Why is there so little observed relationship between environmental factors and LO performance? We recognize that our sample was not perfectly random and thus may not mirror perfectly the whole universe of local organization, but we are reasonably sure it reflects the whole range of experience, if not its distribution. The statistical findings are consistant with those of Gow et al. (1979) and Charlick (1984), whose quantitative analysis along similar lines pointed in the same direction.

One possibility should be considered, though it raises difficult questions about our understanding of social causation. The *interaction* among environmental variables in microenvironments may be so complex that it confounds the ability of our statistical models and techniques to trace effects. The significance and uniqueness of microenvironments is seen persuasively in King's analyses (1975, 1976, 1981) of different communities in northern Nigeria into which cooperatives were introduced by officials,

[24]We found the following differences in summary scores when the sample was divided according to administrative support. LOs where the bureaucracy was unsupportive or even opposed (scores 1 and 2; $N = 20$) had a score of 65, compared with 79 for the rest (scores 3, 4, and 5; $N = 130$). A chi-square test did not show the distributions to be significantly different, however, so one can only say the results point—but not strongly—in a positive direction.

where each locality presented very different physical, economic, social, and political situations. Refining analysis along these lines would require much larger samples and more precise measurements than are thus far available and might produce only modest increments in predictive power.

One possible reason for low statistical associations is that the variables are not as linear as is required for proper use of most available analytical techniques. This would suggest more extensive experimentation with data transformations (log, semilog, and so forth) and with models introducing thresholds and interaction effects. There is good reason not to expect linearity in the effects of most of the variables, even if they can be measured in a linear manner. We have commented previously that the consequences of increments in literacy are not likely to be constant, though establishing this conclusion would be very difficult. We do not think that more sophisticated statistical techniques are likely to produce much more valid knowledge, however, even if higher coefficients or R^2s could be achieved.

Having inspected the data rather extensively, we believe the problem lies less with their distributions than with their contingency. The universe of social relationships we are dealing with is different from the one Boyle observed and measured to formulate his famous law about the predictable interactions of temperature, volume, and pressure in gases.[25] Some tenacious positivists may continue to insist that the problem is one of defining and measuring social variables properly and more precisely, and improvements surely can be made. But the problem appears to be one of misconception more than of misspecification, assuming that there are fixed relationships "out there" waiting to be captured by proper modeling and measurement of what everyone concedes are very complex variables.[26]

A simpler explanation for the low statistical associations is that the variables are not constant in their effects; that is, the effects they have can work in either direction. When we cited contrary hypotheses found in the literature for most of the environmental variables, we suspect that readers wondered which was correct. Would high or low administrative support be more likely to produce effective LO activity? One answer is that *both* hypotheses may be correct—that *either* may produce a given effect, depending on the circumstances. If the real world contains both sets of dynamics, we should expect to find low or nil correlations, with the respective opposite relationships offsetting each other statistically.

This is not to say that there is *no* causal influence between, say, administrative support and LO performance. Depending on how it is provided, a

[25]Johnston and Clark comment with appropriate criticism on the debilitating effects of "physics-envy" in the social sciences (1982:19–20).

[26]This mind-set has been characterized as a form of "neo-Pythagoreanism" for its assumption that the social universe can ultimately be understood in terms of invariant mathematical relationships (Uphoff and Ilchman, 1972:494).

great deal of assistance or little assistance from the bureaucracy may contribute to processes of resource mobilization and management, because incentives for rural people to work together can be produced either by administrative support or by its absence. Conversely, either a great deal of assistance or the lack of any outside aid might discourage local initiative and self-reliance. The discovery that causal influence may run in either direction must be a disturbing conclusion for any budding Boyles within the fraternity of social scientists. One might say that the task, then, is to specify the conditions under which the variable has positive and negative effects. This seems more promising than working on definitional, statistical, or methodological refinements. But it still assumes that social relationships, even if more complex than physical ones, are relatively invariant once truly identified and correctly understood. We would welcome such knowledge but must express agnosticism because of a further possible explanation for the relationships we have observed.

The fact that a number of the significant correlations were in the opposite direction from what was expected does not demonstrate that environmental factors like adverse topography or social stratification "cause" LOs to perform better. Rather it points to the critical link between environments and performance: *what people living in those environments choose to do about them.* It would probably be more accurate to regard environments as settings than as influences. One of the most acute observations on this subject is by Schlomo Eckstein, who concluded from his study of Mexican experience with cooperatives (*ejidos*) that it is very difficult to generalize within the social sphere: "Certain groups have a highly developed sense of social motivation, while others respond only to purely individual inducements. . . . the most important fact is that *neither the positive nor the negative elements are constant. . . . It is precisely in these . . . that men can and must intervene,* including those possible transformations that will bring us close to the final social and economic goals of development" (1971:307; emphasis added).

Our own thinking on this subject was spurred by the Philippine case study referred to above (Simpas et al., 1974), where the good infrastructure and easy access of San Quintin had not contributed to effective local organization; community leaders had chosen to seek individual improvement in a nearby city rather than attempt collective action within their own area. Conversely, the isolated community of Sallapadan, with stimulus from a resident priest and local leaders, was making considerable self-managed progress by means of its LOs. Under other conditions, opposite outcomes might have occurred: with supportive leadership and participation, San Quintin could have flourished; without these, Sallapadan would have languished.

Reflections on such relationships lead perilously close to the challenge-response reasoning of Arnold Toynbee (which is anathema to social scientists because of its unavoidably tautological quality, ascertainable only post hoc): progress comes in response to a challenge, which must be enough to elicit collective action but not so much as to deter such action or to cause it to fail. What was enough or too much is known only after the fact. Actually, the circular quality of such explanations does not make them incorrect so much as unprovable by accepted methods of scientific inquiry.

We are not prepared to propose a challenge-response theory of local organizational success. But we would underscore the importance of the "response" part of that equation, if not the role of "challenge" as a causal factor. One reason why correlations between environment and outcome are so low may be that they are not predetermined. We need to look at relationships with some regard for the purposiveness of people's actions. Certain outcomes may well be probable if there are no new or different initiatives by actors who are part of a given situation. Statistical relationships represent such probabilities and almost always show some distribution rather than a fixed connection, since most social phenomena are stochastic in nature. The variability is a reflection of the fact that various sets of actors, often seeking quite divergent outcomes, become involved in shaping outcomes, investing their own and others' resources to raise the probability of the outcomes they desire most.

What is most to be desired in developmental terms is not necessarily the most probable outcome. Indeed, persons whose preference is for the most probable outcome may not become as actively engaged in shaping the future; they can expect their preference to prevail without expenditure of effort, or they will need to expend less because having that outcome is less difficult than achieving more "deviant" outcomes. But it is likely that some persons will nevertheless desire these less probable outcomes. Whether they will be able and willing to mobilize their own and others' resources to produce some other outcome than the one historically most probable cannot readily be predicted. And even if we could know this, we cannot predict the skill with which persons will work to accomplish some outcomes and to block others.

If the foregoing view correctly represents the nature of the social as contrasted with the physical universe, and we think it does, then Eckstein's observation that neither positive nor negative elements are "constant" makes sense. The implication of this outlook is that persons, individually or collectively, in private or in public roles, affect through their choices and their ensuing actions what the shape of the social universe will be. The emphasis then is on interventions, as we think Eckstein correctly puts the point. The environment of LOs sets the stage, as the extensive hypothesizing reported earlier in this chapter suggests. One can point to a probable "script," based on past patterns of objectives, capabilities, initiatives, interactions, and outcomes. But actors need not necessarily follow this script, and the frequent appearance of disorder in the social universe is attributable to the fact that actors with divergent objectives and capabilities can rewrite their own parts.

The actors with regard to local organization and rural development are many and varied. They include residents of rural communities who may have quite different values and resources; leaders within those communities who may be cooperating or competing with each other and enjoying varying degrees of respect and support; government staff, with diverse technical skills and responsibilities, having allegiance to different ministries and operating at many different levels; staff from private voluntary organizations, foreign donor groups, research institutions, or international agencies. With some consensus among them and some concerting of resources so that activities do not work at cross-purposes, desired outcomes—whether the probable or the deviant ones—may be promoted. It is helpful for those who intervene to know the probabilities of success for alternative courses, whether the odds favor their venture or not, and to take this knowledge into account. This information need not, however, be a straitjacket for those who are prepared to work against the odds, encouraged by the knowledge that environmental factors need not predetermine the fate of LOs. Since an environment usually presents some degree of freedom to actors—whether locals, officials, or outsiders—the next question is what kinds of LOs are most likely to contribute to developmental goals.

APPENDIX TO CHAPTER 4

Associations among Performance Measures

As discussed in Chapter 3, the classification of each case in terms of overall performance reflected the number and quality of performance accomplishments. Data on the various specific accomplishments in each case were entered on a standard form from the documentation available. Such information was reviewed and used to score the case, according to the protocol reproduced in Appendix B. As it turned out, the specific performance indicators were almost all positively correlated, so the overall scoring was not a matter of comparing and reconciling contradictory accomplishments. The coefficients for these intercorrelations are shown in Table 4-2, along with the correlation of each performance variable with the summary measure of overall performance. The average correlation among specific measures was .35, and only three coefficients were negative (with significance levels only .61, .75 and .87).

Table 4-2. Correlations among performance variables and with overall performance

	Agricultural production/ income	Nonagricultural production/ income	Education improvements	Health improvements	Income for poor	Assets for poor	Access to services	Reduced sex discrimination	Reduced social discrimination	Participation in government decision-making	Participation in community decision-making
Overall performance	*.60**	*.45**	*.57**	*.44**	*.51**	*.38**	*.54**	*.36**	*.50**	*.49**	*.63**
Agricultural production/income (106)†	—	.52*	.20	.57	.25*	.13	.40*	.16	.29	.21*	.29*
Nonagricultural production/income (39)		—	−.03	.01	.50*	.45	−.11	.28	.09	.34*	.34*
Education improvements (48)			—	.63*	.16	.19	.52*	.22	−.08	.36*	.62*
Health improvements (41)				—	.16	.29	.35	.33	.11	.11	.24
Income for poor (110)					—	.53*	.78*	.33	.57*	.15	.32*
Assets for poor (40)						—	.24	.11	.29	.16	.48*
Access to services for poor (116)							—	.53*	.46*	.21*	.39*
Reduced sex discrimination (40)								—	.74*	.31*	.50*
Reduced social discrimination (36)									—	.42*	.52*
Participation in government decision-making (146)										—	.47*
Participation in community decision-making (150)											—

*Significant at .05 level.
†N in parentheses.

Although there was no formal weighting of performance variables in the summary indicator, certain dimensions of performance affected it more than others by virtue of the fact that some were attempted by LOs more often than others. The number of cases for which each performance variable could be scored ranged from 150 for participation in community decisionmaking (which could be judged for all cases) to 36 for reduced social discrimination. (In many cases there was no issue of social discrimination to be redressed; in some cases where it existed, we could not ascertain whether or not the LO had contributed to reduction.)[27] Those performance variables for which there was a larger *N* correlated somewhat more highly with overall performance than did those with a lower *N*, but this should be expected statistically and does not invalidate the summary measure.[28] The statistical results of analyzing the overall measure as well as the disaggregated measures have enough consistency and coherence that we believe we are justified in treating primarily with overall performance. We extend the analysis, however, to any specific elements of performance that raise divergent issues or findings.

The associations within the matrix are worth noting in that the correlations among performance indicators within the five sets were higher than the correlations with indicators outside the set:

	Correlation within set	Correlations outside set
Economic gains	.52	.24
Social benefits	.63	.24
Equity effects	.52	.30
Reduced discrimination	.74	.29
Participation in decision-making	.47	.33

The last set, participation, had the lowest correlation *within* its set—between empowerment through participation in government and in community affairs—but also the highest correlation with performance gains *outside* the set.

[27]Several performance measures in the social benefit area had fairly small *N*s because they were not reported, with positive or negative results, for very many cases (nutrition, $N = 16$; water supply, $N = 27$; transportation, $N = 23$; public facilities, $N = 15$). Because of their small numbers and the lack of significant correlation or regression coefficients, they are not reported in the tables. They were figured in the overall performance measure, however, because where they were accomplished, the LO should be given credit for them, and unproductive attempts should be considered a deficiency in overall performance.

[28]The average correlation for those with *N* over 100 was .55, and for those under 50 it was .45, so one can say that the summary measure is slightly weighted toward those outcomes more frequently associated with LO activity. But such outcomes were not necessarily achieved more successfully. The average score for performance measures with *N* over 100 was 3.01, and for those with *N* under 50 it was 3.48. The average standard deviation for the two sets of performance variables was almost the same (0.93 and 0.90 respectively), so the distributions were not different in their skewness.

Regression Analysis of Environmental Factors

Having analyzed the different environmental variables individually, we introduced them into a regression equation in stepwise fashion to see which would best explain statistically the variation in overall performance of LOs. With this set, the problems of collinearity were considerably less than for task variables analyzed in the previous chapter. The "best" equation, with three environmental variables significant at the .05 level, attained an R^2 of only .15, reflecting the limited statistical association between environmental factors and LO performance. The three variables, their beta weights, and significance levels were as follows:

social stratification	community norms	bureaucratic support
.32	.30	.15
(.00)	(.00)	(.05)

Disaggregated Analysis of Environment and Performance

We have made some references in the chapter to associations between environmental factors and specific LO outcomes. Those correlation coefficients that were significant at the .05 level in such disaggregated analysis are shown in Table 4-3. Eight of the environmental variables had no significant correlations with *any* specific outcomes (nor did seven in regression analysis). That group patterns or settlement patterns did not produce significant associations is perhaps not surprising or very important, but the failure to find income or literacy levels associated with specific outcomes is both surprising and significant in substantive terms. We can look at the relationships from the perspective either of environmental factors as independent variables or of performance outcomes as dependent variables.

To begin with environmental factors, we note that all the correlations of topography, resource endowment, and infrastructure that show up as significant at the .05 level are negatively associated with LO performance outcomes. This is consistent with the indication discussed previously, that LO success—at least within our sample of 150 cases—is often inversely related to supposedly favorable physical environments. This pattern holds up generally in regression analysis, except that better resource endowment, infrastructure, and income level are positively associated with agricultural production in the confidence range of .10 to .20. Economic diversification and income level, themselves related, show no correlations significant at the .05 level in either analysis. Unequal income distribution is associated

Table 4-3. Correlations of LO environments with development outcomes

	Agricultural production/ income	Nonagricultural production/ income	Education	Health	Income for poor	Assets for poor	Access to services	Reduced sex discrimination	Reduced social discrimination	Government decision-making	Community decision-making
Topography	—	—	−.29	—	—	—	—	—	—	−.16	−.18
Resource endowment	—	—	−.40	—	—	−.39	—	—	—	—	—
Infrastructure	—	—	—	—	—	−.35	—	—	—	—	—
Economic diversification	—	—	—	—		(none significant)		—	—	—	—
Income level (GNP)	—	—	—	—		(none significant)		—	—	—	—
Income distribution	.22	—	—	—	.26	—	—	—	.36	—	—
Settlement pattern	—	—	—	—		(none significant)		—	—	—	—
Social heterogeneity	—	—	—	—	.28	—	—	—	.37	—	.18
Social stratification	.32	—	.30	—	.26	—	—	—	.35	—	.22
Sex discrimination	—	—	—	—		(none significant)		—	—	—	—
Social discrimination	—	—	—	—		(none significant)		—	—	—	—
Literacy	—	—	—	—		(none significant)		—	—	—	—
Partisanship	—	—	—	—		(none significant)		—	—	—	—
Group patterns	—	—	—	—		(none significant)		—	—	—	—
Community norms	—	—	—	—	—	.40	—	—	—	.26	.29
Societal norms	−.21	—	—	—	—	—	—	—	—	—	—
Political support	—	—	—	—	—	—	—	−.47	—	—	—
Administrative support	.20	—	—	—	—	—	—	−.35	—	—	—
Number of significant coefficients	4	0	3	0	3	3	0	2	3	2	4

with several outcomes, but this is consistent with the observed relationship between "adverse" environments and LO performance.[29]

Settlement pattern is not significantly correlated with any specific outcomes, and with reduced sex discrimination only in regression analysis (reflecting apparently a common South Asian factor). The pair of variables with the highest number of significant correlations in both modes of analysis were social heterogeneity and social stratification. The positive coefficients, also observed in regression analysis, contradict the usual expectation, as discussed in this chapter. That there were no significant correlations, or regression coefficients, for sex discrimination, social discrimination, or literacy was also unexpected.

The one variable for which regression analysis diverged markedly from correlations was partisanship. It had no significant correlations, as seen in Table 4-3, but showed negative beta weights for access to services, −.22 (.01), and participation in community decision-making, −.17 (.03), as well as for agricultural production, −.19 (.06) and income for poor, −.15 (.11). This conforms to the common belief about the negative effects of factionalism. Group patterns, not really a linear variable, did not show any significant correlations.

Community norms also showed a number of positive associations with performance, not surprisingly with equity and participation outcomes, in both correlation and regression analysis. Societal norms, on the other hand, showed less significant association with LO outcomes, perhaps because outcomes were measured always in local contexts. The one significant negative correlation for societal norms was with agricultural production gains, which can be independent of egalitarian ideals and practices. In regression analysis, societal norms had positive beta weights for a number of performance variables at lower levels of confidence, and a negative beta weight for agricultural production.[30]

The only inexplicable statistical relationships emerging from this analysis are the negative correlations between reduced sex discrimination and political or administrative support. In regression analysis, political support had a beta weight of −.52 (.00) for reduced sex discrimination, but there was no relationship even near significance for administrative support. The statistic,

[29]In the study by Gow et al. (1979), skewed income distribution correlated .36 with gains in income, paralleling our statistic, though income distribution did not correlate significantly with their equity or empowerment measures. In multiple regression analysis to account for different "impacts," income distribution was not significant for explaining any of the "outcome" impacts—financial viability, service provision, or leadership (problem solving). Charlick (1984) did not find skewed land distribution significantly correlated with his technical impact measure.

[30]Beta weights and significance levels were as follow: education, .15 (.22); assets for poor, .16 (.28); participation in government decision-making, .13 (.13); participation in community decision-making, .13 (.09); and agricultural production, −.16 (.09).

if correct, suggests that when politicians and administrators back LOs, they are not necessarily assisting the advancement of women, which is attained in spite of officials' attitudes.[31] This is, however, speculative. We can say only that the relationship observed is unexpected and warrants further study.

Looking at the analysis in terms of performance outcomes, we find gains in agricultural production and income significantly associated with economic and social inequality, with nonegalitarian societal norms, and with administrative support. This is borne out by the regression analysis, which added positive associations with better resource endowment, infrastructure, and income level, as noted above. In contrast, none of the environmental measures correlated significantly with nonagricultural gains; the same was true in regression analysis. A similar differentiation showed up with services: educational improvements correlated with adverse topography, resource endowment, and social stratification, yet health improvements correlated with no environmental variables.

The significant correlations of skewed income distribution, social heterogeneity, and stratification with income gains for the poor may be due to the fact that there is more *scope* for such gains under these conditions. But it is encouraging nevertheless, because most analyses have predicted that such conditions would create power disparities and resistance by the rich that would block the gains. Improved asset distribution for the poor is associated with conditions of poor natural resource endowment and underdeveloped infrastructure as well as egalitarian community norms. Perhaps most significant substantively is the observation that no other environmental variables correlated significantly in either direction. None of these variables correlated with increasing access to services by the poor, though regression analysis showed both social stratification and a lack of partisanship significantly correlated with improved access.

We have already discussed the unexpected associations for reduced sex discrimination. Reduced social discrimination was associated with unequal income distribution, social heterogeneity, and social stratification—all adverse conditions. Improvements both in community-level and government-level participation were associated with adverse topographical conditions and egalitarian community norms. Additional factors associated with community-level participation were social heterogeneity and stratification, which possibly provide opportunity for such participation more than they promote it. The only significant correlation (and not a very strong one) that

[31]There were only 40 cases out of the 150 where we could code any change, positive or negative, in sex discrimination, so the size of this subsample is smaller than most others. But the coefficient was great enough to be statistically significant, unlike 88 percent of the coefficients in this table. Since both political and administrative support had high negative correlations with regard to sex discrimination, they could not be simply passed over without comment.

Charlick found with any measure of community-level participation was *less* agricultural diversification. This was also associated with extra-community participation, as was technical competence of the bureaucracy.

There are only small differences between the results of our study and those of Gow et al. and of Charlick, and the results overall are encouragingly similar. The overwhelming proportion of hypothesized relationships between environmental factors and LO outcomes were *not* statistically significant. Since such factors have been conventionally seen as presenting inhibitions or even obstacles to LO success, their apparent minimal influence is a finding to be welcomed. Indeed, two-thirds of the few significant relationships observed are in the *opposite* direction from what is usually expected. These indicate that progress with and through LOs may even be somewhat more likely where social stratification or heterogeneity are higher than average, or where physical disadvantages outweigh advantages.

Environment and Task Performance

The relationships observed between environmental variables and success in the various tasks analyzed in the previous chapter can be quickly reviewed, with no need to present the correlations in a table, since only 18 of the 144 coefficients were significant at the .05 level.

Internal tasks. Planning and goal-setting, and conflict management both correlated with social stratification (.24 and .23) and difficult topography (−.18 and −.20), reinforcing the finding that success may be somewhat more likely under adverse conditions. Planning and goal-setting was also correlated with social heterogeneity (.23), and conflict management was correlated with egalitarian community norms (.29), an understandable association and the highest observed.

Resources. Mobilization was associated with egalitarian community norms (.23) and difficult topography (−.22), again reflecting the pattern discussed, where adverse conditions combined with participatory values are supportive of LO success. It is very significant substantively that resource management had *no* statistically significant environmental correlates, suggesting that this important task variable faces the fewest environmental "constraints."

Services. Both provision and integration of services correlated with social stratification (.23 and .27), but integration correlated also with administrative support (.24), reflecting the need for bureaucratic cooperation in this task—the *only* task such support correlated with significantly. Service integration also correlated with group patterns (.20), though we do not think it was patron-client or associational patterns per se that contributed to better service integration, but rather regional differences. The net scores for integration of services by region were as follow: Latin America, 165; Asia, 138; and Africa, 110.

External tasks. Both control of bureaucracy and claim-making correlated with social heterogeneity (.25 and .21) and with societal norms (.23 and .22). Incidentally, such norms reflecting the ideological orientation of the society at large were significant *only* for external tasks. Control of bureaucracy was also associated with literacy (.23) and community norms (.22), both substantively interesting observations. It is worth adding that the average correlation of literacy with all *other* tasks was only .08.

The five physical-economic factors (excluding topography) had an average magnitude of correlation of only .08, consistent with the stepwise regression that found none of them significant in *ceteris paribus* terms. This disaggregated analysis of environment-task relations further supports our general conclusion that influences from the environment are not very strong or likely in themselves to present major barriers to LO performance.

CHAPTER 5

Structural Factors in Rural Local Organizations

Our finding that environmental factors are not strongly associated with the overall performance of local organizations should direct attention all the more to LO design and structure—factors that can be influenced by sympathetic governments, voluntary agencies, and international donors. Here we shall assess which, if any, structural features show a favorable association with LO performance. The literature is full of propositions about what kind of LOs are most likely to succeed. We shall review such arguments and the conclusions to be drawn from the experience of our 150 cases.

Readers should bear in mind the limitations as well as the strengths of the statistical relations we extract from the data. We are, of course, limited by our sample and cannot know the extent to which the observed relationships apply for all LOs. Our knowledge of the literature and our field experience give us some basis for qualifying and interpreting the statistical analysis of cases and for a critical perspective on the results.

Like environmental factors, structural relations other than the most favorable can be made to work for desired developmental ends. That most correlations, even those that are statistically significant, are not very high indicates that there are substantial *exceptions* to any general rule. A few of the structural variables we find are not related to performance in a linear manner, so correlation coefficients by themselves cannot represent such relationships properly. In such instances, we undertake other modes of analysis.

Various writers, we included, have tried previously on the basis of less extensive and detailed analysis to make suggestions along these lines. Gow and his associates (1979, I:22–24) have identified two schools of thought: advocates of small, informal, homogeneous groups pursuing usually a single function (Hunter, 1976; Mensah, 1977; Tendler, 1976), contrasted with supporters of larger-scale, more formal, multitiered, and multifunctional organizations (Inayatullah, 1972; Uphoff and Esman, 1974; Texier, 1976). They observe correctly that the experience of the first set of authors has been chiefly in Africa and Latin America, whereas the latter have focused mainly on Asia, raising the possibility that differences between the regions account for the variance in prescription. However, an analysis of LO characteristics by region as related to overall performance does not bear out such a pattern of differences.[1]

Structural Options

We will review in this chapter what our data suggest about various structural options raised in the literature. We address each feature initially in correlational terms, as has usually been done in the literature because it was not possible to control for other variables. But where we have found that controlling for other structural or environmental factors enhances understanding of structural options, we introduce these considerations, which our larger sample and more detailed analysis permit. Because even the basic overall relationships provide a great deal of information and complexity, we have reserved discussion of the findings of multivariate and disaggregated analysis for an appendix to this chapter, except where they qualify or contradict the results of simpler analysis.

One variable examined here that is perhaps only quasi-structural is the factor of *initiation:* that is, at whose initiative the LO was established. This, as much as any other factor, appears to affect the structural relationships and their outcomes. We will deal separately with the implications of structural relationships associated with local initiation and with government instigation, as these are important for formulating strategies of support for local organizations as intermediaries in rural development.

[1]The differences in the distributions of LO structural characteristics by region can be compared in terms of the following summary scores. (As noted previously, in our "random walk" through the literature we found only four Middle Eastern cases among the first 150, not enough to analyze separately.)

	Latin America (N = 31)	Africa (N = 50)	Asia (N = 65)
Larger	61	74	75
Formal	190	176	137
Economically heterogeneous	73	54	54
Vertically linked	80	35	22
Multifunctional	125	94	92
Assembly decision-making	32	75	97

We cannot prove that our sample randomly represents these respective regions. The inclusion of a number of Small Farmer Development Program (FAO) cases from Asia may have affected the respective samples by region, but in fact the average size of LOs in Asia is larger

(1) *Single versus multiple functions.* One of the sharpest differences of opinion in the literature concerns the advisability and effectiveness of having LOs focus on one function, or a very few, as opposed to their undertaking a larger number. The most cogent argument in favor of single-function LOs is by Tendler (1976), based on her study of cooperatives in Ecuador and Honduras. She proposes that LOs start with just one task. Indeed, she would plan to circumscribe the organization in its intention, if not its ultimate form: "They [should] organize around a concrete goal, which can be achieved in a limited time period and would not necessarily require the organization thereafter. . . . Though the organization may continue and expand into other activities, once the concrete goal is achieved, the farmers perceive themselves as organizing to achieve this one goal, and not to create an organization" (1976:7).

Our findings (1974) led to the view that multifunctional LOs were more likely to build up both the resources and the commitment from members to operate effectively over time. Particularly when associated with agricultural production, single-function LOs have peaks and troughs of activity that make institutionalization more difficult. Fals Borda (1976), on the other hand, concluded from his study of cooperatives that single-function organizations would be more stable; and Peterson (1982b:136) suggests that the poor are more likely to benefit from single-function LOs that they can operate and control themselves. It could be argued conversely, however, that because the needs of the poor are multiple, improvements on only one front are likely to have little effect. Certainly the Small Farmer Development Program (SFDP) experience with small farms and landless households in Asia shows the value of diversified LO projects when the poor themselves initiate multiple activities (Ghai and Rahman, 1979; Alam, 1979). Again, it can be argued that single-function LOs are less likely to become involved in internal conflicts, as Golladay (1983) suggests.[2] On the other hand, more functions may permit more trade-offs, or "side payments": members who are not happy with one activity might receive compensatory benefits from a different activity. So advantages and disadvantages can be seen on both sides.

Measuring the number of functions an LO performs is not, it turned out,

---

than in Latin America, though no larger than in Africa. The Latin American cases, which we took as randomly as possible from the literature, were more formal, heterogeneous, vertically linked, and multifunctional than the Asian cases and oriented to executive rather than assembly decision-making. Because of sampling limitations, the suggestion of Gow et al. is not disproved by such data, but regional differences do not appear to be an adequate explanation for the different lines of argument.

[2] "Organizations with multiple objectives not only accumulate conflicts peculiar to each objective but also raise problems of allocation of resources among the numerous objectives. For example, a dairy marketing cooperative that extends itself to offer family planning services jeopardizes its original success by increasing the potential for conflict" Golladay (1983:28).

as easy and unambiguous as one might expect. Where the same people with essentially the same skills handle two activities, such as distributing both fertilizer and improved seeds, we counted this as a single function, in contrast to activities like credit or marketing, which require different skills and/or personnel. There was a danger that our measure of overall performance would be biased in favor of multiple functions. But we emphasized quality of performance rather than simply the number of accomplishments.[3] As a check on our findings, we analyzed the cases in the DAI study, correlating the number of functions listed for each LO (Gow et al., 1979, I, tables 10 to 13) with the overall impact score assigned it. Their score was less liable than ours to any circularity with multiple functions. The coefficient was .52 (significance, .00). So we think there is some association between successful performance and multiple functions, though this needs to be interpreted carefully.

Disaggregated analysis shows only positive correlations between multiple functions and LO performance (see Tables 5-16 and 5-17), and the summary scores for this variable show a linear direction: one function, 35; two functions, 44; three functions, 87; four or five functions, 146; six or more functions, 189. The relationship is somewhat more complicated than these numbers suggest, however, because the association is closer between single-function LOs and lack of success than between multiple functions and success. This can be seen from the percentage distributions, by number of functions, for different levels of LO success and vice versa. Table 5-1 shows the breakdowns in both directions, to avoid any inference of causation. The outstanding LOs are fairly well distributed across the spectrum of functions, whereas the very poor ones tend to have a single function. Looked at by number of functions, comprehensive LOs tend to be more successful, though at least one-quarter of the single-function LOs are in the very good or outstanding category.

One obvious inference from the data is that the number of functions is *more likely a consequence than a cause* of LO success. Neither our data nor DAI's could be analyzed longitudinally in a satisfactory manner. But when we read the case studies with an eye to their development over time, we saw much experience that is consistent with the admonition of Tendler and others to *begin* with a single function. The disagreement in the literature may well turn on a time perspective. We would suggest that starting with a single function appears advantageous for initial success, but that such success supports a branching out into more and other activities. These in turn

[3]The observed correlation between our measures of the number of functions undertaken by an LO and overall performance was positive and significant, .30, but not so high as to represent a tautological relationship. The R^2 for such a coefficient implies that the number of functions would account for less than 10 percent of the variance in performance scores. Moreover, there are positive correlations between number of functions and the quality of performance of all LO tasks, where quantitative considerations should be minimal (see Table 5-15).

Table 5-1. Number of functions and LO performance (percentage distributions)

PERFORMANCE	(%)	Single (N = 52)	Two (N = 32)	Multi (N = 32)	Numerous (N = 23)	Comprehensive (N = 11)
Very poor	(13)	55	30	10	5	0
Good	(29)	42	21	16	14	7
Outstanding	(17)	24	16	12	28	20

FUNCTIONS	(%)	Very poor	Poor	Good	Very good	Outstanding
Single	(35)	21	19	35	13	12
Multi	(21)	6	25	22	38	9
Comprehensive	(7)	0	9	27	18	45

can reinforce the LO's capacity and the commitment of its members to its survival.[4] We saw this pattern in the Sukuma cotton cooperative in Tanzania (Lang et al., 1969), which moved into processing once marketing was performed well, and then into agricultural improvement. The cooperatives in Ecuador that Meehan (1978) studied started with consumer goods and then ventured into various lines of production and into literacy programs. Furthermore, there were cases where starting with many functions invited failure. M. Moore (1979), for example, found that the disposition of the government in Sri Lanka to overload the Cultivation Committees with a variety of technical tasks reduced their effectiveness.

Our conclusion after analyzing LO experience supports that of Johnston and Clark (1982:180), that the pursuit of multiple functions carries potential benefits *and* potential liabilities, and of Bottrall (1980), that the effectiveness and solidarity of LOs can be increased by the accumulation of related functions, but the pace must be carefully judged in relation to the group's capacity for self-management. Starting with relatively simple tasks and moving on to more difficult ones seems to be the most appropriate way to proceed. Neither correlation nor regression analysis showed any special advantage to single-function LOs as such. Indeed, the latter mode of analysis suggests that, other things being equal, multiple-function LOs are preferable.

(2) *Informal versus formal organization.* As we have noted, there is considerable support in the literature for more informal modes of organization when it comes to engaging rural people in development tasks. Tendler advocates, for example, "relatively informal, unsophisticated and familiar

[4]Johnston and Clark put the issue well: "Mature organizations have sensibly and naturally diversified their functional base after getting started and gaining experience from a narrow—and often single-function—beginning" (1982:179).

organizational forms" (1976:9). On the other hand, Dore (1971) predicts that such organization, reflecting traditional norms and power structures, will be either unable to perform more modern developmental tasks or unwilling to direct efforts in ways that will help the less advantaged sectors of the community. This gets into the distinction between "traditional" and "modern" organization. We find Hunter (1976:202) suggesting that relatively informal organizations are preferable to more formalized institutions. Yet he argues (in work with Jiggins, 1977) that in most situations it is best to establish new LOs rather than try to work through existing traditional ones. Saunders (1977), in his analysis for the World Bank, concluded the opposite, that one should work as much as possible through traditional organizations because their informal modes of operation are more compatible with people's capabilities and needs. We will examine this controversy more closely in Chapter 8.

There are two dimensions to this issue, related though not identical, that present analytical difficulties. Assuming that one can adequately define "traditional" and "modern," the first is generally associated with informal organizational patterns, and the latter with more formal ones. It is possible to have traditional LOs with a fairly high degree of formality—established, even written rules, with delineated authority roles filled by members' decisions and not by ascriptive criteria. This pattern is evident in the Sidamo *mahabar* associations in Ethiopia (Hamer, 1976). In one of our studies for the RDPP, we found burial societies in Botswana that were entirely "indigenous" yet with almost all the characteristics of formal rather than informal organization (C. Brown, 1982; C. Brown et al., 1982). Conversely, some LOs are "modern" in form but have considerable informality in their actual operations. Though one of the defining characteristics of "modernity" is the extent of formalization—departing from ascriptive, implicit patterns of organization—mixed formal and informal patterns are not uncommon; we classified such LOs as quasi-formal.

It is difficult to specify criteria for coding cases as "traditional" or "modern" that are not parallel to degrees of formalization. Since this latter variable could be spelled out more concretely—in terms of characteristics like a written constitution and by-laws, or government recognition and regulation in contrast to governance by community-established roles and sanctions—we focused on the degree of formalization, recognizing that it contained much of what is involved in the traditional-modern distinction. Cases were classified on a continuum from highly formal to highly informal, as defined in Appendix B.

There was a consistently negative association between the degree of formalization in the LO and its contributions to rural development. Disaggregated performance measures were also negatively correlated with formalization, even regarding income, assets, and access to services for the poor, thus

Table 5-2. Formalization and LO performance (percentage distribution)

PERFORMANCE	(%)	Very informal or informal (*N* = 18)	Mixed (*N* = 41)	Formal or very formal (*N* = 91)
Very poor or poor	(32)	5	15	80
Good	(29)	16	28	56
Very good or outstanding	(39)	16	36	48

FORMALIZATION	(%)	Very poor or poor	Good	Very good or outstanding
Very informal or informal	(12)	11	39	50
Mixed	(27)	20	29	52
Formal or very formal	(61)	43	26	31

countering the assertions of Dore and others.[5] When we calculated summary scores of performance according to the degree of formalization, we found very informal/informal, 135; quasi-formal, 128; formal, 61; and very formal, 21.[6] The consolidated distributions shown in Table 5-2 indicate a very clear tendency for the more formal organizations to be poor or very poor in overall performance and, conversely, for the very good and outstanding LOs to be informal or at most quasi-formal.

There is naturally an association between the size of an LO and the degree of formalization in its internal structuring. De los Reyes (1979) found that 90 percent of the irrigation associations serving areas larger than 100 hectares could be classified as formal, and 60 percent of them were subdivided into subassociations. Only 25 percent of those serving areas under 50 hectares were formal, and only 5 percent were subdivided into two tiers.

In assessing the effects of formalization, we found problems of attribution because of its substantial association with government linkage, which correlates negatively with overall LO performance. In his study of irrigation associations in Indonesia, Duewel (1984) observed that "outsiders" (usually government officials) tend to make local organizations too formal, and thereby to diminish their operational performance.[7] We thus would not

[5]Charlick (1984) used a variable estimating the weakening of traditional political institutions at village level. He found some association between this and participation (empowerment) at the community level, but not at the extra-community level, and he found no significant association of this variable with his measures of efficiency or equity gains. His findings corroborate our inference.

[6]The first two categories were combined because the number of cases in each of them (5 and 13) was too small to give us as much confidence in their distributions as in those of the other categories (*N* = 41, 49, and 42).

[7]We found a substantial correlation (.47) between formalization and government linkage—which correlated −.21 and −.22, respectively, with overall performance. In multiple regression, neither was significant (beta weights of −.08 and .07, with significance levels of .38 and .45, respectively). In effect, both gave way to local initiative, which had a beta weight of .30 (.00).

attribute too much independent influence to formalization; it should be viewed in conjunction with a number of the other factors discussed in this chapter. Having more formal modes of organization was not significantly associated with specific performance measures in either correlation or regression analysis.

One problem in dealing with formality as a variable is that it is difficult to establish *de novo* an informal association. One can hardly set up contemporarily a "traditional" organization, though it may be possible through government channels to stimulate groups that have some action capability without formal roles and rules. In our RDPP work with water users' groups in Sri Lanka, we have seen quite satisfactory progress in starting informal groups at the field channel level and building gradually toward higher-level and more formal organization (Uphoff, 1981, 1982a, 1982c). This was done outside regular government channels by the Agrarian Research and Training Institute (ARTI), not a line agency, though it was paid for and sponsored by the Irrigation Department. Paul (1982) reports a similar and very successful approach in a rural education program (CONAFE) in Mexico.

K. March and Taqqu (1982) have argued, with reference to women's informal organizations, that one should be wary about trying to coopt existing very informal LOs into development programs. Their bases for association and cooperation are likely to be undermined by such a change in the terms of organizational attachment. An alternative would be to recognize the advantages of informal modes of organization and support new, parallel organizations that adopt the familiar modes of operation, rules and roles, incentives, and sanctions of preexisting organizations. We shall return to this issue of working with existing and/or new LOs in Chapter 8.

(3) *Decision-making structure.* The roles established for decision-making within an organization represent one of the more visible structural characteristics. The three main alternatives parallel Aristotle's classification of political systems—rule by one person, by a few persons, or by the many. Thus LOs may be governed by an executive, by a committee (or committees), or by an assembly of all the members. There can be combinations of these, of course. We classified cases as follows: (1) essentially executive decision-making, 14 percent of the sample; (2) an executive plus committee(s), 19 percent; (3) decision-making by committee(s), including an executive committee as one possibility, 30 percent; (4) committees plus an assembly or committee-of-the-whole, 22 percent; and (5) essentially assembly decision-making, 15 percent. There was, understandably, some correlation between the size of the organization and decision-making practices. The coefficient (−.21) indicated a tendency toward governance by assembly in smaller LOs and toward executive rule in larger ones. Committees were clearly the most common form.

There are two lines of argument to be considered. One of these contends that LOs need strong executive leadership to deal decisively with their environments. It may be that the more an LO is oriented to changing the status quo and the more resistance it is likely to face, the more it must "speak with one voice" and pursue a coordinated course of action. The opposite view is that such situations demand the maximum consensus and individual commitment to the enterprise that can be generated only through broad participation in decision-making. This implies greater reliance on assemblies or committees-of-the-whole.

One can argue in favor of executive decision-making in the name of efficiency, as does Chambers (1975:56–59) with regard to water management organizations in Sri Lanka, where the alternatives were appointed irrigation headmen or elected farmer committees.[8] His finding that a majority of cultivators preferred the former as a more authoritative system, however, is not supported by studies in three other districts, where Uphoff found a stronger preference among farmers for elected committees to manage water (Uphoff, Wickramasinghe, and Wijayaratne, 1981).

In a study of rural local organizations in Botswana, Willett (1981:893) concludes that given the mobile nature of the population in its pastoral and agricultural activities, it would be better to vest responsibility in one person than in a committee. This may be an instance where the environment introduces a bias for one kind of structural arrangement over another. The analysis of this environment by Roe and Fortmann (1982) has detailed the kind of constraints likely to affect different modes of decision-making. For example, to have committees making decisions throughout the year would require that they be located in the permanent villages. This would bias participation in favor of the older, richer, male members of the community. Since such environments are not common, however, it is difficult to deal with them comparatively. Our analysis here covers a full range of physical circumstances, biased possibly toward more adverse ones.

The observed correlation (.22) between decision-making structures and overall LO performance signifies greater success with the more participatory arrangements. Moreover, it correlates similarly with success in the various LO tasks, indicating that assembly forms of decision-making are often associated with effective performance of specific tasks. The relationship is not strictly linear, however, as indicated by the following summary scores for different modes of decision-making:

(1) Executive primarily ($N = 20$)	−45
(2) Executive plus committees ($N = 28$)	79
(3) Committees primarily ($N = 43$)	94

[8]When water is scarce, Chambers argues, "there is much work of allocation, arbitration and discipline which it is in the common interest of cultivators as a whole and often in their individual interest that someone should carry out . . . . these functions do not lend themselves to being split up among several individuals. They are most effectively vested in one person with adequate powers" (1975:57).

(4) Committees plus assembly ($N = 32$)	122
(5) Assembly primarily ($N = 21$)	95

A good example of the decision making structure that is most often successful, a combination of committees and assembly, is the Mraru women's organization in Kenya (Kneerim, 1980). Decisions such as whether to invest in a new line of activity are discussed within a nine-member executive committee that formulates alternative courses of action. These are presented to the membership as a whole for final decision. Usually the executive committee recommendation is approved, but only if members are satisfied with it. A similar structure is reported by Wasserstrom (1982) for a women's cooperative in Bolivia. Such a decision-making structure combines the advantages of both modes of deliberation: a smaller group for working out details and a larger group for eliciting divergent views and building consensus through participation.

The purely or primarily executive form of decisionmaking is, on balance, unsuccessful. Executive structures were not significantly associated with any disaggregated performance variables in either correlation or regression analysis. We might have expected to find executive governance advantageous when the government and/or bureaucracy is unsupportive, even hostile. But when we controlled for this political-administrative environmental factor in analyzing the performance of LOs with different decision-making structures, we found the advantage of participatory decision structures even greater, as shown by the summary scores in Table 5-3.[9]

The averages shown in the last line of the table reduce the impact of extreme scores from the smaller samples and correspond fairly closely to the scores for the whole sample. This pattern confirms a general association between more participatory modes of decision-making and LO performance. While not equally strong for all kinds of developmental outcomes, it shows up in the analysis as one of the more important structural variables.

(4) *Size.* Some of the most categorical assertions about LO structure concern the preferred size of an organization, although they are sometimes ambiguous: writers can be referring to base-level organizations—the "building blocks" for larger, federated groups—or to the whole organization. A number have argued in favor of small base-level LOs (Hunter, 1976; Tendler, 1976; Buijs, 1982a:54–55). Golladay (1983) suggests that conflict will be less frequent in smaller LOs, and Doherty and Jodha (1979) believe that

[9]The scoring of political (regime) and administrative (bureaucratic) support for an LO as an environmental factor was as follows: 1, active opposition; 2, non-support; 3, indifference; 4, support, some commitment of resources; 5, active support, significant commitment of resources. The cases falling in categories 1 and 2 were not many, and the difference between 2 and 3 was the hardest to assess, so we looked also at categories 1, 2, and 3 taken together.

Table 5-3. Nonsupportive government and LO decision-making (summary scores)

LO ENVIRONMENT	(*N*)	Executive only	Executive + committee	Committee only	Committee + assembly	Assembly only
Political hostility or nonsupport (categories 1 + 2)	(12)	—	−100	200	160	133
Bureaucratic hostility or nonsupport (categories 1 and 2)	(20)	−200	50	57	125	83
Political hostility, nonsupport, or indifference (categories 1, 2, 3)	(57)	−28	66	78	149	126
Bureaucratic hostility, nonsupport, or indifference (categories 1, 2, 3)	(64)	−122	45	33	112	108
Weighted average		−88	45	66	131	113

domination is less likely in small groups. A report of the U.N. Secretary General on cooperatives (A/36/115, March 4, 1981) expresses disapproval of an observed trend toward larger co-ops because of the probability that control and management will get away from the members. Johnston and Clark (1982:182), on the other hand, suggest that smaller LOs are not necessarily better; they discuss kinds of activities and tasks to which larger LOs are likely to be better suited.

Our own view on this subject (1974) has favored a multitiered system of local organization, with small base-level organizations linked into larger entities. Discussions of size almost invariably get into considerations of vertical linkage as a variable, which we address in the next section. In trying to extrapolate from the anthropological literature some optimum size for task groups, Doherty (1980) concludes that 25 to 75 persons (5 to 15 households) constitute an equilibrium range, to be aggregated into larger groups of about 500 persons (100 households) for tasks of "long-term maintenance." Seligson's analysis of land reform cooperatives in Costa Rica (1982:137–38) indicated that members' participation as a function of LO size was at a fairly constant level up to a membership of 100, after which it dropped off.

Duewel (1984), in studying water users' associations in Indonesia, found that base-level units (in a two-tiered organization) tended to have between 10 and 50 members, cultivating 10 to 15 hectares. (Not coincidentally, he finds the lower organization "very informal.") The definition of "small" used in the Taiwan farmers' associations is rather different. Their base-level organizations, called "small agricultural units," have 150 to 200 members;

interestingly enough, however, these are supplemented by teams of 15 to 20 farmers operating informally below the unit level (Stavis, 1974a).

There is some reason for expecting smaller LOs to be more successful for many tasks because of their informality and the social control they can exercise to insure participation and equitable sharing of responsibilities. Yet where we might have expected a significant negative correlation between size and overall LO performance, we found instead a small positive one (.09), though it is not significant at the .05 level. A more definite positive association between size and performance emerged when we controlled for the effects of other structural variables through regression analysis (see this chapter's appendix). There were significant associations between size and a number of specific kinds of LO performance on a *ceteris paribus* basis. Inferring causation is dubious, but we can say that the presumed association between small size and success does not show up in our analysis.

When we compare summary scores for overall performance according to size, there is little difference below 500 members: LOs under 50 members and those between 50 and 100 members both had 74, while those between 100 and 500 members had 67—only slightly lower. A higher score of 127 for LOs with more than 500 members points to the problem of sorting out cause and effect. From the two-way breakdown of distributions between size and performance shown in Table 5-4, we see that none of the least successful LOs were large. This stands to reason, since large membership could hardly be sustained with poor performance. Thus the positive association observed between size and performance appears to indicate that successful LOs grow in membership rather than that large membership causes success. A qualification on this interpretation is that with regard to certain LO tasks, size correlated significantly with control of bureaucracy and

Table 5-4. LO performance according to size (percentage distributions)

PERFORMANCE	(%)	Under 50 (*N* = 63)	50–100 (*N* = 43)	100–500 (*N* = 27)	Over 500 (*N* = 17)
Very poor	(13)	50	35	15	0
Poor	(19)	32	32	25	11
Good	(29)	49	21	14	16
Very good	(22)	39	30	24	6
Outstanding	(17)	36	36	12	16

SIZE (NO. OF MEMBERS)	(%)	Very poor	Poor	Good	Very good	Outstanding
Under 50	(42)	16	15	34	21	15
50–100	(29)	16	20	20	23	20
100–500	(18)	11	26	22	30	11
Over 500	(11)	0	18	44	13	25

claim-making—externally oriented tasks where the clout of greater size probably makes some difference.

Although we understand the rationale of other analysts in maintaining that smaller size itself contributes to better LO performance, we have not found a basis for this assertion in our data. Indeed, when other factors are controlled for as they are in regression analysis, larger size represents an advantage, *ceteris paribus,* for certain kinds of LO performance. One can say, reversing the implied direction of causation, that poor performance will militate against large size. Moreover, we need to bear in mind that size itself is not necessarily a static structural feature. We observed from case materials that successful LOs (such as the San Martin Jilotepeque cooperative in Guatemala) may grow rapidly, moving from one size category to another.[10] Purely statistical analysis cannot unravel all the strands of cause and effect surrounding the factor of LO size. We are able to take the analysis an illuminating step further, however, when we consider it in relation to LO linkages, both vertical and horizontal.

(5) *Vertical linkage.* One of the main contributions of our 1974 analysis of local organizations was to focus attention on linkages between lower- and higher-level LOs. We defined linkage in terms of interaction and exchange—of information and other resources—on a regular and reliable basis, looking at the extent and effectiveness of communication and influence both downward to and upward from different levels of local organization (Uphoff and Esman, 1974:26). In that analysis we began from a macroperspective, whereas here we have started from the microperspective of the LOs themselves.

To assess vertical linkage, we classified LOs as having (1) no connection to any higher organization, (2) linkage to a higher-level group in a two-tier formation, (3) more complex linkage in a three-tier system, and (4) linkage to a regional federation or (5) a national federation. For purposes of subsample comparison, we combined the 5 percent of the sample in regional federations with the 11 percent in national ones. The cross-tabulations between vertical linkage and overall performance are given in Table 5-5. The pattern can be summarized in the following summary scores: unlinked, 45; two-tier, 84; three-tier, 117; and federation, 143. The great majority (80

[10] The San Martin co-op grew from 32 to 732 members in six years. Fifteen marketwomen's cooperatives in Nicaragua grew from a total of 219 members to 5,530 just seven years later, increasing their average size from 15 to 385 (Bruce, 1980). We are reminded by John D. Montgomery (personal communication) that his studies of local organization and development administration, using a "critical incident" method of data gathering, suggest a tendency for members of a successful organization to want to exclude new members. There can be, however, some advantages to existing members of enlarging their organization's membership and resource base. Tendler (1983) reports on how four successful Bolivian cooperatives have not expanded their membership, but their spread of benefits is quite broad.

Table 5-5. Vertical linkage and LO performance (percentage distributions)

PERFORMANCE	(%)	Unlinked (*N* = 81)	Two tiers (*N* = 33)	Three tiers (*N* = 12)	Federation (*N* = 24)
Very poor	(13)	80	10	5	5
Poor	(19)	52	24	10	14
Good	(29)	52	29	5	14
Very good	(22)	61	21	3	15
Outstanding	(17)	29	17	21	33

VERTICAL LINKAGE	(%)	Very poor	Poor	Good	Very good	Outstanding
Unlinked	(54)	20	19	27	25	9
Two tiers	(22)	6	22	37	22	13
Three tiers	(8)	8	25	17	8	42
Federation	(16)	4	17	25	21	33

percent) of LOs that registered very poor performance were unlinked, whereas less than one-third of the outstanding LOs had no vertical linkage.

That regionally or nationally federated LOs tend to be found in the more successful categories probably reflects certain evolutionary processes. Relative success in performance can strengthen LOs' affiliation with higher-level bodies or give them reason and resources to form such bodies. At the same time, the process is interactive, since higher-level bodies can contribute to the further success of their constituent organizations. The data distributed in this way produced a correlation of .22—positive and significant, though not as great as might have been expected.

When we undertook to analyze the effects of a *combination* of size and vertical-linkage variables, we found a more substantial and important relationship. It will be recalled that LO size and overall performance were positively if not always significantly correlated. When we partitioned the total sample to compare vertical linkage in small LOs (membership under 100) and large ones (over 100), vertically unlinked small LOs had lower summary scores for overall performance than did large unlinked ones. But small LOs linked beyond two tiers did *better* than large ones similarly linked, even though small LOs in general had a lower summary score than did large ones. There was little difference according to size for two-tier organizations, both sets performing well above the average score (77) for the whole sample.

	Small (under 100)		Large (over 100)	
No vertical linkage	(*N* = 35)	31	(*N* = 18)	61
Two tiers	(*N* = 15)	106	(*N* = 10)	110
Three or more tiers	(*N* = 12)	134	(*N* = 15)	108
Total	(*N* = 62)	73	(*N* = 43)	87

These scores support the conclusion in our 1974 study that there are advantages in multiple tiers of organization with smaller units at the base. These have the benefits of solidarity, augmented by the advantages of scale and specialized services that a higher level of organization can provide.

The advantage is not automatic. King describes two-tier cooperatives in northern Nigeria in which, once certain members were elected as buyers and given loans for their task, "their connection with the village cooperative came to an end and their relationship was entirely with the cooperative union [at the higher level] under the control of government officials" (1981:260–61). If the smaller base is not an effective unit of operation, the benefits of solidarity and collective responsibility are lost. This is one of the limitations on the effectiveness of the "training and visit" system of agricultural extension as often implemented (Compton, 1982; Howell, 1982).

The implication of our analysis is that vertical linkage, up to and including possible federation at the regional or national level, is a desirable structural feature. But as with other characteristics of LOs, the role of "outside" actors must be handled with care, so that the linkages proposed and assisted are well understood and supported by both the membership and the leadership of the organizations involved.

(6) *Horizontal linkage.* Our previous work in Asia underscored the importance also of horizontal linkages between and among LOs at the same (usually village) level. Tendler (1976), on the other hand, as a result of her study of Latin American cooperatives, has argued for "unconnected" organizations. Johnston and Clark take a middle position, saying they would not recommend forsaking horizontally structured groups, but suggesting that for many purposes simple vertical organization would accomplish development objectives at less cost, particularly for the poor, who must bear the costs of their local organizations. They suggest that certain functions, such as health care or agricultural extension, require little in the way of horizontal organization. "The essential need is for organizational techniques that tap local knowledge, experience and judgment," channeling these to higher-level decision-makers (1982:188).[11]

We would be more willing to accept this view if the results of our data analysis did not point precisely toward the importance of horizontal linkage. Its correlation with overall LO performance was .29, one of the highest for any structural feature, and the summary scores showed some clear ad-

[11]Johnston and Clark (1982) refer to the arguments of Chayanov, who offered a concept of vertical cooperation whereby local outputs and needs are handled by major commercial or administrative organizations at a regional or higher level without requiring any horizontal grouping of the poor. This can be taken to mean that no LOs at all are necessary, only services provided commercially or administratively from above. But we read Johnston and Clark as arguing not for doing without LOs but for keeping the costs of organization to participants as low as possible. They are speaking more of the efficiency than of the equity or empowerment goals of local organization.

Table 5-6. Horizontal linkage and LO performance (percentage distributions)

PERFORMANCE	(%)	Unlinked (*N* = 113)	Occasional cooperation (*N* = 28)	Regular cooperation (*N* = 9)
Very poor	(13)	100	0	0
Poor	(19)	83	17	0
Good	(29)	72	21	7
Very good	(22)	72	19	9
Outstanding	(17)	50	36	14

HORIZONTAL LINKAGE	(%)	Very poor	Poor	Good	Very good	Outstanding
Unlinked	(75)	18	22	28	21	10
Occasional cooperation	(19)	0	18	32	21	29
Regular cooperation	(6)	0	0	33	33	33

vantages: 41 for horizontally unlinked LOs, 143 for occasionally linked LOs, and 200 for regularly linked LOs, calculated from the distributions shown in Table 5–6. Even if one combines the second and third categories (the last is quite small: N = 9), the summary score is 151 for all horizontally linked LOs, compared to 41 for all unlinked ones. This is one of the strongest structural associations with overall performance that we found. Horizontal linkage was not significantly correlated, however, specifically with agricultural, health, or educational gains or with improvements for the poor (see the appendix to this chapter), so part of Johnston and Clark's argument is not contradicted by our data.

One of the most interactive relationships we found in analyzing the data is the reinforcement between horizontal and vertical linkage. Our conclusions cannot be very firm in light of the small size of some of the subsamples. But when we calculated summary scores for LOs having neither, either, or both kinds of linkage, we could see a marked difference in overall performance for LOs having both, with horizontal linkage the more influential of the two.

	No horizontal linkage		Horizontal linkage	
No vertical linkage	(*N* = 74)	36	(*N* = 5)	140
Two tiers vertical	(*N* = 18)	34	(*N* = 14)	146
Three or more vertical	(*N* = 16)	75	(*N* = 18)	166
Total	(*N* = 108)	41	(*N* = 37)	151

LOs with neither kind of linkage averaged a performance score only half that for the sample as a whole. Those with substantial vertical linkage but no horizontal linkage did as well as the average for the whole sample, while vertically unlinked LOs that were horizontally linked had a performance score almost twice as high as the average. Adding substantial vertical linkage boosts the summary score even higher. Thus while both forms of linkage are able to make positive contributions to performance, horizontal linkage seems particularly valuable for effective local organization. The advantages of horizontal linkage also appear in disaggregated analysis of performance, as shown in the appendix of this chapter.

(7) *Government linkage.* The horizontal and vertical linkage we have been considering refers to structural relations within and among LOs. We considered also linkage to government agencies and staff, looking at the frequency of communication and cooperation with officials as well as government control over LO resources. The degrees of linkage went from (1) *autonomy,* with effectively no interaction and no government control over LO resources, to (5) *direction,* where there was heavy interaction controlled by the government. In between were (2) *low* linkage with little interaction, (3) *moderate* linkage with some but not regular interaction, and (4) *high* linkage with much interaction but some reciprocity (that is, LOs were able to maintain some control over the flow of resources to and from themselves).

There is little systematic discussion of this variable in the literature, though there are partisans as well as opponents of government linkage. The "Marxist" and "liberationist" schools of thought discussed in Chapter 2 presume a negative correlation between such linkage and LO contributions to developmental outcomes, at least in terms of equity and empowerment, while the "technocratic" school would expect a positive correlation, presuming that government's intentions and effects are basically benign. Our 1974 analysis suggested a positive association in that we did not find effective development performance associated with "autonomy." But this would not necessarily mean that government linkage should be maximized.

As formulated, the government linkage variable appears linear, with degrees of linkage stated in rank order. But the corollary variable of "reciprocity" between LO and government is not so ordered, because (1) and (5) are both nil in this regard, and (4) represents a maximum. As soon as we began analyzing this variable, its nonlinear nature became clear, as indicated by the following summary scores of performance: autonomy, 107; low linkage, 162; moderate linkage, 63; high linkage, 99; and direction, 7. This presents an unusual bimodal relationship, with low linkage better than lowest linkage (autonomy), high linkage better than highest linkage (direction), and moderate linkage better only than the extremes. How these patterns differ can be seen from the distributions given in Table 5–7, with the mode for each distribution in bold face.

Table 5-7. Government linkage and LO performance (percentage distributions)

PERFORMANCE	(%)	Autonomy (*N* = 19)	Low linkage (*N* = 19)	Moderate linkage (*N* = 28)	High linkage (*N* = 43)	Direction (*N* = 41)
Very poor	(13)	15	0	15	30	**40**
Poor	(19)	3	10	24	21	**42**
Good	(29)	14	7	16	**33**	30
Very good	(22)	18	24	**27**	15	15
Outstanding	(17)	12	20	8	**48**	12

OFFICIAL INVOLVEMENT	(%)	Very poor	Poor	Good	Very good	Outstanding
Autonomy	(13)	16	5	**32**	**32**	16
Low linkage	(13)	0	16	16	**42**	26
Moderate linkage	(19)	11	25	25	**32**	7
High linkage	(29)	14	14	**33**	12	28
Direction	(27)	20	30	**32**	12	7

When we discussed governmental and bureaucratic support in the preceding chapter as part of the political-administrative environment of local organization, we noted the importance of qualitative factors. *How* government linkage with LOs is managed is apparently more significant than *how much* involvement there is. Whether government linkage is beneficial or not depends on its nature and extent, on the kind of stimulus that official resource contributions, suggestions, rules, or guidelines represent—whether they induce greater and more productive activity by LO members or discourage it. King's description (1981:260–61, 276–77) of how government officers related to agricultural cooperatives in northern Nigeria exemplifies the kind of linkage that is unlikely to produce developmental benefits of any kind. How to achieve positive-sum outcomes is one of the themes of our discussion in the rest of this book, having noted statistically the negative effects of excessive government linkage.

A local organization with no government linkage (autonomy) and with no vertical or horizontal linkage represents the "unlinked" extreme. It contrasts with the majority of LOs, which have some kind and degree of linkage with other organizations and/or with government. In our sample the number of LOs in this extreme position was 13, or 9 percent. The summary performance score for this group was only 46, compared with the sample average of 77. The scores for other LOs that had no vertical or horizontal linkage were, according to their degree of government involvement: low linkage, 138; moderate linkage, 49; high linkage, 33; and direction, −22. Although some of the subsamples were fairly small, the differences in scores were substantial enough to support the conclusion that the best situation

Table 5-8. Government/vertical linkage and LO performance (summary scores)

GOVERNMENT LINKAGE	VERTICAL LINKAGE (*N* in parentheses)					
	None		Two tiers		Three or more	
Autonomy or low linkage	(23)	90	(8)	163	(6)	232
Moderate or high linkage	(38)	42	(13)	95	(19)	157
Direction	(19)	−6	(11)	0	(11)	45

for otherwise unlinked LOs is to have at least *some* government linkage.[12] Government linkage at a low or high level, short of direction, produced favorable summary scores when combined with at least some vertical linkage, as shown in Table 5-8.

Unfortunately, the subsamples were too small to permit controlling further for degrees of horizontal linkage, but this would differentiate the observed performance of LOs even more. Our conclusion is that all three kinds of linkage can offer some advantages to LOs, with horizontal connections clearly the most important. Vertical linkage, especially in conjunction with horizontal linkage and small base-level groups, is certainly valuable. Government linkage warrants only qualified endorsement; local organizations are better off with none at all than with too much, although some official involvement appears to be the most favorable relationship. Even considerable government linkage can be quite desirable as long as it does not become directive, and as long as local leaders and members do not lose control of their organization or their feeling of responsibility for it. This may seem a complex prescription, and it is—but the data are quite persuasive on the point.

(8) *Incentives.* The rewards and sanctions associated with membership and with performance or nonperformance of membership roles will structure the way members relate to their organization. We analyzed this variable in terms of the extent to which becoming *and remaining* a member was voluntary or compulsory. On the one hand, attitudes of members toward the LO will presumably be more positive if voluntarism is the basis for membership, with free entry and exit. On the other hand, the LO's resource base and strength may well be greater if it can enforce enough discipline on members so that they cannot simply contribute or not as they please; the "free rider" effect, discussed below, can be a serious problem for LOs.

This variable of incentives reflects the degree of control that the organization—in effect, its leadership—can have over the membership. It represents

[12]If the three middle degrees of linkage (low, moderate and high) are combined to get a subsample of 43, the summary score is 57, compared with 46 for autonomy ($N = 89$) and −22 for direction ($N = 18$). The size of the low, moderate and high categories, respectively, was 8, 14, and 21.

the converse of accountability, the control that members can exercise over their leaders. It is an important issue in the design of LOs because opinions differ as to whether LOs should be purely voluntary, in the Western tradition of "private voluntary organizations," or able to impose some discipline in order to act more cohesively in pursuit of their objectives.

One of the important observations that Akhter Hameed Khan made about the development of what is now known as the Comilla model in Bangladesh was that some discipline was necessary to make small farmer groups viable and effective (Owens and Shaw, 1972:93–98). All members of these groups, if they wished to remain in good standing, had to attend weekly meetings, pay membership dues, and make weekly deposits in their savings accounts. Attendance was important because representatives were going to weekly training sessions at the subdistrict (*thana*) training center and bringing information back to the community. Mobilization of funds gave the groups a basis for making loans and thus reducing members' dependence on private moneylenders. Once the groups were functioning, they could also get government funds to increase the amount lendable to members. The fact that the members had some of their own money in the pool increased their care and responsibility in making loans. Indeed, it was found that once outside funds dwarfed the farmers' own balances, "overdues" mounted rapidly (Blair, 1974).

This variable was not easy to score, because voluntarism and compulsion are in some ways subjective, and the continuum is not unidimensional.[13] The scale we used was as follows: (5) purely voluntary membership, with no restrictions on entry or exit from the organization; (4) quasi-voluntary, with free entry accompanied by some obligations in order to remain a member in good standing (dues, attendance, savings, or other duties); (3) mixed, with some sanctions but a number of voluntary features; (2) quasi-compulsory, with strong leadership sanctions over members (expulsion and exclusion from benefits, for example); (1) compulsory, with membership in the organization obligatory and compliance with the rules enforced by strong sanctions. Since compulsory conditions go against the common conception of LOs, it is not surprising that the literature included very few with such features. The last two categories (containing 19 cases, or 13 percent of the sample) were combined for purposes of analysis. The distributions between performance and incentives are shown in Table 5-9.

More than half of the LOs that performed very poorly were purely voluntary whereas nearly half of those that were outstanding or very good were quasi-voluntary, indicating that this variable—like government linkage—is

[13]Cohen and Uphoff (1977:90–94) identified a scale ranging from volunteered participation initiated from below to enforced participation initiated from above, with rewarded (induced) participation representing intermediate motivation. Unfortunately, the nine combinations of motivation and impetus for participation do not form clearly ordinal steps.

Table 5-9. Incentives and LO performance (percentage distributions)

PERFORMANCE	(%)	Compulsory/ quasi-compulsory (N = 14)	Mixed voluntary and compulsory (N = 19)	Quasi-voluntary (N = 50)	Voluntary (N = 62)
Very poor	(13)	15	10	20	55
Poor	(19)	10	28	17	45
Good	(29)	16	12	35	37
Very good	(22)	6	6	46	42
Outstanding	(17)	12	12	44	32

INCENTIVES	(%)	Very poor	Poor	Good	Very good	Outstanding
Compulsory/quasi-compulsory	(13)	17	17	39	11	17
Mixed (voluntary and compulsory)	(13)	10	40	25	10	15
Quasi-voluntary	(33)	8	10	30	30	22
Voluntary	(41)	18	21	26	33	13

not linear. When we consider summary scores for the four incentive categories, we find compulsory/quasi-compulsory LOs (score, 63) generally performing better than those that are purely voluntary (52) or have mixed incentives (30). By far the best score, however, is for quasi-voluntary LOs (128).

Such figures support the contention of A. H. Khan and others that some discipline is important for success. They do not, however, tell us, what the requirements should be or who should set them. We cannot test statistically the following observation, based on our reading of the case studies, but it is consistent with the overall pattern of results emerging from our analysis: one of the most important areas for local participation is in determining *what* will be the obligations of membership and *how* they will be enforced. There can be outside suggestions, but unless the members agree on the rationale and substance of sanctions, external efforts to obtain discipline are likely to be unsuccessful.

The implication for outside agencies which seek to foster or assist LOs is that pure voluntarism with easy entry and exit is not advisable. The problem of "free riders" has been persuasively commented upon by Olson (1965) and Popkin (1981), among others. Organizations have a difficult time surviving unless all who benefit from them share in the costs. Even if an LO can manage financially without contributions from all members, the feeling that burdens are unfairly shared will undermine its effectiveness.

Some organizations, such as those for processing milk (like the AMUL milk cooperative in India) or tea (the Kenya Tea Development Authority), require a degree of compulsion by the higher-level organization to ensure the quality of the product and the promptness of its delivery for processing;

otherwise, all members suffer.[14] Even harsh penalties against adulteration or delay appear quite reasonable to the majority of members, and those not willing to accept such discipline have to be excluded for the sake of the others. In such instances, even if the rules are handed down from a higher organizational level, there is definite benefit to members, and those who do not wish to comply have the option of exit.

Ordinarily, requirements should be decided by the members themselves, though some restrictions or obligations based on experience elsewhere may be suggested by outside agencies, government or private. It is also important that members be involved in the enforcement of the rules, though we note that their willingness to invoke sanctions against friends and neighbors may be strengthened by some outside involvement. The degree to which rules are enforced becomes an indicator of the value of the LO to its members and a sign of organizational viability (or decline)—provided, of course, that the rules are seen as reasonable.[15] Efforts by outside agencies to maintain high levels of attendance or dues payment are likely to be misdirected. If not achieved willingly by members, such indicators represent hollow progress and may even hasten the LO's decline.

(9) *Participatory orientation*. An organization's rules and values regarding participation and egalitarian outcomes can be structured so as to encourage and reinforce such outcomes, or can be indifferent or even unsupportive. This variable represents in summary form an estimate of how participatory a local organization was in its internal proceedings. It correlates positively with assembly structures of decision-making (.43), but not so much as to be duplicative. The way organizations actually operate can be somewhat different from the way they are formally structured, as we found in our sample. The summary scores for LOs in the different categories were as follow: (5) actively participatory and egalitarian, 108; (4) participatory and egalitarian, 113; (3) neutral, 25; and (2 + 1) nonparticipatory and ineg-

[14]On the Amul cooperative, see Hunt (1974), Somjee and Somjee (1978), Franda (1979:55–63), D. Korten (1980:485–86), and Paul (1982:15–36); on the KTDA, see Steeves (1975, 1978) and Paul (1982:51–62).

[15]In judging enforcement of rules, one must take *seasonality* into account; there may be no reason to maintain certain forms of discipline all year round. This became clear in our fieldwork with groups supposed to manage small dams in Botswana (Roe and Fortmann, 1982). Officials in the capital city thought that these groups were performing poorly, since no regular meetings were held, and user fees were seldom collected from herdsmen to pay for maintenance. However, once the cycle of receding water supply during the dry season made the small dams important to the herdsmen, the groups began to function visibly, though because maintenance was fairly simple and inexpensive (it could be done by voluntary labor), no fees were collected. The user charge per beast that had been set by officials was several times higher than necessary to maintain the dams, giving evidence of the value of involving group members in setting requirements and sanctions. The same caveat applies to the enforcement of irrigation rules by water users' associations when there are differences between wet and dry seasons for cultivation (Duewel, 1984:33).

alitarian, 0.[16] The correlations between participatory orientation and effective performance of specific LO tasks were all relatively high. As shown in this chapter's appendix, the only task for which the correlation was not significant was integration of services.

The general finding that LOs contribute more effectively to rural development goals, broadly defined, when they are oriented to operating in a participatory way should not be surprising. One might ask whether this relationship is simply circular, because of the methodology employed. But one sees the same connection when considering the association of participatory orientation with specific developmental outcomes. None of the correlation or regression coefficients was significantly associated with the nonparticipative end of the scale.

An interesting question of theoretical and practical significance is what happens when the orientation of an LO is at variance with that ot its environment. Since we also scored the immediate community and the broader societal environments in terms of their orientation to participatory and egalitarian outcomes, we can see whether more participatory LOs were more or less successful when "deviant" in their mode of operation. It might be that LOs operating under adverse social conditions should adapt themselves to these circumstances and become more authoritarian in response or downplay their developmental objectives. To investigate this possibility, we examined separately those LOs operating in community or societal environments that were hostile or indifferent to participatory/egalitarian outcomes. The summary scores for the different kinds of LOs in such unfavorable circumstances are given in Table 5-10 and compared with those for the whole sample.[17]

We see that LOs operating under indifferent or adverse normative conditions have less success than the average *unless* they are highly participatory and egalitarian in their operations—and then they showed much more success than the average. This should be a clear signal to any organizations, government or private, seeking to promote development in unpromising situations. Even under indifferent or adverse conditions, LOs that are highly participatory and egalitarian can register an impressive degree of success. The summary scores for such organizations (140 and 192) compare very

[16]We found only one LO in the actively nonparticipatory and inegalitarian category (a mothers' club in South Korea: see Kincaid et al., 1976), so we included it with category (2). The percentage distribution among the categories was (5) 14%, (4) 41%, (3) 35%, and (2 + 1) 10%. Thus almost half were not clearly participatory and egalitarian. Although it is possible in principle for LOs to be participatory but not egalitarian, or vice versa, such a separation did not appear in practice.

[17]We combined the indifferent (3) and adverse (2 and 1) categories of environment because there were not enough in the latter to break down into sub-categories for reliable comparative analysis. The subgroups are fairly small as it is, but the differences are large and consistent enough to be taken seriously.

Table 5-10. LO orientation and normative environment (summary scores)

	Indifferent/adverse community norms (*N* = 54)		Indifferent/adverse societal norms (*N* = 75)		All LOs (*N* = 150)	
	(*N*)	Score	(*N*)	Score	(*N*)	Score
Inegalitarian/ nonparticipatory	(7)	−13	(5)	0	(15)	0
Neutral	(20)	−5	(36)	15	(52)	25
Egalitarian/ participatory	(17)	65	(21)	41	(62)	113
Actively egalitarian/ participatory	(10)	140	(13)	192	(21)	108

favorably with the average success score for the whole sample (77). It takes unusually effective local leadership and sophisticated provision of outside support, however, to operate under negative conditions (we consider such factors in Chapters 8 and 9).

(10–12) *Economic, social, and sex composition.* The composition of LO membership was one aspect of organizational structure for which we expected to find more definite associations, especially since our overall performance measure reflected possible gains in income, assets, or access to services for the poor. The consensus in the literature is that homogeneous groups are more likely to be more effective, particularly for the poorer sectors (e.g., Morss et al., 1976; Buijs, 1982a:55–56; Hunter, 1980b). Golladay cites the experience of the Bangladesh Rural Advancement Committee (BRAC) in this regard:

> If the participants in an organization share common problems and capacities for resolving them, then the likelihood of serious differences in priorities emerging is smaller. The early failures of the BRAC illustrate this problem. Under the initial design both landowners and landless laborers were expected to collaborate in resolving community problems. The experiment encountered serious conflicts among members because the concerns and interests of the landowners were competitive with those of the landless. After the scheme was reorganized to include only the landless, it developed into an effective organization. [1983:27–28; see also Ahmed, 1980]

We would note here again, however, the qualification that Grijpstra (1982a) has formulated (discussed in Chapter 4) in connection with social stratification. How socioeconomic differences are perceived and acted upon can vary. The poor may see inequality as exploitative, or regard it as natural and as offering opportunities for clientage or social mobility. The social meaning of the fact of heterogeneity is not everywhere the same.

Heterogeneity can be found in various aspects of membership, as discussed in Chapter 3. We looked at this variable in both economic and social terms, since LOs can have members with essentially the same economic status but diverse social backgrounds, or vice versa. We classified LOs on a scale from (1) relative homogeneity to (5) substantial heterogeneity in terms of differences within the LO in income and control over assets, and in ethnic, caste, religious, or other characteristics. If an LO's environment is relatively homogeneous, of course, its membership will be also, so composition of membership is not entirely independent of the socioeconomic circumstances. We were also concerned to know of any differences associated with composition of membership according to sex. Simply looking at homogeneity versus heterogeneity in this regard, however, would not be as significant as considering *which* sex, if either, was predominant in the membership, so we classified LOs as (1) all male: more than 95 percent; (2) mostly male: 60–95 percent; (3) mixed; (4) mostly female: 60–95 percent; or (5) all female: more than 95 percent.

An argument in favor of heterogeneous LOs could be that larger ones, if more effective, would benefit poor members more than would smaller homogeneous groups because the larger organizations have more clout and draw on more established leadership. As we have seen, size per se does not appear to be a major determinant of LO performance (correlation only .09). Size, understandably, does correlate with the heterogeneity of LO membership economically (.41) and socially (.18), though not by sex (.00).[18]

We did not find any general association between membership composition and overall LO performance; the correlations were .07, .05 and .01 for economic, social, and sex differentiation. Given the views expressed in the literature, we had expected economic heterogeneity—and quite probably social heterogeneity as well—to be negative and significant. As shown in the appendix to this chapter, economic homogeneity correlates significantly with several specific LO outcomes, including income and service access for the poor. But economic heterogeneity is just as strongly correlated with gains in health and in participation in government decision-making. So disaggregated analysis does not negate the overall correlation. Regression analysis, making *ceteris paribus* assessments, shows economic homogeneity as favoring a few performance outcomes (consistent with the literature) but heterogeneity as preferable for some other objectives. Social heterogeneity seems generally more supportive of rural development outcomes in disaggregated analysis. Only reduced social discrimination is correlated with socially homogeneous LOs. Regression analysis shows definite advantages for social heterogeneity, though only *ceteris paribus,* when other structural factors are

[18]This last correlation is an encouraging confirmation of the methodology, since there was no reason to expect sex composition as scored to have any association with size.

Table 5-11. Homogeneity/heterogeneity and LO performance

Composition	Economic (*N*)	Economic Score	Social (*N*)	Social Score	Sex composition	(*N*)	Score
Relative homogeneity	(23)	35	(62)	72	All male	(25)	64
Low heterogeneity	(31)	142	(42)	50	Mostly male	(51)	116
Moderate heterogeneity	(50)	72	(26)	108	Mixed	(50)	56
Relative heterogeneity	(21)	100	(10)	140	Mostly female	(4)	125
Substantial heterogeneity	(25)	76	(10)	50	All female	(11)	99

held constant. Divergent patterns for LO composition, definitely not linear, can be seen in the summary performance scores comparing different levels and kinds of homogeneity-heterogeneity (Table 5-11).[19]

Since one of our concerns was to know how local organizations would perform in inhospitable economic or social environments, we examined all LOs where the environmental variables had extreme scores (4 or 5) for income distribution, social heterogeneity, social stratification, social discrimination, and sex discrimination. These ranged from one-quarter to one-third of the total sample. Some of the subsamples were rather small, so the reliability of the scores shown in Table 5-12 is less than ideal. Still, they give an idea of what appear to be the relationships between LO composition and success when negative socioeconomic conditions are controlled for.[20]

The data suggest that under inegalitarian conditions, economically and socially heterogeneous LOs are somewhat more likely than homogeneous organizations to achieve a range of success. The smallest subsample of LOs operating under adverse socioeconomic conditions was made up of all-female organizations (shown in the bottom row of Table 5-12). With such a small sample one cannot prove any assertion, but the consistency of high scores suggests that women's organizations can be very effective. The basic finding is that virtually all the sets of LOs under adverse socioeconomic circumstances performed better than the average for the whole sample. One could qualify this by saying that the potential for improvement may have been greater there. But by the same token, the progress achieved may have been more beneficial and is indicative of considerable leadership skill and membership commitment.

[19]Since nine cases could not be scored for sex composition, the *N* for analysis was 141. Only ten percent of the total number of cases were all or mostly women's LOs. Some of the subsamples are fairly small, but the pattern (or lack of one) is consistent enough that larger subsamples would not be likely to produce a more definite correlation in either direction.

[20]Since some subsamples on a five-interval scale were small, the calculations were done for three categories (1 + 2, 3, and 4 + 5). Only 7 out of the 45 subsamples are more than 30, so specific scores are likely to have some margin of error in measurement. This means we need to see whether there is any pattern discernible in the data.

Table 5-12. LO composition and negative socioeconomic conditions

Composition of membership	Unequal income distribution (*N* = 53)		Greater social hetero-geneity (*N* = 43)		Greater social stratification (*N* = 60)		More social discrimi-nation (*N* = 50)		More sex discrimi-nation (*N* = 68)	
	(*N*)	Score	(*N*)	Score	(*N*)	Score	(*N*)	Score	(*N*)	Score
ECONOMIC										
Homogeneous	(19)	104	(17)	92	(22)	88	(21)	93	(28)	70
Intermediate	(15)	86	(10)	150	(21)	111	(19)	136	(22)	86
Heterogeneous	(19)	100	(13)	178	(17)	123	(10)	20	(18)	110
SOCIAL										
Homogeneous	(32)	100	(19)	109	(40)	98	(34)	106	(47)	73
Intermediate	(11)	100	(7)	186	(10)	120	(6)	150	(10)	110
Heterogeneous	(10)	90	(14)	150	(10)	118	(10)	110	(11)	126
SEX										
All/most male	(32)	96	(23)	147	(36)	100	(28)	116	(42)	100
Mixed	(16)	81	(13)	107	(18)	118	(16)	112	(17)	94
All/most female	(5)	180	(4)	150	(5)	180	(5)	160	(7)	73

Structural Correlates of Local and Outside Initiative in Establishing LOs

Consideration of initiative as a variable gets us into the debate over the merits of top-down versus bottom-up approaches to development (Stöhr and Taylor, 1981). These days the most frequently heard views are in support of the latter. On the basis of her extensive observation and analysis of Asian experience, Hollnsteiner (1977) concludes that officially-initiated LOs are unlikely to help the poor or even to be very successful. Hutupea and his associates (1978:174) claim specifically that "viable local irrigator groups are less likely to be established where government plays a dominant role in developing water resources." On the other hand, Buijs (1982b:15) concludes that participatory LOs "are rarely, if ever, founded spontaneously by the participants. Usually they are initiated by some outsider." Further, Johnston and Clark argue that organization from below "often results in both inefficiencies and grave inequities; the organizers, after all, are more likely to be the local elite than the poor. Conversely, initiatives from above have produced some of the most successful and participative local organizations in development history" (1982:174–75). All of these views are stated as probabilities, recognizing that there can be exceptions. We were interested in finding out what probabilities of success were associated with top-down and bottom-up initiative.

We considered what could be regarded as a continuum ranging from "local" to "outside" initiative: (1) local residents, or some number of them, acting mostly on their own without previously recognized leaders; (2) local leaders, including the much castigated "local elites"; (3) shared initiative between locals and outsiders; (4) government agencies; (5) nongovernment agencies. The last category includes international agencies like the Food and Agriculture Organization (FAO), which has sponsored the interesting and often very successful Small Farmer Development Programs in Nepal, Bangladesh, and the Philippines, religious groups like the Catholic Church; private voluntary organizations like OXFAM; and others like universities or research institutes.

In some respects, initiative from members of the last category is more exogenous to the community than that of regular government agencies, especially if they are international bodies. Even a university is in many ways more distant from a rural community than the government staff assigned to the area. But there can be factors of empathy and identification with the community that bridge the gap, especially when the outside agency is playing a catalytic role, as described by Meehan (1978) and Lassen (1980). "Catalyst agents," as distinguished from more conventionally conceived "change agents," are supposed to live with rural people and help them initiate problem-solving, resource-mobilizing, and interest-articulating activities (Grijpstra, 1982a; Buijs, 1982a; Volken et al., 1982; Gran, 1983:165ff; Rabibhadana, 1983).[21]

When a government had set up a special agency to work with rural communities in such a catalytic mode, according to the procedures and philosophy pioneered by nongovernmental agencies, we classified the LOs as being initiated from outside (5). This was done to avoid lumping together LOs that had activist origins with those started under more routine bureaucratic auspices. It means, however, that the variable for initiative is no longer a simple continuum; it reflects not just the initiators' *distance* from the community but also their *style* of initiation.

The summary scores for different kinds of LO initiation turned out to be definitely nonlinear: local residents, 153; local leaders, 138; shared initiative, 50; government agencies, 16; and outside agencies (catalysts), 114. As shown by the distributions in Table 5-13, LOs initiated by catalysts—although some appear to fail—are on the average much more successful than those inspired by regular government programs. Locally initiated LOs tend, however, to be more successful than those started from *either* outside source, (4) or (5). This accounts for the negative correlations between initiative and performance measures in Table 5-15 in this chapter's appendix.

It may be disappointing to many readers that purely governmental approaches to LO formation are so often unsuccessful.[22] Actually, all the

[21]We are using the term "catalyst" in the same way that Grijpstra and Buijs refer to "promoters." This term is similar to *promotores* in Spanish and *animateurs* in French. The "catalyst approach" is dealt with at more length in Chapter 8.

[22]In the regression analysis reported in the appendix to this chapter, outside initiative had the strongest—and most negative—association with overall LO performance.

Table 5-13. Initiative and LO performance (percentage distributions)

PERFORMANCE	(%)	Local residents (*N* = 14)	Local leaders (*N* = 26)	Shared (*N* = 24)	Government (*N* = 53)	Catalysts (*N* = 33)
Very poor	(13)	5	10	10	50	25
Poor	(19)	3	10	24	52	10
Good	(29)	9	14	19	40	19
Very good	(22)	12	24	18	21	24
Outstanding	(17)	16	28	4	16	36

INITIATIVE	(%)	Very poor	Poor	Good	Very good	Outstanding
Local residents	(9)	7	7	29	29	29
Local leaders	(17)	8	12	23	31	27
Shared	(16)	8	29	33	25	4
Government	(35)	19	28	32	13	8
Catalysts	(22)	15	9	24	24	27

positions cited above are in some ways compatible with the results of our analysis. Hollnsteiner and Hutupea are correct in that locally-initiated LOs have a greater probability of success. But Buijs's observation points to the value of catalyst-style initiation that is reflected in the data. Johnston and Clark are right that *some* of the most successful LOs have been initiated by government, and not necessarily in the catalyst mode: for example, the Malawi self-help water supply committees (Glennie, 1979, 1982; Liebenow, 1981), the Taiwan farmers' associations (Kwoh, 1964; Stavis, 1974a), and the growers' committees of the Kenya Tea Development Authority (Steeves, 1975; Paul, 1982:51–62). In such cases, government initiative resulted in LOs that were not only productive but also relatively equitable, though the degree of empowerment that resulted was perhaps lower.

One feature common to more successful government-initiated cases was their structure of small base-level groups linked to larger district, regional, or even national organizations. This is the same principle of vertical linkage found in some of the more successful locally-initiated LOs, such as the *zanjera* irrigation associations in the Philippines (H. Lewis, 1971; Coward, 1979a; Siy, 1982) and the bridge-building committees in the Baglung district of Nepal (P. Pradhan, 1980). Guidelines, procedures, and resources (technical and financial) are provided by government, but the responsibility for making things happen in the locality is conferred upon the committee (or small agricultural unit within the Taiwan FAs). The higher-level organization concentrates on working with those groups that are proceeding well, relying primarily on "demonstration effect" to motivate the laggards. Thus the paternalism that often manifests itself in government-initiated activities is reduced, and local groups must exhibit a degree of self-reliance even within an officially-sponsored scheme. As officials discussing water projects with communities in Malawi told their listeners, "This is not the government's water scheme, it is yours. It will work if you are willing to work. And it is you, rather than government, that will make the decision on whether to proceed, on organizing yourselves into committees, and on deciding the order in which various villages would participate" (Liebenow, 1981:5).

A similar strategy by a nongovernment agency, the Fundacion del Centavo (the Penny Foundation) in Guatemala, is described by Peterson (1982b:131–35) as one that reaches and assists the rural poor. The foundation plays a catalyst role and provides only minimum services to local groups, after making field visits to assess their viability. The organizations are responsible for their own structure and must assume joint responsibility for loans. Unlike most paternalistic government programs "the Fundacion does not seek to ensure the institutional continuity of its primary societies. About one-third of the groups do not request further assistance; some groups affiliate with cooperatives; others find alternative funding sources; and some disband. The Fundacion does not have the resources to encourage dependency and thus allows the dissolution of weak local organizations" (Peterson, 1982b:133).

One relevant variable may be the normative environment in which LO initiation is undertaken. In environments that are indifferent or hostile toward participatory and egalitarian outcomes locally-initiated LOs have somewhat lower summary scores, and catalyst-sponsored LOs have relatively if not absolutely higher scores. Shared initiative seems to work poorly if community norms are unfavorable, and government initiative shows the poorest results—not surprisingly—where societal (regime) norms are adverse. The implication of these scores is that outside agents are likely to be needed more when local norms are adverse or indifferent, and it is encouraging to note that they have reasonable prospects of starting successful LOs under these unfavorable conditions.

In case readers suspect that government linkage will "spoil" locally-initiated LOs, we should note that such a result is not evident from the data.[23] LOs that start and remain entirely without interaction and cooperation with outside agencies do not perform better than the average of those that have

[23]Because only 40 of our cases were started by local residents or local leaders, some of the subsamples for degrees of government linkage after establishment are fairly small, but they can be combined into a larger, more reliable subsample as shown below. Here are the *N*s and summary scores for locally founded organizations, distinguished according to degrees of government linkage.

Autonomy	(15)	140		
Low linkage	(9)	230	}	
Moderate linkage	(11)	36	(25)	144
High linkage	(5)	220	}	
Direction	(0)	—		
	(40)	142		

some involvement with government agencies. Further, none of the LOs locally established reached the extreme of total direction by officials. So we include that government linkage, even to a substantial degree, can be positive for locally-initiated organizations.

When we analyzed the relationships among structural variables including initiative, we found that two alternative patterns, associated respectively with local initiative and outside initiative, could be identified from the correlations shown in Table 5-14. A discussion of them makes clearer why the overall statistical associations reported here are not higher—because there are a number of countervailing relationships. When we deal with correlations, inferences can be made in either direction. If formalization in LOs correlates with their initiation, as it does, we can say either that locally-initiated LOs have the advantage of informality or that outside-instigated LOs have the disadvantage of formality, since formalization correlates negatively with overall performance. In the discussion that follows, we use the significance interval of .05 not to test any hypothesis but rather to help determine which correlations should be given attention. Those that are not large enough to be significant offer less certain guidance.

Locally-initiated LOs appear to have the advantages not only of being more informal in their operations but also of using assembly modes of decision-making, which correlates positively with performance. (Conversely one could say that outside-initiated LO's have the disadvantages of executive decision-making, which goes along with their greater degree of formalization.) Locally-initiated LOs should also benefit from having more horizontal but less government linkage. The association of local initiation with smaller LOs and outside initiation with larger ones implies no clear-cut advantage either way, because size per se has little correlation with performance. Only in *ceteris paribus* terms is larger size an advantage, and then it may be as much an effect as a cause of success.

One reason size is so ambivalent in its relation to performance is that smaller LOs are associated with assembly decision-making and more informal procedures, while larger ones more frequently have horizontal and vertical linkages (the correlations are shown in Table 5-14). So smaller and larger groups are correlated with *different sets* of favorable features. As we have seen, the performance of small base LOs improved remarkably when they had both vertical and horizontal linkage. Conversely, larger LOs might be strengthened by more informal modes of operation, though these might be difficult to manage—as would assembly decision-making—in organizations with many members.[24]

[24]We would note additionally that the structural feature most likely to be positively associated with both LO performance and outside initiation is multiple functions. Yet, as we have said, it may not be advantageous for LOs to begin by taking on many activities. The implication is that LOs instigated from outside should seek to build on whatever strength multiple functions can give them, while locally initiated LOs should be encouraged and assisted to

Table 5-14. Correlations among structural variables

	Functions	Formalization	Decision-making	Size	Vertical linkage	Horizontal linkage	Government linkage	Incentives	Participatory orientation	Economic composition	Social composition	Sex composition
Initiative (outside vs. local)	.28*	.36*	−.17*	−.16	−.03	−.22*	.53*	.09	−.08	−.12	.05	−.10
Functions (multiple vs. single)	—	.06	.03	.10	.30*	.17*	.14	.05	.13	.09	.04	.06
Formalization (formal vs. informal)		—	−.34*	.30*	.21*	−.04	.47*	−.02	−.30*	.24*	.22*	−.03
Decision-making (assembly vs. executive)			—	−.21*	−.14	.02	−.37*	.20*	.43*	−.32*	−.10	−.12
Size of base organization (large vs. small)				—	.15	.17*	.03	−.07	.04	.41*	.18*	.00
Vertical linkage (federation vs. none)					—	.35*	.14	−.11	.02	.32*	.26*	.04
Horizontal linkage (cooperation vs. none)						—	−.16	−.05	.05	.15	.11	−.05
Government linkage (direction vs. autonomy)							—	−.21*	−.23*	.13	.04	−.10
Incentives (voluntary vs. compulsory)								—	−.04	−.14	.05	.19*
Participatory orientation (egalitarian vs. inegalitarian)									—	−.22*	−.15	−.03
Economic composition (heterogeneous vs. homogeneous)										—	.40*	−.04
Social composition (heterogeneous vs. homogeneous)											—	−.09
Sex composition (female vs. male)												—

*Significant at .05 level.

Because there are so many correlations among structural factors, it is difficult to sort out their respective effects in bivariate analysis. We approached multivariate analysis with some caution. The specific results of stepwise regression analysis, using structural variables to account for LO performance, are discussed in the appendix to this chapter. Initiation by government is not significantly associated with *any* disaggregated performance outcomes. But this reflects probable rather than necessary relationships, since it is possible for government-initiated LOs to be quite successful. Looking at all the data, we conclude that *how* the initiation is attempted and *what kind* of LO emerges are more important than the simple fact of *who* established it.

APPENDIX TO CHAPTER 5

LO Structural Characteristics and Performance of Functions

We first review the way in which various structural characteristics of LOs appear to be associated with the performance of tasks described as "functions" in Chapter 3. Table 5-15 presents the correlations between LO structural features and performance of functions. Before commenting on specific relationships, we would note, first, that the correlations of structural features with performance of the four *sets* of tasks (internal, resource, service, and external tasks) are generally quite similar; for example, the number of functions correlates .13 with planning and goal-setting, and .17 with conflict management. The average difference between correlations within sets was only .07. Second, the *average* correlation for any particular structural feature with the various functions taken together, shown in the next to last column, is generally quite similar to the correlation of that feature with overall performance, shown in the last column. For example, the average correlation of formalization with performance of all functions is −.22, and

---

evolve toward multiple functions. The structural feature which seems least "determined" in advance by the source of LO initiation is vertical linkage (see Table 5-14). Since this is positively associated with performance, it means that an LO should aim at developing connections from base-level organization up to district, regional, or national levels.

The associations that different compositions of membership have with other features further point up some of the complexity of structural relationships. Local organizations that were economically and socially homogeneous are also smaller—or, in other words, larger LOs are more likely to be heterogeneous. Small, homogeneous, LOs have the advantages of more informality and more participatory relationships, while large, heterogeneous LOs are more likely to have vertical and horizontal linkage, an aid to better performance. This helps to explain why economic or social composition per se had little correlation with overall performance. The only significant associations for composition by sex were women's LOs with more voluntaristic, less compulsory incentives and men's LOs with more coercive ones.

Table 5-15. Correlations of LO structure with functions and performance

	Planning/ goal-setting	Conflict management	Resource mobilization	Resource management	Service provision	Service integration	Control of bureaucracy	Claim-making	Average for functions	Overall performance
Functions (multiple vs. single)	.13	.17	.19*	.12	.25*	.19	.24*	.29*	.20	.30*
Formalization (formal vs. informal)	−.30*	−.28*	−.24*	−.24*	−.22*	−.15	−.13	−.18	−.22	−.21*
Decision-making (assembly vs. executive)	.25*	.21*	.19*	.21*	.15	.15	.31*	.17	.21	.22*
Size of base organization (large vs. small)	.06	.05	.14	.01	−.05	.10	.28*	.20*	.09	.09
Vertical linkage (federation vs. none)	.03	.15	.16	.07	.21*	.04	.11	.22*	.12	.22*
Horizontal linkage (cooperation vs. none)	.26*	.27*	.24*	.18*	.13	.21	.31*	.27*	.23	.29*
Government linkage† (direction vs. autonomy)	−.35*	−.25*	−.24*	−.20*	−.15	−.09	−.35*	−.37*	−.25	−.22*
Incentives† (voluntary vs. compulsory)	.12	−.12	−.05	−.03	−.04	−.06	.16	.17	.03	−.02
Participatory orientation (egalitarian vs. inegalitarian)	.29*	.32*	.24*	.30*	.25*	.17	.26*	.28*	.26	.28*
Economic composition (heterogeneous vs. homogeneous)	−.05	.04	.03	−.09	.05	−.10	−.05	.00	−.01	.07
Social composition (heterogeneous vs. homogeneous)	.06	−.04	.01	−.08	.00	−.03	.06	.08	.01	.05
Sex composition† (female vs. male)	.00	−.02	.11	−.04	.03	−.09	−.02	.09	.01	.01
Initiative† (outside vs. local)	−.20*	−.15	−.13	−.17	−.04	−.01	−.08	−.19	−.11	−.16

*Significant at .05 level of confidence.
†Structural variables that are nonlinear; coefficients do not state relationship well.

with overall performance −.21. The consistency of association of structural features with LO out*puts* and out*comes* (task performance compared with developmental improvements) enhances the confidence one may place in either set of correlations. The average difference between correlations in the two right-hand columns in Table 5-15 is only .04.

We offer this kind of comparison, which does not simply and directly test hypotheses, as a way of probing the data for what patterns may be evident. We find, for example, that the number of functions correlates more with overall performance (.30) than with the average for all tasks (.20). This might reflect a bias in scoring overall performance so that LOs with multiple functions come out higher; success in performing generic LO tasks should be less subject to such a bias. The fact that we see a consistent, positive association between number of functions and performance of each task indicates some apparently favorable connection between multiple functions and LO performance, however measured. It gives us more confidence in the .30 correlation between number of functions and overall performance. Also, as shown in Table 5-16, the number of functions correlates positively with the degree of success in *all* disaggregated performance measures, further supporting the conclusion about the generally positive value of multiple functions, subject to the usual caveats about inferring causation.

As seen in Table 5-15, multiple functions appear more supportive of success in external than internal tasks. It is not clear from the data that conflict is more likely to be a problem with multiple functions (Golladay, 1983), since conflict management seems to be performed somewhat more successfully when there are more functions (confidence level .10). Formalization in LO operations seems to present more difficulties in internal and resource tasks than in service and external tasks, as indicated by the more negative correlations. (The average correlation for the first four tasks is −.27, compared with −.17 for the last four tasks.) Assembly as compared with executive decisionmaking shows a positive association with successful performance of tasks, especially with internal and resource tasks though also with control of bureaucracy. That larger size of LOs at the base is associated only with external task performance is an interesting and understandable finding.

Vertical linkage is significantly associated only with service provision and with claim-making (not, interestingly, with service integration), whereas horizontal linkage is definitely associated with all but the service tasks.[25] Government linkage is negatively associated with all tasks, though less so with the service tasks. As this is not a linear variable, correlation does not

[25]Our reference to "significance" in this discussion departs from the notion of statistical significance (how likely it is that the observed correlation is different from zero) to introduce some threshold of substantive significance (how seriously the coefficient should be taken). Since we are not dealing with a properly random sample, the significance interval is better seen as providing a check as to which correlations, if any, we attribute "significance."

reflect well its relationship with performance.[26] Incentives analyzed according to the five-interval scale we have described do not correlate significantly with any of the tasks or with overall performance, because that variable is also nonlinear (we saw previously that quasi-voluntary LOs had the best performance on average).[27] A participatory and egalitarian LO orientation, on the other hand, correlates consistently and positively with overall performance and with all tasks. None of the membership composition measures correlated significantly with tasks or overall performance.

The most interesting association observed is that of initiative, shown at the bottom of Table 5-15. The fact that all the coefficients with tasks are negative, though the association with service tasks is essentially nil, indicates more favorable outputs and outcomes from LOs established by *local* than by outside initiative. However, this is a complicated variable; when LOs are set up from outside in a "catalyst agent" mode, the results are generally more favorable than in those established under conventional bureaucratic auspices. Thus we find one kind of outside initiative fruitful, and the other usually not.

Structural Features and Specific Developmental Outcomes

Our analysis in this chapter focused on performance in terms of an overall assessment of successful contributions to developmental objectives. To have tried to deal with all the components of performance as we went along would have made the exposition encyclopedic and difficult. We do, however, want to consider what associations emerge when specific kinds of LO performance are compared with structural variations. The correlation coefficients for this are shown in Table 5-16 along a continuum of outcomes from efficiency to equity to empowerment. A consideration of the same relationships analyzed by multiple regression follows. The results can be quickly reviewed.

Economic gains. Few structural features are significantly associated with either agricultural or nonagricultural gains. We find horizontal linkage correlated positively with the latter (though with an N of 38, the significance level is .07. Otherwise the average correlation is .06 and .08 for the other

[26]Combining autonomy and low linkage to represent the lowest value on the scale, keeping moderate linkage as the middle value, and combining high linkage and direction to constitute the highest value makes this a more linear variable, with summary scores of 134, 63, and 54. When such a three-interval scale is used, the average correlation with tasks becomes only −.11; with overall performance it is −.06, not significant.

[27]If this variable is also analyzed according to a three-interval instead of five-interval scale, to contrast voluntarism generally with more compulsory approaches, the correlation with overall performance is .23 and the average correlation with tasks is .16, indicating a positive association between voluntarism and performance. Charlick (1984) found coercion significantly correlated with none of the performance measures except participation in implementation activities.

Table 5-16. Correlations of structural features with developmental outcomes

	Agricultural production/ income	Nonagricultural production/ income	Education	Health	Income for poor	Assets for poor	Access to services for poor	Less sex discrimination	Less social discrimination	Participation in government decision-making	Participation in community decision-making
Functions (multiple vs. single)	.08	.33*	.44*	.13	.28*	.17	.31*	.18	.17	.19*	.33*
Formalization (formal vs. informal)	−.33*	−.02	−.27	−.26	−.31*	−.18	−.34*	−.05	−.23	−.08	−.08
Decision-making (assembly vs. executive)	.03	−.15	.17	−.08	.26*	−.08	.20*	.08	.19	.02	.27*
Size of base organization (large vs. small)	−.04	.12	−.17	.10	−.08	.06	−.09	.23	.12	.17*	.13
Vertical linkage (federation vs. none)	.07	.10	.16	.09	.10	.00	.26*	.07	.11	.34*	.18*
Horizontal linkage (cooperation vs. none)	.10	.30	.06	−.14	.13	−.01	.08	.27	.41*	.28*	.29*
Government linkage (direction vs. autonomy)	−.05	.00	−.09	−.07	−.23*	−.01	−.11	−.42*	−.18	−.22*	−.20*
Incentives (voluntary vs. compulsory)	.10	−.15	.21	.15	.05	−.27	.12	.41*	.15	.01	−.02
Participatory orientation (egalitarian vs. inegalitarian)	.12	.02	−.08	−.05	.39*	.08	.36*	.12	.17	.07	.23*
Economic composition (heterogeneous vs. homogeneous)	−.07	.03	.11	.37*	−.34*	.04	−.22*	−.12	−.26	.22*	.14
Social composition (heterogeneous vs. homogeneous)	−.07	.20	.20	.23	−.13	.23	−.03	−.11	−.38*	.17*	.25*
Sex composition (female vs. male)	−.01	−.05	−.05	.08	.15	.11	.31*	.54*	.47*	.01	.07
Initiative (outside vs. local)	.01	.01	.23	.04	.00	−.11	.02	−.35*	−.37*	−.18*	−.08

*Significant at .05 level; level varies with number of cases for which degree of success for that development outcome could be scored.

features, indicating that good results can be achieved with a variety of structural arrangements.

Social benefits. These are not much associated with any particular structural feature, either, though we see literacy and enrollment gains correlated with multifunctionality, and health improvements with economic heterogeneity.

Equity effects. This is an area of performance where structural arrangements seem to have more importance, though none of the features is significantly correlated with increasing assets for the poor. Increased income and access to services for the poor are correlated with multifunctionality, informal modes of operation, assembly decision-making, participatory orientations, and economic homogeneity (the first significant association we find for this latter variable). Lack of government linkage is associated with improved income for the poor but not so much with their access to services. Better access to services results more often in all-female organizations than in all-male ones.

Reduced discrimination. This is associated with horizontal linkage and with less government linkage, as well as with all-female LOs. The more socially homogeneous organizations appear better able to reduce social discrimination, just as the more economically homogeneous LOs more readily reduce economic disparities.

Participation in decision-making. In this area we also found a number of significant associations. Multifunctional, larger, vertically and horizontally linked, governmentally unlinked (that is, more autonomous), and economically and socially *heterogeneous* LOs were more able to increase the voice of rural people in governmental levels of decision-making. The pattern for increased empowerment at the community level was slightly different. Multifunctionality appeared important, but assembly modes of decision-making emerged more definitely. Also quite important was participatory orientation, and social heterogeneity showed more influence than economic heterogeneity. These are all interesting findings, not counterintuitive but suggesting distinctions in relative importance that would otherwise be impossible to discern.

If we look at the different structural features across all the disaggregated performance measures, we find that *multifunctionality* is generally positively associated, just as *formalization* is negatively associated.

There are a few significant correlations for decision-making. *Assembly structures* appear less important for nonagricultural improvements, for health gains, and for redistributing assets to the poor, as well as for participation at the government level; they appear reasonably important for participation at the community level and for increasing income and services for the poor.

Size of base-level organization appears to make some contribution to empowerment, but not to efficiency gains. Larger LOs may help in reducing discrimination, but their relation to other equity gains is not consistent.

Vertical and horizontal linkage are not especially important for efficiency or productivity improvements, and their relationship to equity is mixed. Vertical ties do appear to assist in improving services for the poor, and horizontal ties are associated with reducing discrimination. Both kinds of linkage are positively associated with our measures of empowerment.

Government linkage is generally negatively associated with specific gains, though not clearly with efficiency gains or with redistributing assets for the poor. It is reasonable that better official ties could be helpful in these specific areas of performance. We do not find that such involvement has been useful in reducing discrimination or increasing people's empowerment. (The negative correlation with reduced discrimination is consistent with the relationship observed in Chapter 4 on environment, where political and administrative support were *not* associated with reduced discrimination.)

Few significant correlations are observed for voluntary or compulsory *incentives*. The former are associated with reduced sex discrimination, and the latter with redistributed assets for the poor. (With N = 40, this coefficient is significant at the .10 level.)

The most interesting finding in this disaggregated analysis is that *economic homogeneity* does show up as a contributing factor to improving income and services for the poor, as many have argued. Economic and, to a lesser extent, social heterogeneity are more important for empowerment measures, though homogeneity is better for reducing discrimination, as it is for gains in education and health. *All-female composition* is more favorable for equity and reduced discrimination, but not for empowerment or for economic and social benefits.

Results of Regression Analysis

Correlations are not qualified by any controls and represent "average" relationships. With multiple regression, we can get relationships at the "margin" by controlling for other variables statistically. The beta weights (and their significance levels in parentheses) are shown in Table 5-17, along with the R^2 for the "best" regression equation, containing only structural variables significant at the .05 level. Because we are interested in the *direction* of association (positive or negative) of structural features with specific performance outcomes, to see whether this disaggregated analysis produces any relationships different from those that emerged when we addressed

Table 5-17. Beta weights of structural features in multiple regression with development outcomes

	R^2	Functions	Formalization	Decision-making	Size	Vertical linkage	Horizontal linkage	Government linkage	Initiative	Normative orientation	Economic heterogeneity	Social heterogeneity	Female membership
Overall performance	.26	**.26** (.00)					**.20** (.01)		**−.30** (.00)	**.20** (.01)			
Agricultural production	.12		**−.34** (.00)		.15 (.16)			.17 (.11)					
Nonagricultural production	.15	**.39** (.02)					.18 (.30)					.18 (.27)	
Education improvements	.30	**.40** (.00)		.19 (.17)	−.23 (.10)				**−.33** (.02)			.22 (.10)	
Health improvements	.17		**−.31** (.05)			−.18 (.28)	−.20 (.23)				.17 (.32)	**.36** (.03)	.19 (.26)
Income for poor	.32	**.25** (.01)	−.10 (.29)						−.15 (.10)	**.26** (.01)	**−.33** (.00)		.12 (.16)
Assets for poor	.20	.20 (.26)	−.20 (.24)		.17 (.29)	−.30 (.11)				.32 (.06)		**.47** (.02)	
Access to services for poor	.39	.15 (.10)	**−.25** (.01)			**.33** (.00)			−.11 (.21)	**.17** (.05)	**−.28** (.00)		**.28** (.00)
Reduced sex discrimination	.37	**.28** (.05)			.15 (.33)		.15 (.29)		−.18 (.20)		−.17 (.23)	.21 (.12)	**.59** (.00)
Reduced social discrimination	.38			**.51** (.01)	**.36** (.05)		.18 (.27)	.26 (.11)			−.24 (.18)		**.43** (.01)
Participation in government decision-making	.15				.11 (.18)	**.34** (.00)	.09 (.30)	**−.21** (.01)				.13 (.13)	
Participation in community decision-making	.27	**.24** (.00)		**.28** (.00)	.14 (.08)		**.21** (.01)			.11 (.20)	.09 (.33)	**.26** (.00)	.10 (.20)

R^2 shown is for "best" regression equation, having largest number of independent variables significant at .05 level; beta weights for these variables are boldface; beta weights for variables having significance level up to .33 are also shown; significance levels are given in parentheses.

overall performance, we show coefficients up to a confidence level of .33. That is, we consider coefficients that have at least a two-in-three probability of being correct according to statistical inference.

Because regression analysis proceeds with the assumption of linearity in all variables, the structural features of government linkage and initiative were recoded as follows (along the lines discussed in this chapter): government linkage—(1) autonomy plus low linkage, (2) moderate linkage, (3) high linkage plus direction; initiative—(1) local residents plus local leaders, (2) catalyst agents, (3) shared initiative, (4) government. These alignments represent more approximately linear continua. The variable of incentives could not be recoded in a way that was linear and theoretically valid, so it was dropped from the regression analysis.

For most of the variables there is essentially no difference between the results of correlation and regression analysis, or between the relationships observed for structural factors with overall performance and with disaggregated performance measures. This can be seen, for example, in a comparison of the respective correlation coefficients, regression beta weights, and significance levels (up to .33) for the number of *functions:*

	Correlation		Regression	
Overall performance	.30	(.00)	.26	(.00)
Nonagricultural production	.33	(.04)	.26	(.00)
Education improvement	.44	(.00)	.40	(.00)
Income for poor	.28	(.00)	.25	(.01)
Assets for poor	.17	(.29)	.20	(.26)
Access to services	.31	(.00)	.15	(.10)
Reduced sex discrimination	.18	(.26)	.28	(.05)
Reduced social discrimination	.17	(.33)		
Participation in government decision-making	.19	(.02)		
Participation in community decision-making	.33	(.00)	.24	(.00)

Neither correlation nor regression analysis showed any relation approaching significance between number of functions and health improvement. Only two of the beta weights for functions regressed on specific performance outcomes were negative: agricultural production, −.02 (.82), and reduced social discrimination, −.01 (.96). Only one of the correlation coefficients was negative: transportation improvements, −.43 (.04; N = 23).[28] This indicates that with the possible exception of agricultural improvement and the probable exception of transportation, single-function LOs did not have an advantage on average or at the margin.

To review the results of regression analysis, some of which have been reported in the chapter already, there are no favorable associations between

[28]The other service outcomes for which the N was too small to produce many statistically reliable results were nutrition .25 (.34), water supply .07 (.71), and public facilities .46 (.08). These point in the same direction as the rest of the analysis, though they might have been expected to benefit from single-function organization.

formal LOs and specific outcomes even at the .33 level; on the other hand, there were a number of associations between informal organization and specific performance gains—agriculture, health, and all three equity outcomes. These are very similar to the associations observed in correlation analysis.

There were no associations between executive decision-making and specific development outcomes up to the .33 level in multiple regression analysis, whereas educational gains, reduced social discrimination, and participation in community decision-making were associated with assembly structures. This is interesting because in simple correlation analysis there had been negative correlations between assembly decision-making and nonagricultural gains, health improvements, and assets for the poor (significance levels of .38, .62 and .64). The inferred connection between more participatory decision-making structures and development outcomes holds up in this disaggregated and controlled analysis.

One variable that shows a different marginal relationship from its average is *size*. Larger LO base organizations are associated at the .05 confidence level only with reduced social discrimination, but up to the .33 level we find agriculture, assets for the poor, reduced sex discrimination, and both kinds of increased participation positively associated with size—while educational gains are associated with smaller LOs in such analysis. This suggests, as stated in the chapter, that although the average relationship is not very great (the average correlation for size with performance outcomes in Table 5-16 was relatively low), larger size is an advantage if other factors are equal. This may, of course, be a reflection of effective performance rather than a cause of it.

Whereas there are no significant correlations of development outcomes with the *absence* of vertical or horizontal linkage, or with government linkage, in regression analysis the picture is not quite so sharp if we go up (or down) to the .33 level of confidence. Health gains may be associated with no vertical or horizontal linkage, and assets for the poor with no vertical linkage. A good number of performance measures are positively associated with horizontal linkage and also with vertical linkage. In regression analysis, increased participation in government decision-making is significantly associated (at the .01 level) with a greater degree of independence (autonomy), whereas agricultural gains and reduced social discrimination are associated with government linkage (at the .11 level).

The association between egalitarian norms in LO structure and performance, observed for overall performance and in correlation analysis, holds up in the multiple regressions, as does the association between local initiative and performance.

Disaggregated performance analysis produced its most interesting results with regard to LO membership structure. Correlation analysis showed no significant overall associations between overall performance and membership composition. But we found some significant correlations with specific performance outcomes. Gains in health and in participation were correlated with economic *hetero*geneity, while increases in income and access to services for the poor were correlated with economic *homo*geneity. These same results emerged from the regression analysis with the addition of reductions in discrimination associated with economic homogeneity (at confidence levels between .18 and .23).

Social composition had no association with overall performance, but social *hetero*geneity generally correlated with specific outcomes. The exception was that social *homo*geneity in LO membership correlated with achieving reductions in social discrimination. In regression analysis, the significance of social heterogeneity stands out more clearly and strongly.

Overall performance had the least correlation with sex composition, indicating that in general it made no difference whether LO membership was predominantly male or female. But female membership was significantly correlated with several specific LO performance gains (access to services, reduced sex discrimination, and reduced social discrimination). In regression analysis this relationship holds up, and other development gains are also associated with greater female membership—health improvements, income for the poor, and participation in community decision-making.

The consistency of results between correlation and regression analysis, and between analysis of overall and disaggregated performance measures, can be seen in Table 5-18, which compares correlation coefficients and beta weights both for direction of relationship (sign) and levels of significance. In the latter regard, significance at the .05 level or better is indicated by a capital letter; significance up to the .33 level is shown with lower-case. In only two instances were there differences in sign *even up to the .33 level* i.e., in the relation between government linkage and reduced social discrimination, and between initiative and education improvements. These small discrepancies could be due simply to our recoding, described above, for government linkage and initiative.

The discussion in this chapter and particularly in this appendix will probably have been too detailed for many readers, yet not detailed enough for a few. We have checked the relationships in many ways and are confident that the patterns of association—in terms of magnitudes, signs, and statistical significance levels—support our conclusions and generalizations. We are not relying on any particular correlation or coefficient in isolation. We recognize, however, that such use of statistical analysis crosses the bounds of many conventions and must be treated as descriptive rather than conclusive. No firm hypotheses were formulated and then tested according to the strict canons of statistical demonstration. As reviewers of the manuscript commented, our work was bound to leave some statistical purists dissatisfied. But we think we have been able to illuminate the terrain of local organization structure and experience, even if some shadows remain.

Table 5-18. Comparison of results of correlation and multiple regression analyses

	Functions	Formalization	Decision-making	Size	Vertical linkage	Horizontal linkage	Government linkage	Initiative	Normative orientation	Economic heterogeneity	Social heterogeneity	Female membership	Voluntary incentives
Overall performance	R B	−R	R	r	R	R B	−R	−R −B	R B				
Agricultural production		−R −B		b		r	b		r				r
Nonagricultural production	R B					r b					r b		
Education improvements	R B	−r	r b	−r −b				r −B			r b		r
Health improvements		−r −B			−b	−b				R b	r B	b	
Income for poor	R B	−R −b	R		r	r	−R	−b	R B	−R −B	−r	r b	
Assets for poor	r b	−r −b		b	b				b		r B		−r
Access to services for poor	R b	−R −B	R	−r	R B		−r	−b	R B	−R −B		R B	r
Reduced sex discrimination	r B			r b		r b	−R	−R −b		−b	b	R B	R
Reduced social discrimination	r	−r	r B	B		R b	−r b	−R	r	−r −b	−R	R B	
Participation in government decision-making	R			R b	R B	R b	−R −B	−R		R	R b		
Participation in community decision-making	R B	−r	R B	r b	R	R B	−R	−r	R b	r b	R B	b	

R = correlation coefficient sig. .00–.05 r = correlation coefficient sig. .06–.33
B = regression beta weight sig. .00–.05 b = regression beta weight sig. .06–.33

CHAPTER 6

Vulnerabilities of Rural Local Organizations

The ability of rural residents to advance their common and individual interests depends substantially, as we have seen, on their success in sustaining local organizations beyond the immediate tasks that precipitated them. The extension of the state and of commercial exchanges into rural areas—trends that are well underway in most countries—present rural organizations with opportunities, but also with threats to their ability to defend and promote the interests of their members. Whether they are small-scale, informal, face-to-face associations; larger-scale, more formal, traditional structures; or modern organizations conceived and introduced by governments or voluntary agencies—all face many of the same difficulties.

One may ask why, if rural local organizations are able to enhance the productivity and well-being of their members, they are not more often effective and not more widespread. In this chapter, we shall assess the major obstacles to the establishment, maintenance, and effectiveness of local organizations, particularly those that represent and serve the rural poor. These vulnerabilities can be grouped generally into five categories: resistance, subordination, internal division, ineffectiveness, and malpractices.

Some of these were discussed at length by Michels (1915) when he examined the performance and weaknesses of political organizations and unions established in Europe at the turn of the century to further the interests of what was then the poor majority in *those* "underdeveloped" countries. While such organizations were necessary to advance those interests, they were vulnerable to various kinds of external manipulation and self-inflicted disabilities. The now-famous "iron law of oligarchy"—the subordination of an organization to its leadership—was not then the only hazard facing organized efforts to improve the situation of disadvantaged groups, nor is it now.

In our review of case studies, we have been able to glean considerable information about the maladies, and their antidotes, of local organizations.

This material does not lend itself to statistical analysis, however, because it is incomplete. When problems such as internal division or financial mismanagement are not reported for a particular case, we have no way of knowing from the literature whether these are nonexistent, were simply not observed by the author, or have been screened out. We have noted an inverse relationship between the number of problems an organization acknowledges and the forthrightness with which it tackles them. Those who have the most problems often say the least about them.

We have been impressed, for example, with the way Akhter Hameed Khan and others working at Comilla in the Pakistan Academy for Rural Development (later the Bangladesh Academy for Rural Development), while making more progress than most other programs, were self-critical in their reports on their efforts to establish farmers' committees (Raper, 1970:210–26). Similarly, when the director of marketing for the farmers' associations (FAs) in Taiwan visited Cornell in 1975, she detailed for us many problems the FAs had encountered with participation and management. Yet one of our colleagues, who had studied FAs firsthand shortly before that (Stavis, 1974a), had found that despite such problems, their performance and contribution to rural development were well above the average for Third World countries. So we start our analysis knowing that virtually all LOs will have some problems if they are attempting to accomplish worthwhile developmental change, but also suspecting that outsiders' and even insiders' reports will not be representative or complete. Indeed, the most critical accounts may be those of the most successful programs.

This "bias" means that standard quantitative modes of analysis are not appropriate. Yet the case materials offer much evidence of the kinds of problems that LOs encounter. We shall review that evidence in this chapter to characterize the obstacles that LOs and their supporters face. The materials also provide numerous examples of measures that can counter the negative factors. These provisions—treated in Chapter 7—may not be effective in all situations, but they can give guidance for strategies to improve the effectiveness of rural LOs.

Resistance

Local organizations, particularly of the rural poor, can encounter active or passive resistance from many sources. These include local and regional elites, national government leaders and administrators at all levels, rural people themselves, and other organizations with conflicting interests. Resistance, it should be said, is not always harmful to local organizations. Some thrive on adversity, as did the Sukuma cotton cooperatives in Tanzania, where the resistance of the colonial government in the 1950s led to solidarity among the growers; they withheld their cotton until granted official recognition of the cooperative, which went on to grow and prosper (Lang et al., 1969).

The most usual and most intractable opponents of rural associations are local and regional elites—such as substantial landowners, or merchants and their retainers—who also dominate local government and maintain commercial and political links with regional and even national centers of power. They tend to regard organizations of tenants, laborers, or marginal farmers—however benign their stated purpose—with as much enthusiasm as that with which mill and factory owners in the United States and Europe greeted the appearance of labor unions a century ago. They look upon such organizations as a threat to their economic and political control (especially over the terms and conditions of tenancy, employment, credit, and marketing) and often as the work of subversive "outside agitators." The Farmech mechanization project in Lesotho in 1960 offers an example. As often happens in Africa when new organizations are formed, traditional chiefs, sensing that their power was in danger, sought to undermine the newly elected leaders (Wallman, 1969). The Bangladesh Rural Advancement Committee (BRAC) was similarly resisted by landed farmers when its activities began to emphasize the interests of the poor (Ahmed, 1980).

Resistance by local elites takes many forms. The MODECBO cooperative in Haiti found that landlords raised rents when members wanted to start communal cultivation of vegetables (Maguire, 1979). More serious measures, such as economic sanctions and even violence, may be used to discourage tenants, workers, or customers from participating in mutual assistance associations. The Henna project supported by the Evangelical Church in Ethiopia in the early 1970s encountered such resistance, but was nevertheless able to grow to 15,000 members prior to the revolution against that country's feudal regime (Ståhl, 1974). Rumors may be circulated that the organization is subversive, as was done against the Tambon Yokkrabat organization in Thailand, even though it was funded by the Rockefeller Foundation (Rabibhadana, 1983). Such rumors are designed to alarm the government and frighten potential members.[1]

More subtle means of discouragement include undercutting the prices offered by co-ops or offering lower interest rates on credit. Because many local elites are in positions of influence, they may resort to the courts and legal measures to prevent their tenants or laborers from organizing for any

[1]Rumors were used against organizers helping to start the farmers' water management groups we worked with in Sri Lanka. One influential farmer, who was drawing an unfair share of water, first said the organizers were spreading *ganja* (marijuana); when that was not believed, he said they were working to increase production in order to get farmers on their side and then make a revolution. Subsequently, however, the resisting farmer came around to support the farmers' organizations actively (Uphoff, 1981, 1982a).

UNIVERSITY LIBRARIES
CARNEGIE-MELLON UNIVERSITY
PITTSBURGH, PENNSYLVANIA 15213

purpose. If such measures do not succeed, employers or landlords may employ stronger tactics, even the use of goon squads to intimidate members or leaders, beat them up, or burn their houses. In Nepal some group leaders for the Small Farmer Development Program were subjected to harassment. The first year, it involved the theft of their cattle and home grain stores; the third year, a rape charge was engineered against one of the emerging leaders (Clark et al., 1979). Further, there was an attempt to poison one of the organizers (Ghai and Rahman, 1981:25–26). In extreme situations, we may find the leaders of local organizations assassinated, as witnessed in the case of the Farmer Federation of Thailand (Morell and Samudavanija, 1981: 205–30). Established and led by farmers themselves, the FFT emerged in 1973 to promote land reform and got a national law enacted, reaching a membership of 10,000 families in Chiengmai province alone. At least 21 FFT leaders were killed between March 1974 and August 1975, contributing to the demise of the organization.

We would note that this sort of resistance is found most often in situations where land ownership and wealth are most unequally distributed. Where resources are more equitably distributed, the resistance of local elites may take the form of opposition from community "elders," which can usually be reduced or neutralized by judicious efforts (e.g., Kincaid et al., 1976; Gow et al., 1979, II:3–40; Sheppard, 1981).

The hostility of local elites is often reinforced by the hostility of government, national or provincial, and of administrative bureaucracies. Or, as in the case of DESEC in Bolivia, political parties may fear competition from an autonomous organization (Morss et al., 1976, II:G1–G15). Many authoritarian governments are unwilling to tolerate organizations that are not directly controlled by the state; they find atomized publics easier to deal with. In addition to resistance at the national level, local bureaucrats on their own may attempt to quash an organization that is forming in their jurisdiction. In Ethiopia, the *mahabar* local development committees had problems with the governor of their area and with the courts. When members sought to settle disputes among themselves through their own courts, this was construed by local administrators as a challenge to their authority (Hamer, 1976, 1980). And in both Kenya and Tanzania, as Holmquist (1979: 137) observed, "The bureaucracy. . . has tried to muzzle, if not curtail, self-help organization because the entire self-help process threatens the bureaucracy's managerial functions, its ideology, and ultimately its means of survival."

Some civil servants may believe that local organizations introduce "politics" into administrative operations, resulting in loss of control and lower technical standards, less uniformity in implementation and results, and possibly corruption. Such fears were manifested but quite unjustified in the case of local efforts by committees to build suspended bridges in the

Baglung district of Nepal (P. Pradhan, 1980). The style of public administration prevailing in most developing countries is threatened and complicated by assertive publics. So civil servants, preferring to deliver prepackaged services to a waiting and receptive community, may see little benefit and much potential trouble in organized groups, especially among the poor.

An active, effective program like that of Tambon Yokkrabat in Thailand can indeed be an embarrassment to indifferent civil servants. Their opposition may be stronger if they are in collusion with local elites or private interests. When members of the San Jose credit union in the Philippines tried to oust dishonest leaders, local officials first refused them permission to use the town auditorium for a meeting, then declared meetings of the dissidents illegal and spread rumors that the military would arrest any persons attending the meeting (Hollnsteiner, 1979). Nevertheless, members persisted and were able to persuade higher government officials to audit the books. The first audit was unsatisfactory, and they had to insist on a second before the truth was documented—after which they were finally able to remove the officers. When a tribal population in the Indian state of Uttar Pradesh resisted the destructive felling of trees in their area, they found officials backing the contractors and had to resort to nonviolent direct action (De, 1979).

Governments and administrators may resist local organizations in entirely legal ways. The implementation of policies can be slowed greatly by the foot dragging of unenthusiastic bureaucrats. The most obvious indication of government attitudes toward organizations of the disadvantaged is its allocation of resources, which does not always benefit and sometimes even impedes the rural poor. The authorities may go even further, if they see the local organization as a threat to the established order, to dissolve it and intimidate or imprison its leaders. The hostility and suspicion of government, reinforcing the intractability of local elites, can doom most local organizations of the poor. Without any channels to express their needs and seek redress of their grievances, the rural poor may withdraw completely from contact with government, or turn to extralegal methods of expressing their frustration and anger at an indifferent or hostile government.

Local organizations may also encounter resistance from within the ranks of the rural poor themselves. Since local societies are seldom homogeneous, divisions based on kinship networks, on residence patterns, on ethnic or religious affiliations, or on political party allegiance can inhibit effective organization. For example, efforts to establish a vegetable-growing cooperative among the poor in Barpali, India, were resisted by members of a very low caste (gardeners) who felt they should have a monopoly on such activity (F. Thomas, 1968). Social groups may share so little trust that they are unable to cooperate, especially if the organization appears to be unduly influenced or controlled by one faction at the apparent expense of the others. These divisions may be independent of class alignment, but they represent vulnerabilities that can be exploited by local elites. Where generalized distrust, hostility, and dissatisfaction are very high, as in Barangay Liberty in the Philippines (Lopez, 1979), the task of forming and maintaining associations among the poor becomes extremely difficult.

In many rural societies, cross-class social structures continue to function. Low-income people rely on links with kinfolk, patrons, employers, officials, or others to protect themselves and ensure their survival. Until such networks have been rendered unreliable or nonfunctional by population growth or the commercialization of agriculture, people often find these particularistic but inequitable ties more reliable and productive, and far less risky, than efforts to satisfy their needs through association with people in similar circumstances.[2] In the latter—more impersonal—relationships, bonds of trust have not been built, and fellow members may be competitors for employment, tenancy, credit, or public services. Community members may also resist if they believe the organization is being promoted by untrustworthy outsiders, especially government officials. Among the Sukuma in Tanzania the community's suspicions of the government organizers nearly doomed a cattle cooperative at the beginning (Lang et al., 1969). Fear of "politics" and the consequences of being involved in controversial activity may also inhibit participation in local organizations, as we have observed in the operation of rural health committees in Guatemala (Colburn, 1981:21).

Resistance by the poor is frequently passive. Potential participants may decline to affiliate, because they prefer to rely on their individual or kinship resources and are unwilling to risk the wrath of landowners, moneylenders, merchants, or officials on whom they depend for vital economic resources and even for physical protection. Passive resistance may turn to active hostility, not excluding violence, when an organization apparently dominated by one group begins to claim resources or to encroach on opportunities that threaten to injure an opposing faction. In the case of the Ujamaa villagization program in Tanzania, rural people sabotaged cooperative efforts that were imposed by the heavy-handed tactics of bureaucrats and party functionaries: peasants did nothing, made only token efforts, or simply dropped out. Sometimes unfortunate "accidents" occurred, or peasants "misunderstood" instructions (Fortmann, 1980; Hyden, 1981a). In North Cauca, Colombia, a program to introduce improved varieties of maize ran into passive resistance by farmers for whom maize was not traditionally a cash crop and who therefore were unwilling to invest cash in "modern" inputs for it (Morss et al., 1976, II:H23–H34). In other parts of the large Caqueza project, however, when farmers were approached in a more consultative

[2]This has been frequently emphasized in the literature on peasants; see, for example, Hollnsteiner (1963), Galjart (1967), Lemarchand (1972), and Scott (1976), and also the critique by Popkin (1979).

manner, better results were achieved for both farmers and technicians (Zandstra et al., 1979).

Some of the most striking and unfortunate resistance was reported for women's organizations within rural communities where male dominance was strong and unaccustomed to any challenge. An organizer for a mothers' club at Wae Am Li in South Korea was beaten by village elders when she talked about contraception, and the establishment of a club was opposed by the elders until the government ordered them to accept it. They quoted a Korean folk saying, "Misfortune will fall upon the house where the hen crows like a rooster" (Kincaid et al., 1976). In Bangladesh, an SFDP group, set up so that women at Digharkanda could earn income by processing paddy, was stubbornly opposed by husbands, who complained that their wives were becoming "uncontrollable." One man divorced his wife, and one woman had to leave the village without repaying her share of the group loan, presumably under pressure from her husband. Enough husbands interfered with their wives' use of proceeds from their work that the loan went into default and the group collapsed (Raha, 1979). In Jamaica (Antrobus, 1980), Kenya (Kneerim, 1980), and Bolivia (Wasserstrom, 1982), on the other hand, no similar resistance was reported to women's organizations. The opposition to groups seeking to advance women's interests may be as deep-seated as to those seeking to improve the condition of the rural poor. If the poor are also women, the resistance may be greater still—unless they are part of a movement by and for the poor of both sexes, such as the Chipko and Bhoomi Sena movements in India (De, 1979; Rahman, 1981). An exception is the Dhulia women's movement in India, where tribal women by mass action were able to improve their economic well-being and personal security (Kanhare, 1980).

Resistance may also come from organizations that compete with one another for members or for the authority to operate in certain areas. The Federation of Free Farmers (FFF) in the Philippines has experienced continuing difficulties in its relations with plantation workers' unions. The FFF has enrolled few plantation workers, although many of its services would be of benefit to them. These interorganizational disputes can lead to flurries of hostile propaganda among potential members, to the sabotage of the efforts of each organization, and even to violence (Hodsdon, 1979). It is, however, an indication of the lack of organization among the rural poor that we found few instances of competition between organizations for the same membership.

Subordination

Local organizations are in chronic danger of losing their freedom of action and falling under the control of more powerful outsiders. Subordination or dependency may be the result of intrigue; it may come from the exercise of superior power by local elites or officials; or it may reflect the price of relying on the services or resources provided by government agencies or voluntary organizations. In general, we find four kinds of subordination: to government, to local elites, to local leaders (even if they come from among the organization's target membership), and to outside agencies assisting the organization. These will be considered in turn, though they are not always separable. Government may co-opt local leaders, outside agencies may in effect be co-opted by local elites, and still other combinations are possible.

Local organizations often face a common dilemma. If they remain small and informal, they avoid the attention of politicians, officials, and local influentials—but their impact may be limited. On the other hand, if they expand in scope and scale with more consequential benefits to their members, they invite the attention of outsiders. Local membership organizations are the inevitable targets of ambitious politicians wishing to build constituencies, of bureaucratic agencies hoping to enlarge their scope, of governments jealous of their political control in rural areas, and of local landowners and merchants concerned with their domination of the local economy. Each has its own purposes in relating to rural organizations. The danger of a local organization's becoming subject to powerful outside influences is still greater when the initiative for organization has come from the government or from a source close to government, such as an officially sanctioned political party. Experience has often taught rural people to be wary of government and its agents. While complaining of neglect, they expect little good to come their way from the state, which generally represents and acts in the interests of more powerful people.

Local organizations should represent the interests of their members as the members define those interests. Once government intervenes, even with beneficent intentions, this self-determination may be at risk. This is one of the major dilemmas for rural local organizations: government assistance may be helpful and often necessary, yet that same assistance may lead to dependency and control that undermine the organization. In Costa Rica it has been noted that assistance provided by a supportive government often usurped the influence of local leaders and, consequently, the goals and objectives of their organizations (Seligson, 1980). The literature on village development committees is replete with instances in which governments sought to use local organizations as retailers of services, initially giving them some autonomy and authority but gradually imposing limitations and controls. The terms *demobilization* and *departicipation* are indicative of the direction that governments may take once they sense a loss of control of local membership organizations (Kasfir, 1977). This occurred with the village development committees in Botswana, Kenya, Tanzania, and Zambia

(Chambers, 1974; Vengroff, 1974; NIPA, 1976). In Madagascar, among the Bara people, village development committees were set up in order to provide some government presence in the countryside. Participation by the members was limited, however, because the LOs were only the symbolic gesture of a government seeking to gain the support of peasants (R. Huntington, 1980).

The price of government-derived benefits is inevitably some loss of local control. There are always strings attached to loans, grants, and services. At a minimum, records must be kept in certain ways, reports must be written and filed, and officials will enforce regulations and policies enacted not by the local members but by a remote and impersonal presence called government. The benefits may be worth the limits they impose on self-determination, but they are controls nonetheless and may become so irrational or burdensome that the organization funneling these services loses its value to local people.

Some governments proscribe local organizations, except of the most informal kind, and are not inclined to help them in any way. Any services provided to the poor in rural areas come through government agencies directly to individuals. Increasingly, however, governments are recognizing that they lack the administrative resources to provide services directly to individual rural households; they are compelled to rely on intermediary organizations, but they usually insist on exercising some influence on the organizations they patronize. The motives may vary: they may desire to use local organizations as objects of patronage or as "transmission belts," to preclude their capture by opponents of the regime, to ensure that resources are used honestly and for the purposes intended by government, or to promote uniformity as an aid to good administration. In some cases an important motive seems to be to please a foreign donor agency (Blustain, 1982a). A mixture of these motives was found in government dealings with an association of agricultural credit users (AUCA) in Paraguay (Morss et al., 1976, II:K3–K12) and with the Farmers' Associations and Irrigation Associations in Taiwan (Stavis, 1974a; Abel, 1975).

Senior politicians and administrators often assume that because of their superior education they have an accurate idea of what local people need and what will benefit them, and that there is one best way for local publics to help themselves and interact with government.[3] In Ethiopia, for example, the "minimum package program" was implanted in rural areas by the government and a foreign donor, without consulting the people who would be

[3]"The sad fact is that analysts, planners and politicians simply *do not know* what kind of local organization is actually in the poor's interests. The delusion that sufficient cognition can overcome this ignorance—that the "newest direction" will finally be the right direction—may be a greater obstacle than ignorance itself to designing better organizational programs" (Johnston and Clark, 1982:169; emphasis theirs).

affected and whose participation would be vital to its success (Lele, 1975). The results of even well-meaning top-down prescriptions can undermine the ability of local organizations to represent and act on behalf of the self-defined needs and interests of members and potential members.

Government regulations can suffocate local organizations with paternalism, so that even good intentions may produce undesirable results. This fact contributed to the negative correlation we observed (see the previous chapter) between government linkage and LO performance. Standard rules for the conduct of business, conditions for the use of resources, and prescriptions for the activities to be undertaken by the organization may be closely supervised by resident bureaucrats and technicians. Services are provided on terms that the supervisors set, rather than according to the priorities and convenience of the members. In Thailand, officials of the Department of Cooperative Promotion virtually operated the cooperatives themselves through detailed enforcement of regulations (Thai Khadi Research Institute, 1980); in India, government auditors refused to accept a *panchayat*'s expenditure of one rupee to replace a drinking glass because the pieces of the broken glass were not submitted with the voucher (Haragopal, 1980:45). Officials may also undermine the self-confidence and sense of responsibility of local organizations:

> Some of the policies of the Banco Ejidal carried out through the Taretan Sugar Mill appear to be based in part, at least, on the assumption that ejidos never perform well and, therefore, should not be held responsible for their mistakes. The bank does not, for example, insist on the repayment of debts by ejido members. The same mentality is visible in the Department of Agrarian Affairs, where frequently no attempt is made to straighten out financial irregularities. [Landsberger and de Alcantara, 1971]

The kind of paternalism demonstrated in this Mexican example breeds dependency and prevents local organizations from representing the interests of their members. As they lose the capacity to act on their own initiative, they become less effective in mobilizing self-help, in managing local resources, in articulating the needs of their members, and in providing reliable feedback on government programs. The consequence of external control is usually a passive, dependent, even alienated constituency.

The loss of self-determination may be reflected in decisions about what the organizations should do. Once they are subordinated to government, the latter—rather than LO members, or even their leaders—makes program decisions. Government assumes that members will naturally agree, since they are to benefit. The consequence is that the priorities of local organizations become distorted and no longer reflect the preferences of their members or their knowledge of local conditions. As they lose their sense of control, their participation diminishes, and the organization comes to be regarded as the instrument and responsibility of the state. Its failures can be imputed to government, and government is expected to rectify them. Having lost control, the members abandon their initiative and responsibility. Moreover, the imposition of numerous—often extraneous—tasks on the organization may overwhelm its managerial capacity. Even successful multipurpose local organizations reach a threshold that threatens their effectiveness and viability. If they survive, it is because the government is then propping them up with outside resources, including government personnel. But the efficiency of local organization as a channel of services declines, since member contributions fall off. Such consequences were evident among the Multi-Purpose Cooperative Societies in Sri Lanka (Uphoff and Wanigaratne, 1982).

Local organizations that manage to avoid becoming subordinate to government sometimes fall victim to local elites—the more prosperous farmers, merchants, moneylenders, and resident government officials—whose status, education, economic power, and organizational skills enable them to assume effective control. Where this occurs, they are free to preempt for their own use the resources provided by government (such as subsidized credit) and deny them to the smaller farmers for whom they were intended. The organization is thus converted to the service of rural elites. Small and marginal cultivators drop out, become inactive, or remain only to receive whatever small benefits that they may claim once major resources have been appropriated by more powerful members. In the Swanirvar movement in Bangladesh, representatives of the local elite insinuated themselves into the leadership of the organization and gradually monopolized the benefits of membership (Hossain et al., 1979), repeating a pattern observed often in Bangladesh—in the Thana Irrigation Program and other rural development efforts (Blair, 1974). Many organizations are set up in such a way that only the more prosperous farmers are able to participate, as in the North Clarendon Processing Company in Jamaica and similar cooperatives, where participation and benefits were based on the number of shares purchased (Gow et al., 1979, II:313–39).

An active role for local elites is not necessarily harmful. To the extent that the local organization can mobilize resources from government and provide services that are widely diffused and shared ("public goods")—such as roads, health clinics, religious centers, or cattle dips—control and leadership by established elites may spread benefits and access widely, if not equitably (Leonard, 1982a:12–18). Where extended kinship and other forms of clientage or community solidarity continue to operate and promote the basic interests of all members, elite leadership of the local organizations may represent a valuable collective resource. This is consistent with our statistical findings (Chapter 5) that economic heterogeneity of membership is positively associated with favorable performance of several LO functions. Among the most impressive LO performances were those of the self-help water supply organizations in Malawi, operating under the leadership of traditional chiefs and elders (Robertson, 1978; Glennie, 1979; Liebenow, 1981). In Chapter 8 we shall consider other examples of the role that local elites can play in promoting the activities of LOs. Where such structures of solidarity no longer operate, however, the subordination of local organizations to local elites is likely to exclude the majority and represent a form of exploitation.

Even when the leadership of local organizations comes from among the more typical members or the poorer strata, there is no assurance that a form of subordination will not result. This was, after all, the experience that prompted Michels (1915) to propose "the iron law of oligarchy." Even democratic organizations intended to champion the interests of the poor tend to be taken over by their leadership. The Small Farmer Development Program in Nepal has noted an incipient problem of "leaderitis," where those first chosen to head the small groups tend to remain in control and to take advantage of their offices, if only in small ways. We discuss in Chapter 7 some of the means employed to offset this problem. It can be severe, as in the case of an SFDP group of landless fishermen at Fakirakanda, Bangladesh, which failed as a consequence of such a leadership takeover (Chowdhury, 1979).

Well-intentioned outside agencies can also contribute to the domination of local organizations. A classic case was the Cornell experiment at Vicos, Peru, to introduce participatory development in a patron-dominated community (Dobyns et al., 1971; Brinkerhoff, 1980, ch. 5; Lynch, 1982). The very process of intervening in a stagnant, inegalitarian situation drew outsiders into a paternalistic role (Lynch, 1982:94–96). On the other hand, as the process continued, there was some evidence of local assertiveness. For example, the Vicosinos expelled Peace Corps volunteers in 1964 and also Peruvian teachers whom they believed were not behaving in the interests of the students. The Sarvodaya Shramadana movement in Sri Lanka has had some problems of overcentralization in dealing with local initiatives and organization (Ratnapala, 1980; C. Moore, 1981), fostered in part by the very strength that charismatic leadership at the national level has brought to the movement. A deliberate decision was made to decentralize the organization, to try to put more authority back into the district organizations and communities. Implementing this decision has been complicated by the extent to which the organization now receives foreign funding for its activities.

A different kind of subordination to outside interests may occur when the organizations of the rural poor are promoted by a political party, particularly a radical one. The Khet Mazdoor Union, sponsored by the Communist Party of India, lost momentum when party officials changed their strategy of organization, without consulting LO leaders or members, and decided to enroll small cultivators as well as landless laborers—after previously arguing that this was undesirable (Pandey, 1975). Alexander (1980) also documents some of the problems arising in tenant and laborer organizations backed by Communist parties in Kerala and Tamil Nadu. That kind of outside support was deemed necessary to get LOs started among the very poorest of the poor, but it did not come without some costs. Political party influence was found to be adverse in the Farmech project in Lesotho, already mentioned (Wallman, 1969); on the other hand, it seems to have been positive for the land reform cooperatives in Egypt (Harik, 1974) and the peasant leagues in Venezuela (Powell, 1971). Such experiences suggest that outside support may be valuable, even essential, for the initiation and success of rural local organizations, particularly those that include the poorest strata; but the dangers of subordination are real and potentially debilitating.

Internal Division

The fact that few collectivities are free of factionalism and internal politics is another common source of vulnerability for local organizations in rural areas. The larger the organization, indeed, the more likely it is that cleavages will emerge and manifest themselves in competitive claims for control of the organization and its benefits. So bitter are some social divisions that they impair the functioning of the organization. When village centers (Gonokendro) were being initiated for the Bangladesh Rural Advancement Committee, factionalism was so great that in at least one-third of the villages it was not even possible to agree on the sites for the centers (Ahmed, 1980:438).

The basis of rural social cleavages may be communal—ethnic, racial, religious, or caste.[4] The level of trust, the ability to communicate, and the willingness to share obligations and benefits may be so tenuous that cooperation is not possible. The Morningside/Delightful Buying Cooperative in Jamaica split in half over the issue of doing business on Saturday, the holy day of the Seventh Day Adventists in the organization (Gow et al., 1979, II:327–40); ironically, the breakaway cooperative was named the Unity Buying Club. The Ngok Dinka of the Sudan had great difficulty maintaining a cooperative because of the intense conflict among members from different subethnic groups. When mechanized farms were set up, the two that were ethnically homogeneous seemed viable, but the two that crossed social group boundaries were riven with problems (R. Huntington, 1980).[5]

[4]Whyte and Alberti (1976:238–40) list the following types of cleavages that gave rise to conflict in the Peruvian villages they studied: age grading or generational divisions; differences in wealth; differences in economic activities; sociocultural differences; neighborhood divisions; religion and religious organizations; and political party membership. Conflict is greater to the extent these cleavages are overlapping rather than cross-cutting.

[5]Actually, because of bureaucratic suffocation and economic infeasibility, none of the groups succeeded. "A large and talented support system managed to serve these fields so that at best their yields were half that of the traditional fields and the cost per acre very high." (R. Huntington, 1980:13).

On the other hand, we did find examples of ethnically heterogeneous LOs that managed reasonably effective performance, such as the Manyu Oil Palm Cooperative groups in Cameroon (Gow et al., 1979, II:53–66) and the Bagoy Irrigation groups in the Philippines (de los Reyes and Viado, 1979). Indeed, in the case of the *mahabar* in Ethiopia, heterogeneity of kin groups was found to strengthen the committees because member families which themselves had strong solidarity competed to contribute most to group goals (Hamer, 1976). That both heterogeneous and homogeneous organizations registered both success and failures was reflected in the results of our statistical analysis.

Cleavages may also be based on political affiliation; members of different parties may be unable to coexist and work for common goals within the same organization. Such was the case with the Farmech project in Lesotho. Political parties may attempt to use LOs to promote party interests, provoking conflict and risking the participation of large numbers of members and prospective members. The co-ops in Zaire (Janzen, 1969) were ethnically homogeneous but linked to political parties, so that when the multiparty system collapsed after independence, many of the co-ops did too. Even within an ethnic group, there may be factionalism among patrikin and matrikin groups in the same organization. Janzen reports that this tension, when exacerbated by nepotism, led to conflict and dissolution (though sometimes to transformation of the co-op into a corporation of nonkin investors). In contrast, with active political leadership and partisan tolerance, the peasant federation in Venezuela (FCV) productively absorbed members of competing party allegiance and different economic status (Powell, 1971).

Economic diversity can, of course, produce serious cleavages. Alexander (1980) describes how tension between landed farmers and their tenants or laborers in South India debilitated agricultural LOs. A common problem found in irrigation groups is the division between upstream and downstream farmers. The Banes irrigation group in the Philippines was unable to deal with this problem and had to post guards to protect individuals' water supply. Water stealing was a major problem despite local sanctions such as ostracism, witchcraft, and public beating (de los Reyes and Viado, 1979:18). On the other hand, the *zanjera* irrigation organizations in the northern Philippines have been quite capable of dealing with upstream-downstream problems, both within groups and between groups (H. Lewis, 1971; Coward, 1979a). As a rule, local organizations that contain persons with competing social or economic interests are inherently vulnerable to conflict and to the politicization of decision-making. Energies consumed in political struggle—in the assertion and defense of uncompromising positions—will lead to immobilization rather than be invested in performance. If one group wins control, tbe others are likely to feel excluded, drop out, or become inactive.

Closely associated with internal divisions is differential participation in local organizations, usually at the expense of socially and economically marginal groups that have much at stake in rural development. One such group is women. In most rural societies, women in low-income households not only raise children and maintain the home; they also perform essential economic functions, including farming, tending livestock, marketing, and producing handicrafts. Yet they are often excluded or consigned to minor or inconspicuous roles in local organizations. When it was suggested that women be included in the management committee for the Sukhomajri water users' association in India, village elders rejected the idea, saying they wanted the committee to be "a serious body, not a modernistic frill" (Franda, 1981). Similar treatment is frequently accorded low-caste and low-status persons, who may be permitted to provide labor but not to participate in the decision-making or management of local organizations. Their distinctive needs are consequently given little weight. Residents of remote and isolated settlements may also be underrepresented, less because they are excluded than because their opportunities for participation are limited by the costs involved in transportation and communication.

Since unorganized members of the public are severely disadvantaged, those whose participation is limited by discrimination and social taboos share inequitably in the benefits of organization. Organization is invariably more problematical for the socially and economically marginal than for those with more resources. Most cooperatives that are established to provide credit and production inputs exclude the poorest peasants, the landless. This was the case with the Federation of Regional Agricultural Cooperatives (FECOAR) in Guatemala, yet in the same country, membership in the cooperative at San Martin Jilotepeque did include the rural poor (Gow et al., 1979, II:141–52).

Among the poor in rural India, "cooperative" has become a dirty word (Moulik, 1980:6). The same is true in much of Africa and Latin America, and our data confirmed that co-ops are generally less successful than other types of LOs (see Chapter 3).[6] Still, if an organization conveys benefits to

[6]With regard specifically to improvements for the poor, the following summary scores can be reported for the three types of LO (the number of cases for which the outcome could be scored is shown in parentheses):

	LDAs		Co-ops		IAs	
Increased income for the poor	(18)	74	(61)	61	(10)	124
More equitable asset distribution	(11)	216	(14)	184	(15)	161
Better access to services for poor	(25)	132	(33)	33	(58)	128

Not many LOs undertook or could be scored on asset redistribution, but most that did were reasonably or very successful. Co-ops as a type of local organization show up as particularly unsuccessful on relative improvements in access to services for the poor.

the disadvantaged, even though distribution is unequal and opportunities for participation are circumscribed, it may be of net advantage to them. Eventually, however, they must either gain a greater stake in existing organizations or form their own. The first alternative may be possible for women, but the second is more likely to be necessary for low-status ethnic and economic groups. This in turn is likely to be critically influenced by government decisions to orient programs to these especially vulnerable and disadvantaged publics.

Ineffectiveness

Political skills, organizational skills, and technical skills are often in short supply among disadvantaged groups in rural areas. The underdevelopment of these skills, due to low levels of education and inexperience with formal association, can reinforce low self-esteem and lack of confidence. These deficiencies inhibit organization, even when it is encouraged by government. The smaller and more informal an organization and the less it encounters officialdom and local elites, the less important are these scarce and untested skills and the less their absence inhibits the participation of lower-status, less educated rural people. Unfortunately, government practices can contribute to ineffectiveness by setting formalistic requirements for LO operations that are difficult for members and leaders to meet, or by working with them in arbitrary ways. Harbeson (1972), in his case study of Kenya cooperatives, found that government efforts to make leaders carry out unpopular functions imposed from above led to a high turnover rate in the leadership, which exacerbated the problem of inexperience. This, in turn, was seen by officials as justifying their control over the cooperatives.

The most obvious shortage of skills is usually in the area of business management, though the shortage may be more relative than absolute, depending on what is expected or required of LOs.[7] However, in various situations even simple bookkeeping skills—the ability to manage and account for the funds the organization is responsible for and against which members have legitimate personal claims—may be lacking. Every report on the Federation of Free Farmers in the Philippines, a large organization with substantial sums of money flowing through its coffers, mentions its inadequate accounting methods.

[7]Tendler has observed that government-sponsored credit cooperatives in Ecuador and Honduras required farmers to use complicated auditing and bookkeeping procedures which they could not perform adequately, and administrators were not willing to pay to have such services performed (1976:17). This created problems of ineffectiveness where simpler procedures, as discussed in the next chapter, could have been introduced. King (1981:262) reports that 14 types of written records were required of cooperatives in northern Nigeria. This would have been a burden even if members had not been mostly illiterate, as they were.

The literature is replete with accounts of small cooperatives whose failures are imputed to incompetent recordkeeping and management of funds, often linked to corruption. The failure of the Sukuma cattle cooperative in Tanzania was due in part to the inability of members to understand the accounting system and the resulting mistrust of bookkeepers (Lang et al., 1969). Even when clerical and record-keeping ability is available, a lack of skill in basic business management can lead to imprudent investments in equipment, heavy debts, or the launching of ventures that fail and discredit the organization among its members. The need to overcome these common shortages of business skills through training and to avoid unwise ventures through legal regulations, tutelage, and other forms of technical assistance provides a major rationale for government intervention and assistance to local organizations.

In order to attract membership, promoters and activists are often tempted to promise more benefits at less cost than the organization is subsequently able to deliver, thereby encouraging a perception of ineffectiveness. In reality, benefits are often delayed, while costs are immediate. Failure to realize optimistic expectations strains the credibility of the organization, creates disillusionment and dissension among the members, and leads to the erosion of their participation. The memory of failure, in turn, restrains future efforts at organization. Leaders and members with little experience in formal organization are likely to have some difficulty in sustaining LOs through such setbacks and in managing the interpersonal and intergroup conflicts that invariably arise. They are also likely to feel handicapped in dealing with people of higher status, including local elites and officials. Their limited resources, opposition from officials and local elites, and lack of self-confidence are negatively reinforced by their limited skills in presenting and arguing their case, in bargaining for group benefits, and in exerting pressure on officialdom. Initial rebuffs and disappointments may discourage members, presuade them that the organization is doomed to ineffectiveness, and convince them that they are better advised to pursue their interests through particularistic, rather than collective, channels.[8]

One of the resource management skills that may be overlooked is maintenance, as demonstrated by the case of the rickshaw pullers' group in Boyra village, sponsored by the SFDP in Bangladesh. That group used its credit to purchase five rickshaws for its ten members. All ran smoothly for the first

[8]Possibly even relatively effective organizations such as the Sidamo *mahabar* associations can have problems because of private desires to advance more rapidly than possible through group channels. Hamer (1976) reports that some members, out of self-interest, set up shops competing with *mahabar* co-ops; he concludes that since many members have only short-term goals, associations have to provide immediate gains in order to retain support. In this case, however, the effectiveness of the *mahabar* in conducting local courts and keeping disputes away from the costly and corrupt government legal system helped them survive the competition. The *mahabar* might not have lasted if they had been single-purpose organizations.

four to five months—until the rickshaws began to need frequent and costly repairs. Loan repayment funds were diverted to repairs, and each team was made responsible for its rickshaw, but still the situation worsened as members accused each other of being irresponsible with equipment. Loan repayments became overdue, and several members dropped out. To recoup their losses, the group went into a part-time fish culture project—which also failed, adding to the debt burden. The organization's survival had become perilous at the time the report on it was written (Islam, 1979).

In contrast, a successful SFDP rickshaw group in Digharkanda, apart from the exemplary leadership of its president and secretary, had one member who could do all the maintenance work in a group garage (Raha, 1979). Perhaps the Boyra group was "fragile" anyway, and the maintenance issue only triggered its demise. Still, it might otherwise have lasted long enough to establish member solidarity and develop experienced leadership; conversely, the Digharkanda group might not have survived if its initial rickshaw enterprise had foundered on maintenance. As it was, the group went on to purchase additional rickshaws, a plot of land for its garage, and a cowshed. It also started a fish-raising project and an education program for members and their dependents.

One of the most widespread and consequential problems confronting local organizations is the lack of authority to make binding decisions and to ensure compliance on the part of all members. The absence of such authority weakens the ability of organizations to deal with problems of "free riders," noncompliance, and malpractices, thus weakening interest and confidence among members and undermining credibility among outsiders. The lack of authority in many local organizations increases the possibility of internal disputes and makes their resolution more difficult. Not surprisingly, the problems caused by the absence of authority are especially prevalent when members are called upon to make resource contributions to projects and the benefits turn out to be zero-sum, as in the Banes Irrigation Association in the Philippines (de los Reyes and Viado, 1979).

It has become a truism in the study of rural poverty that marginal farmers tend to be risk averters, because they are too close to the margin of subsistance to sustain failures.[9] By the same token, the rural poor can ill afford the luxury of investing their time and resources in organizations that seem unlikely to provide valued benefits for their family and community. There are risks and costs of membership in organizations that may be perceived as greater than the potential benefits. While the "culture of poverty" notion

[9]How risk averse they are and what they do as a consequence is a matter of debate between Scott (1976) and Popkin (1979), but the general conclusion that they need to be risk averse because of meager assets and uncertain income is not at issue.

has lost much of its appeal to students of rural development, it is nevertheless clear that repeated experience with disappointment seriously inhibits initiative and reinforces a sense of powerlessness and low expectations among the rural poor. Ineffectiveness in LOs may be the consequence neither of leadership nor of membership failures, but rather of the uncertainties surrounding rural life, the difficulties of maintaining cohesive social organization in severely resource-poor environments, and the high cost of organizational learning when the margin of resources available to an LO's members is meager and their self-confidence limited by the experience of failure. Ineffectiveness and the fear of ineffectiveness thus can be a serious impediment to the organization of the rural poor.

It is for this reason that in any attempt to start or improve LOs, care must be taken that there be some evident success fairly quickly. The corollary is that programs may need to start fairly small and on an intensive basis, so that there can be a visible "demonstration effect" and rural people can see for themselves the benefits of cooperative effort. This process has been described concisely by Glennie (1979:141–46) in his report on the successful Malawi self-help water program (see also Liebenow, 1981). But we see it at work also in the Ayni Ruway associations in Bolivia (Lassen, 1980; Healy, 1980; Breslin, 1982) and the Baglung bridge construction committees in Nepal (P. Pradhan, 1980). Such success, demonstrating what people can accomplish with organized effort, is the best antidote to feelings of ineffectiveness in the local population.

Malpractices

Individuals frequently use organizations to pursue personal goals, which violate the goals of the organization and the collective interests of its members. Those who achieve positions of leadership may appropriate the resources of the organization for their personal use, or for the benefit of family, friends, or factions. They may even betray the interests of the membership for personal profit, colluding with politicians, officials, or local elites. Funds may be diverted from organizational to personal purposes, ranging from subsidized credit to contracts for the provision of equipment and services. Such corruption and betrayal is not uncommon, even in organizations that survive because benefits to members still exceed their costs of membership. Frequently, however, the recognition of corrupt practices discredits the organization, destroys the morale of members, and results in its demise.

Outright dishonesty is often a greater problem for LOs than simple lack of skills. The Thana Irrigation Program in Bangladesh became notorious because larger landowners submitted bogus membership lists in order to set up co-ops that would be eligible to get and operate tube wells at highly subsidized prices (Blair, 1974). The cotton marketing cooperative in Gondo, Uganda, was certainly not unique in its corrupt handling of loans and repayment, but it demonstrates the problem very clearly (Vincent, 1976). One of the most graphic accounts comes from a sociologist working with local infrastructure projects in El Salvador for the American Friends Service Committee (Huizer, 1963). He found that a local administrator, having reported that a road project had "run out of money," was actually pocketing the remaining balance himself. The rural people working on the roads understood what was going on but were unwilling to take the case to higher authorities because of their fear and distrust of officialdom. Only when an outsider, the sociologist, threatened to raise the issue himself was the problem settled and work resumed.

One should keep some perspective on the problem of corruption. In Taiwan it is reported that a certain amount of nepotism and petty corruption goes on within the Farmers' Associations and Irrigation Associations (Stavis, 1974a). Some influence is often involved in the awarding of contracts or jobs. But these practices do not necessarily result in unfair distribution of services or water. The important question is what adverse consequences malpractices have on local organizations. The negative effects can be a serious obstacle to LO performance. In our field studies of Local Development Associations in Yemen, we found that although the objective level of corruption was not demonstrably severe, the prevailing opinion that it was rampant—summarized in the expression that "all LDA officers are thieves"—interfered with the functioning of LDAs at Maghlaf and Bani Awwam (Swanson and Hebert, 1982). In particular it became difficult to mobilize local funds for improvement projects, even though such funds were abundant, thanks to repatriated earnings from the Persian Gulf.

Corrupt practices are not confined to organizational leaders. The temptation to pocket LO funds, especially if they have come from the government or a foreign donor and are therefore not really local money, is very great among the poor. Often the members are so little committed to the collectivity that despite the best efforts of leaders, they invade organizational resources. They may fail to repay loans or to contribute their share of labor; they may cut trees in community forests or divert a disproportionate share of irrigation water to their own land. These antisocial practices by rank-and-file members reflect weak institutionalization of the norms of cooperation, which undermines organizational performance. Preventing or correcting these abuses, or controlling free ridership—the tendency of persons to benefit without contributing to the organization—may require sanctions and social controls that the members are unwilling or unable to apply. The organization then ceases to function because of corruption, free ridership, and the inability of the majority of members to enforce organizational discipline.

In the Digharkanda paddy-processing group already cited (Raha, 1979), women members misued their earnings until their doing so destroyed the organization. The successful richshaw pullers' group in Digharkanda faced a problem when members started taking, without cause, interest-free sickness and emergency loans from the group treasury. In this instance, however, with the encouragement of the SFDP's group organizer, the problem was taken up by the group and all such loans were stopped until members had agreed to use the money only as originally intended.[10] Similarly, when an SFDP marketing group in Fakirkanda found two members using loan funds for unauthorized purposes, the other members disciplined them (Chowdhury, 1979).

The imposition of penalties for malpractice may produce unexpected consequences. In the case of the Ngok Dinka consumer cooperative noted above, beset as it was with factionalism based on tribal differences and personalistic allegiances, the jailing of an embezzler led to a split in the co-op (R. Huntington, 1980). On the other hand, when one of the farmers' groups sponsored by the Puebla project in Mexico gave up trying to convince three irresponsible members to repay their loans in 1972 and took the "extreme" measure of having them put in jail, membership in that group climbed from 111 to 200 (CIMMYT, 1974). The majority of members and residents in the community may appreciate strict control of malpractices.

As previously indicated, corrupt and self-seeking leaders may betray their trust for personal benefit (e.g., see Hyden, 1978–79). For those who refuse to do so, what are the rewards of leadership? Respect, the opportunity for service, community power, a step toward future opportunities are all regarded as more or less legitimate returns on the leader's efforts and costs. Often, members will acquiesce in more tangible rewards—the location of community facilities near the leader's house, small jobs for his or her relatives, preferential access to loans—but not beyond some threshold where benefits transgress what is considered "fair" compensation for the burdens of leadership. Past this uncertain point comes conflict, which may split the organization between those who support the leader and those who seek to displace him.

Given the vulnerabilities of organizations that represent and serve the rural poor, it is remarkable that so many of them do manage to survive and

[10] One kind of malpractice the group could do nothing about occurred when the bank that provided the initial loan to purchase rickshaws (so as to free the pullers from control by rich rickshaw owners) went ahead and bought the new rickshaws itself at a higher price (and lower quality) then could be found in the open market (Raha 1979).

benefit their members. The various combinations of resistance, subordination, internal cleavage, ineffectiveness, and malpractice that local organizations can encounter add up to a daunting catalogue of obstacles. In the next chapter we consider some of the measures that have served to control or contain these adverse conditions.

CHAPTER 7

Innovations and Practices to Improve Organizational Performance

Various measures, tactics, or principles may be supportive of better LO performance. Their effectiveness depends on the context in which they are invoked as well as the skill and persistence with which they are pursued. In our review of case studies, we looked for social inventions and organizational techniques that might reduce the likelihood or seriousness of the failings to which LOs are vulnerable. These cannot always be relied on to work by themselves. The policy framework of government, the quality of local leadership, and the strategy of outside support, if adverse, can negate almost any innovation or practice that might be introduced to offset elite capture or internal cleavage. But before turning to broader issues of strategy for working with local organizations, we wish to take note of some practical approaches to improving LO operations, since these should be helpful within almost any strategy to promote rural development.

Any measure to assist local organizations may produce unintended adverse consequences. Smaller, informal groups may be internally stronger but externally less effective, as shown by the data in Chapter 5. Existing groups may be closer to their members than new LOs but perhaps more easily dominated by traditional elites. "Public goods" that cannot be monopolized by the rich may nevertheless not meet the most pressing needs of the rural poor. Organizing the latter in special groups may alienate them from the rest of the community and focus antagonism on their organizations. Government or other outside involvement can undercut the self-reliance and long-run effectiveness of LOs.

There are no foolproof remedies for the shortcomings of local organizations; each solution is likely to create its own problems. Solutions need to be continually assessed to see whether they are achieving their intended purposes, and modified to deal with undesired side effects. More important, they should probably be introduced in combinations so that there are mutually reinforcing effects, such as government audits *and* regular open meetings to curb financial malpractices. Measures to hold officers accountable to members should reduce ineffectiveness and malpractices and also make it easier for the LO to resist subordination from any source.

That several dozen LOs in our sample have succeeded under varying circumstances in contributing impressively to rural development indicates that it is possible for organizations to overcome resistance and operate reasonably independently, to avoid debilitating conflict and achieve fairly high standards of integrity and effectiveness. Even some LOs whose overall performance has not been stellar have been quite successful in certain respects, and we can learn from their experience also.

Overcoming Resistance

What measures are effective will depend on what kind of opposition the LO is encountering—from what source and how intense. If it comes from the public whose participation is sought, it may be debilitating, but so long as a new technology or opportunity is truly advantageous, resistance can be circumvented. Perhaps the best example of a planned effort to overcome public opposition was observed with the Banki water project in India (Misra, 1975). The Planning, Research, and Action Institute of Lucknow worked with the head of the *panchayat* in Banki to organize a piped water supply, but at first there was little response from the villagers. The public had misconceptions about the cost of the program and about the quality of the water they were using and they favored continued use of the existing wells. By making house-to-house surveys on sickness and water use and then discussing the findings in informal "evening sittings" in the village, the organizers persuaded the people that they would benefit from a clean supply of water and that the cost would be reasonable.[1] The *panchayat* also set up a waterworks executive committee, consisting of one member from each village, that evolved into an elected committee responsible for collecting water charges and managing the system. After seven years there were no overdue charges, and a substantial savings fund had been built up. Most impressive, there was a definite reduction in diarrhea, dysentery, and typhoid within three years, reinforcing the message of the educational campaign on the connection between clean water and better health.[2] Not all development efforts produce such dramatic results, but the approach used by the Luck-

[1]An innovative design feature of the project was that it included 42 public standpipes as well as individual connections to houses, for which the householders had to pay an extra fee. In this way, everyone could be served, and more resources could be raised from the better-off members of the community.

[2]Instances of diarrhea declined from 83.8 per thousand in 1965 to 19.0 in 1968, dysentery from 12 to 3 per thousand, and typhoid from 3.3 to zero (Misra, 1975:59).

now Institute in conjunction with the support of local leaders is exemplary as a technique for winning the public acceptance necessary for any such project to move ahead.

Where the activity is broadly beneficial, public resistance is seldom reported in the literature.[3] More often the innovation is resisted by local notables such as elders or chiefs, who fear it will upset the community and their prerogatives. If such local elites are not particularly exploitative, overcoming their opposition is likely to be possible with some combination of patience and co-optation. The village elders at Pelebo in Liberia at first opposed formation of a committee to carry out a program of health education and preventive and curative health care. With patient contact, however, they were won over in a year's time, and the village herbalist was sent for six weeks' training as a health paraprofessional. Once the initial activities were launched, the group also undertook the construction of a school and the digging of a well (Sheppard, 1981). Similar patient, persistent, and successful efforts by LO organizers were reported for DESEC in Bolivia (Morss et al., 1976, II:G2–G14) and the Integrated Rural Development Project at Tambon Yokkrabat in Thailand (Rabibhadana, 1983). The strategy of co-opting resistant village elders was also effective for some of the mothers' clubs in South Korea (Kincaid et al., 1976). Most important was their ability to show results that improved the conditions of member families and the village at large. Some of the benefits can be produced by external assistance, as was the case when the installation of a village water supply system with private funding ended opposition by elders to a women's oil palm cooperative in Manyu, Cameroon (Gow et al., 1979, II:53–66).

Resistance is less easily overcome when it emanates from advantaged class or caste interests, but methods of combating it need not be extreme if the opposition can be neutralized by public confrontation. When the SFDP marketing group of petty vendors in Fakirakanda, Bangladesh, was set up with group credit to escape the control of local moneylenders, the latter tried to defeat it. In this case, open criticism by members in public meetings—an act bolder than individuals had previously dared but encouraged by their coming together as a group—neutralized the resistance of the moneylenders (Chowhury, 1979). The Sarvodaya Shramadana movement in Sri Lanka, attempting to curb liquor vending and gambling in the

[3]We cited in the previous chapter the resistance from farmers to a maize project in North Cauca, Colombia (Morss et al., 1976, II:H17–H27). But there were two problems in the project design itself: first, the crop to be improved was traditionally a subsistence one in which farmers did not make any cash investments (these went into crops they sold in the market); in addition, the project was introducing entirely new organizations when farmers would have been happier working through their existing, less formal channels. Eventually the Caqueza project, of which this LO was a part, did make progress as farmers came to see advantages in improved maize yields and, after initial misunderstandings, to view the organizations as "theirs" (Zandstra et al., 1979).

villages, has usually succeeded in shutting down these operations, its efforts bolstered by prevailing Buddhist norms (Ratnapala, 1980). But these are not examples of overcoming truly hard-core resistance.

Such resistance has been encountered and overcome in a number of cases—some of them already cited in Chapter 6. One key element is group solidarity, often based on common ethnic identity. We have previously mentioned how the tribal leaders and members of the Bhoomi Sena in India successfully challenged exploitation by the *sawkar* caste. This was only one round in a protracted struggle for tribals' rights, but it was won so decisively that the momentum shifted in their favor (de Silva et al., 1979; Rahman, 1981). The Chipko movement among tribal people in India required a similar willingness to go to jail, in this case to stop the clear-cutting of forests on which the people depended for their livelihood and the loss of which contributed to flooding. Having halted the deforestation (by clinging to trees marked for felling), the cooperative's members also undertook to replant vulnerable slopes, earning legitimacy by their positive acts (De, 1979).

Two examples in the Phillipines indicate similar success in confrontation with government agencies, even under martial law (Hollnsteiner, 1979). The San Jose credit cooperative threw out corrupt officers who were colluding with local officials, and the Kagawasam movement successfully protected traditional claims to forest land against commercial encroachment. The Henna cooperatives' perseverance in the face of local elite repression in Ethiopia is also impressive (Ståhl, 1974), as is that of the Sukuma cotton cooperative in Tanzania, which had to threaten a boycott of cotton production and sales in order to gain official recognition (Lang et al., 1969). In all these cases, social solidarity emanated in part from common ethnic identity. Interestingly, we did not find as much documentation of similar confrontational experience in Latin America, though we understand that informal, mass peasant movements in Bolivia, drawing on Ayamara and Quechua solidarity, have been able to conduct nationwide blockades of road traffic and thus to put significant pressure on the government. This is one of the most interesting contemporary situations of peasant organization and participation (Devine, 1981). Generally, the willingness and ability of Latin American governments and elites to use violence discourages confrontation, leaving rural populations the alternatives of submission or rebellion. The latter course takes participants out of the local organization mode and into more political and even revolutionary movements.

It should be noted that the degree of militancy shown in the Phillipine and Ethiopian cases just cited was greatly encouraged, by strong support from religious leaders (priests or lay workers) and organizations (the Catholic church in the Philippines, the Evangelical Church in Ethiopia). In Latin America, too, activism among disadvantaged sectors has often been dependent on church support (e.g., see Sharpe, 1977). To some extent, when such outside support provides legitimacy and other resources, it may take the place of ethnic solidarity. Seldom do local groups venture into sharp confrontation with officials or local elites unless there is the kind of trust and cohesiveness that comes from either common ethnicity or religious commitment.

We would also note a more conventional method, which requires that the group in question be at least a substantial minority: that is, the use of the electoral system to counter elite or governmental resistance. We recognize that elections are liable to manipulation by elites to reinforce their interests (Johnston and Clark, 1982:164), but they can also lead to egalitarian outcomes that are less likely in the brokered, unmobilized mode of nonelectoral politics. The Small Farmer Development Program (SFDP) in Nepal initially encountered strong opposition in the Nuwakot district from landed families who controlled the local government (*panchayat*) system. Their control was broken, however, by the use of electoral power. Shrestha (1980:93) explains the success in one of the district's communities:

> The primary reason for the creation of such a conducive atmosphere is the leadership pattern of the panchayat itself. Although only four out of eleven members of the [Tupche] village panchayat are small farmers and members of small farmers groups, the entire [panchayat] leadership has a favourable attitude towards the project because their continuation in the leadership position is directly dependent on the support of the small farmers. Before the inception of this project in the village the inhabitants were the victims of the tyranny of a rich man in the village. When his power base was threatened, he tried to have them punished by complaining to the district authorities that public property, e.g., the fallow land in the village, had been illegally appropriated by the so-called small farmers. However, with the district administration in favour of the small farmers, not only did his bid to frustrate them fail, but a permanent transformation in the power structure of the village was also achieved.

In many of the SFDP communities such a change was not possible, partly because the social structure is divided according to caste. But even in these the situation has been mitigated at the lower levels under the impact of the program: some of the small groups with higher-caste majorities have elected low-caste leaders (Shrestha, personal communication). The factor of solidarity thus shows up even in the use of electoral means to advance LO interests. The significance of voting power is seen in the Bhoomi Sena and Chipko cases; politicians now court the support of these organizations because the tribal populations, though not a majority, constitute major blocs that can "turn out the vote."

In some circumstances, there may be conflict between the government and local elites, or the government may—for its own reasons—choose to side with the poor majority in LOs. We saw this in the case of the SFDP group in Nepal, just described. In Korea the mothers' clubs enjoyed strong government support, which helped to overcome local elite resistance. Where local elite groups are in competition with the incumbent regime, "outside" support from the latter may be useful in backing up challenges to the local elite. In Chapter 6 we discussed cases in India where communist cadres undertook to assist tenant and landless laborer unions (Pandey, 1975; Alexander, 1980). This may be the only way such groups can get off the ground, but as a means of countering elite or government resistance it is a mixed blessing. Those who provide external support may attempt to dominate the LO, and such support attracts opposition from powerful sources.

Finally, we would note that sometimes resistance is affected by events beyond anyone's control. The Barpali vegetable growers' cooperative, opposed by a gardening caste, found the opposition diminished when an unexpected drought occurred, legitimating any and all efforts to grow more food (F. Thomas, 1968). Crises and sudden external problems have been discussed more generally in this regard by Solomon (1972). In our own project, helping to establish farmers' groups for water management in Sri Lanka, a failure of the rains created a situation where farmers accepted group action to conserve water much more readily than had been expected (Uphoff, 1981). When introducing or assisting LOs, one needs to be prepared to take advantage of whatever events occur to gain support and to counter opposition. This is discussed in Chapter 8.

The tactics that LOs adopt to deal with resistance reflect their perceptions of opportunities or constraints in their environment and of the external support on which they can depend. Where the environment is hostile and there is no external help, LOs tend to withdraw, avoid provocation, and rely on their own resources. When the environment is hostile but assistance is available from a voluntary organization, political group, or government, LOs may resort to confrontational tactics, strikes, withdrawal of patronage, demonstrations, or petitions to achieve their objectives against hostile local elites or officials. When the environment offers them opportunities to exert pressure legitimately, and especially if they are encouraged by an outside ally, LOs can employ participative tactics through elections, bargaining with bureaucratic agencies or local merchants, and accommodating to procedures and rules established by governments to regulate access to resources or influence. Since the rural poor cannot afford high risks, the tactics they employ—withdrawal, confrontation, participation—tend to be prudent expressions of what they can reasonably expect to accomplish.

Overcoming resistance that would otherwise prevent LOs from taking root can give impetus to their growth once the initial obstacle is circumvented. Gow et al. (1979, I:152–53) recount how a Samahang Nayon organization in the Phillipines was opposed by landlords seeking to avoid land transfer under the agrarian reform. They claimed that SN members were not tenants and charged them with trespassing. When a court ruled in the members' favor, the LO doubled in size within a year. Similar LO gains in India followed a Bhoomi Sena victory when its land claims were upheld by a judge (Rahman, 1981). One would not necessarily prescribe resistance as an advantage (in the way that Hyden (1981b) sees value in an "obstructive power"), but it should not be regarded as an unmitigated problem.

Avoiding Subordination

The measures to be used to avoid or minimize domination of LOs depend largely on its source. The most direct and powerful challenges come from government officials and/or local elites. As we saw in Chapter 5, government linkage to the point of domination has negative consequences for LO performance. Capture by more privileged elements of the local community will usually be equally unfortunate. Yet beneficent government assistance is not impossible; indeed, one of the most certain ways to prevent or offset dominance by local elites is government support. The Taiwan farmers' associations are more able and disposed than most other LOs to distribute benefits widely and to limit exploitation by local elites of the resources and opportunities the LOs create; this outcome, as Stavis (1974a) has shown, is largely due to government policies and efforts.

One of the most interesting studies in this regard is by Montgomery (1972), who assessed the achievement of land reform goals in a wide range of countries. He found that the degree to which rural people benefited from land reform—in terms of income, tenure security, and increased political participation—depended more on whether local organizations were involved in the implementation process than on the regime's ideology, its motivation, or the extent of land inequality. What is surprising about this finding is that such devolution of responsibility for land reform did not simply produce capture and distortion of benefits by local elites. The process of engaging rural people through committees, local government, peasant leagues, special courts, and similar means produced less subordination and distortion of benefits than under more bureaucratic or politicized approaches. Even a "progressive" bureaucracy acting in a centralized manner is likely to be more subject to the blandishments and influence of the rural elite than to pressures from illiterate, poor peasants who make claims or pleas on an individual basis. They lack the money and time, the knowledge and manners to be able to meet with distant officials, to understand how paper can be manipulated to their benefit, to turn legal loopholes to their advantage, and (perhaps most important) to gain a continuing foothold in the political system—unless they are involved in an organized way.

Subordination—possibly in the benign form of paternalism—is a risk for any LO that has a heterogeneous membership or that undertakes cooperation with government. Yet there are benefits from large membership or government cooperation that should not be overlooked. These include gaining more educated, experienced, and influential members and more access to technical expertise and outside resources. The small, autonomous, unlinked organization has less capacity to accomplish much on behalf of its members. A pathetic example among the cases we studied involved village patrols set up by the young men in a Senegalese village; they proved to be utterly inept, partly because they were not linked even to the village elders (Snyder, 1978). In Chapter 5 we saw that linkage, even when it involves some degree of government linkage, is likely to be instrumental to organizational success.

The main implication is that the risks of subordination are probably worth running, except in highly stratified socioeconomic environments. The question is how to reduce the risks and keep external influences within a range that is productive for the LO members. One particularly striking common feature of the most successful LOs in our study is that almost all in some way or other were able to reduce the effects of "the iron law of oligarchy," keeping leadership accountable or turning it to local advantage.

When a government is determined to dominate or subordinate an LO, the latter usually succumbs or disappears. We have discussed already how a number of LOs resisted government opposition to their establishment or continuation. But these cases show only that such resistance is possible, not that it is likely to succeed. We have seen that subordination to government is less likely where the rural community, or at least the subset that constitutes the LO's membership, has firm egalitarian values and traditions, like the Sidamo in Ethiopia (Hamer, 1976, 1980), or the people in Daudzai, Pakistan (Bhatty, 1979), or the Indians in the Bolivian highlands (Healy, 1980). One policy that apparently helped preserve the independence of the Subak irrigation associations in Bali was to prohibit government officials from holding office, even though they might be cultivators within the association's jurisdiction (Birkelbach, 1973). This provision has been formally adopted with new irrigation organizations in Java (Duewel, 1984). If government domination becomes particularly oppressive, as with the sugar plantation cooperatives in Peru, a strike may be the best and only remedy (Alberti, 1976). In that case, members of the cooperative regained control over their organization by instituting elections; previously, a majority of the delegates to their general assembly had been appointed by the government. Overall, we found it unlikely that LOs could resist the embrace of a government determined to control them. The probable consequence was that the LO would become an extension of government and lose the support and participation of its members.

Perhaps the most significant measure used to curtail the influence of local elites is barring them from membership. The Taiwan Farmers' Associations, for example, give only only cultivating farmers a full vote and the right to serve on the board of directors (Stavis, 1974a). The Small Farmer Development Program in Nepal, Bangladesh, and the Phillipines have permitted only households with incomes below the average for the rural population and with less than some maximum size of owned or rented landholding to become members.[4] The SFDP has some examples of unusual success with small investment of funds, though with considerable investment of supervision and training through group organizers.[5] Similar results have been obtained in Thailand by the efforts of catalyst agents from the Thai Khadi Research Institute, who worked with small farmer and other poor households in Tambon Yokkrabat (Rabibhadana, 1983). The program of the Agrarian Research and Training Institute in Sri Lanka to establish farmers' organizations for improved water management in the Gal Oya Project has found it possible to "convert" the two major local elite figures, who initially opposed the program of organization for equitable water distribution (Uphoff, 1982a).

A more bureaucratically initiated and centrally managed program is that of the Kenya Tea Development Authority (KTDA). It aims at assisting only smallholders to produce the lucrative crop that was previously thought to be feasible only for plantation agriculture. Tea growing was considered to be technically too demanding and the quality requirements too high for small growers—until it was tried within a well-conceived organizational structure. To restrict benefits to smallholders, tea plants were initially provided on a subsidized basis in quantities sufficient for no more than one acre (though that has since become the minimum and three acres the maximum for new holdings). Impressive improvements for small farmers have been achieved—even though a minority of larger growers, eager to get more

[4]In the Nepal SFDP, for example, persons who own more than 0.75 hectares of irrigated land or 1.25 of unirrigated land in the hill districts, or more than 1.3 hectares of irrigated or 6 of unirrigated land in the plains districts are not supposed to be admitted as members.

[5]An evaluation by the Agricultural Projects Service Centre after two years found the following kinds of differences between SFDP member households and a random sample of nonmember (control) households: household income nearly double; livestock output double (livestock raising was one of the major group activities); 97 percent with improved latrines compared with 50 percent; 51 percent school attendance by children compared with 30 percent; 26 percent acceptance of family planning compared with 14 percent (one group agreed that all members with three or more children would have vasectomies); 45 small farmer members elected to village *panchayats* (B. Pradhan, 1980). Shrestha (1980:93) reports that the groups decided to observe the ceremony "marking the entry of a boy to adulthood on a community basis instead of the traditional individual basis which would have involved the parents in very heavy expense in feasting their relatives and fellow villagers. As it turned out it was accomplished with a token cost of only five to ten rupees—a benefit of which not only about 30 small farmer households availed themselves but also two bigger ones." As households became better off, however, many went back to individual ceremonies.

income from tea, have managed by subterfuge to operate bigger holdings within the scheme (Steeves, 1975, 1978; Morss et al., 1976, II:D31–D42). The average tea plot of a member is less than one acre and yields an annual income of $185, three times the average cash income for agriculturalists there (Paul, 1982:61). The number of small farmers in the program has expanded from 1,000 in 1959 to more than 130,000, and their share of tea exports has risen from 6 percent to over half.

The system of grower committees begins with elected committees at the local level, which are run on an informal basis with a minimum of official representation. At the district level there is a committee of growers appointed by KTDA officials on the basis of being good tea farmers, having leadership qualities and having been cooperative with the Administration, the Department of Agriculture and the KTDA. This brings influential members into the representational system, strengthening the committees vis-à-vis the KTDA. There is also a committee at the tea factory level, with farmer representation. The role of committees at all levels, apart from allocating the plants within their area, is mostly advisory; they recommend the number of new growers to be taken in each year, select sites for buying centers, determine necessary road development and maintenance, and enforce regulations. Those members of the local elite who have been co-opted by the bureaucracy help to articulate local needs that smallholders share (roads, convenient buying centers, better prices).

A similar structure of organization is found in the AMUL dairy cooperative in India, where village-level cooperatives form the basis for daily purchases and payment. More than 850 village societies are amalgamated into a large apex organization with almost 300,000 members, most of them owning only one or two cows or buffalos (Somjee and Somjee, 1978; Paul, 1982:15–36). Participation in operating the organization is not much greater for AMUL than for KTDA, but the distribution of benefits within an organizational framework set up from above is impressively broad. Low-caste and even untouchable households have been able to participate in the economic benefits. The poorest households can participate in a cash-earning activity, and the premium paid for milk compared to the price given by private buyers represents a substantial improvement for them. Landless member households get 65–70 percent of their income from the sale of milk, and even small farmer households get 25–30 percent of their income this way from the sale of two or three liters per day. In addition, with its profits from large-scale processing of cheese and butter, AMUL has been able to provide veterinary and artificial insemination services; road improvement; and even scholarships for qualified children of members. The philosophy of AMUL's founder, A. L. Kurien, has contributed to this bias in favor of the poor in distributing benefits from milk production through this multitiered cooperative mechanism (Dorsey, 1978). The staff continue to be oriented to helping poor rather than rich households.[6]

Generally speaking, it may be possible and desirable to exclude the rich from organizations intended to benefit primarily the poor. Reducing the likelihood or extent of LO domination by local elites depends to a large extent on what attitude the higher authorities take toward the objective of benefiting the poor. One alternative is to establish heterogeneous local organizations, such as the Farmers' Associations in Taiwan, which include virtually all agricultural households. Membership is not compulsory, but the advantages of membership are so great—cheaper credit, lower prices for agricultural inputs, free advice from technicians employed by the FA—that farmers have good reasons to join. The local elite are in many ways prominent in the FAs, and in the similarly constituted Irrigation Associations. But government regulations on distribution of profits and on ceilings for credit preserve access to credit, inputs, and advice for *all* members. Even if benefits are not equal, they are accessible, thanks to government policy and supervision. This government effort, along with the land reform that created a reasonably equitable social structure, was a reaction to the failure of the Nationalist regime to maintain peasant support while on the mainland. For at least their first 25 years, the FAs and IAs assisted a broad spectrum of rural households and avoided the kind of elite exploitation so common among LOs in the Third World (Stavis, 1974a).

We have noted efforts to overcome the "iron law of oligarchy," the

[6]Steeves reports a similar effort by KTDA staff to protect opportunities for smaller growers (personal communication, based on fieldwork in 1979). There are numerous similarities between KTDA and AMUL which we cannot examine fully here. One is that tea and milk both provide certain incentives and impose a certain discipline, making such an organization more viable. Both involve daily operations (plucking and milking) and require quick, high-quality processing to get the best return for the labor (the best tea is produced within 12 hours of plucking; milk must be processed within a few hours unless refrigerated). The tea leaves and milk should not be adulterated with stones, water or other material if quality is to be preserved. A substantially higher price will be paid by buyers for a high-quality product, so a financial reward (incentive) can be paid to producers for careful handling and processing. Purchasing in small-group lots, holding the whole group responsible for any degradations of the tea leaves or milk, and paying a premium for high quality (good butterfat content, for example) can generate social pressures within the group to guard against watering the milk or bulking up the tea delivery with nonprime leaves, for example. There is no zero-sum conflict between large and small producers of tea or milk: both produce for large markets, and they have no price effect on one another. Indeed, larger ones may even gain somewhat from the greater efficiency of large volume in pickup and processing when smallholders increase total production in an area.

The AMUL "model" is now being extended to oilseed production and processing, with Government of India and World Bank assistance. We have not been able to do any study of this, but our expectation, based on a cursory analysis of the way in which the activity (product) affects the efficacy of an organizational mode, is that the oilseed effort will encounter many difficulties. There are already reports from India of violent opposition by "oil kings" to the new organization (*India Today,* 15 Jan. 1982, 113–15). This should not be seen as discrediting the model but as demonstrating how organization and task need to be matched to one another.

domination of an organization by a self-perpetuating leadership. Merely holding elections for LO officers does not assure responsive and accountable leadership, but the absence of elections makes this outcome even less likely. In the case of the Bangladesh SFDP group of landless laborers who went into fish farming at Fakirakanda, the absence of regular procedures for elections made it difficult for members to dismiss a leader who was divisive and irresponsible (Chowdhury, 1979). The Multi-Purpose Cooperative Societies in Sri Lanka had never been fully participatory organizations, being subject to the competing claims of party adherents in each community. But after the government in 1970 abrogated elections for their boards of directors (by having 9 of the 15 members appointed by the Minister of Cooperatives), oligarchic tendencies within the co-ops became stifling.

In favor of elections, we would cite the case of the Mmankgodi Farmers Association in Botswana. It elected officers the first year on the basis of social position and influence but replaced seven of eight the second year, choosing leaders of proven competence and commitment (Kloppenburg, 1983). While not a perfect instrument, elections are an important means of limiting the subordination of LOs to vested interests.

One of the frequent proposals for controlling the accumulated power of leaders is rotation in office. Buijs (1982a:56) cites this as one of several mechanisms that may be used when it is not possible to exclude richer persons from membership. We have found a number of examples where this seems to have had desirable results, though there is the contrary argument that it is unwise to replace effective leaders. There are good arguments for rotation. One of the most elaborate but apparently workable systems is that of the market women's cooperatives set up in Nicaragua (despite opposition from the traditional moneylenders, who spread rumors that the co-op would steal its members' money). Eleven directors are elected by the membership, with five composing a committee of administration, three a credit committee, and three an oversight committee to assure the correctness and honesty of operation in the other two committees. Membership on the credit committee rotates in order to minimize the chance of favoritism (Bruce, 1980). These co-ops, backed by an independent foundation (not the government), grew rapidly; there were more than 12,000 members in 58 co-ops by 1979. Savings grew from $75,000 in 1975 to $1.6 million four years later, and total assets from $125,000 to $625,000, an indication of the vitality and member participation these LOs achieved.

Although the national peasant confederation in Mexico (CNC) has not been noted for membership participation, in the Taretan area of the state of Michoacan, the CNC is vital. There, following the tradition of the Mexican revolution that there be "no reelection," officers are rotated with good results. There is, however, a special attachment in this area to revolutionary traditions and a fairly egalitarian distribution of land and income. According to Landsberger and de Alcantara (1971), "the belief that leaders cannot act without the approval of the majority of their constituents is widespread, among both leaders and followers, and seems quite often to be put into practice." They note that perhaps the single most important cause of the eventual success of the movement was the example set by "E. L.," a charismatic figure who led the earlier agrarian movement and helped to establish almost every peasant organization in Taretan, and by General Lazaro Cardénas, who was governor of the state before becoming president of Mexico, 1934–40. This underscores the value of an ethic of responsive leadership and egalitarian outcomes underlying whatever mechanisms may exist for rotation, for elections, or for other means of officer selection.

The very successful *zanjera* irrigation groups in the Philippines practice rotation of officers. Their explanation is that being an officer is a lot of work, and it is only fair to rotate the responsibility (Coward, 1979a). However, there is also a strong egalitarian norm, a belief that all persons should contribute labor and funds in proportion to the benefit they derive from the irrigation system (which is proportional, in turn, to the amount of land they cultivate that is served by the system; see Siy, 1982). This factor of community norms supports the point just made about rotation of officers in Mexican experience.

When the Korean government set out to create the Saemaul Undong movement for rural uplift, it gave considerable thought to structuring leadership roles. SU leaders are elected by their peers at the village level and, to impress upon villagers that the leaders are local representatives, are not compensated financially by the government, although they do receive such perquisites as free travel by rail and special scholarships for their children. The guidelines provide for rotating the leaders out of office usually every few years as a means of ensuring fresh incoming ideas, as well as preventing the leadership from becoming an independent political force. Incumbents can be rewarded for good performance, however, by reelection. Competition between the new, younger SU leaders and the older village leadership has brought about some conflict but has also had the effect of improving the performance of the Saemaul Undong as well as of the "traditional" village headmen. This effort of local organizations and leadership is reinforced by frequent visits from government staff to the villages to promote the directive that development be made more equitable (Lee, 1979; Goldsmith, 1981).

All of these examples of experience, ranging from Nicaragua and Mexico to the Philippines and South Korea, indicate that rotation can provide a useful procedural check on leader domination of local organizations. We are nevertheless reluctant to propose it as a universal remedy to be prescribed by outside agencies. If members themselves wish to institute a requirement of rotation in office, there is no reason not to welcome it. But if members are actively participating in the LO and are concerned enough about its performance to remove any leader not serving their interests, that may be sufficient; enforced rotation could deprive them of experienced talent. Also, the responsibility that members feel for their LO may be reduced by such intervention in their internal affairs.

One means of reducing the risk of subordination, fairly obvious but not often enough attempted, is *training*—for members as well as local leaders—to spread skills of administration, accounting, and communication, as well as technical knowledge, throughout the organization.[7] One of the factors that Michels noted as leading to oligarchy was a widening of the skills gap between leaders and followers as the former gained expertise, mostly through experience. The German Catholic development agency assisting the Manyu Oil Palm Cooperative in Cameroon undertook to provide organizational and technical training specifically to reduce the probability of its being dominated by government or its own leaders (Gow et al., 1979, II:53–66). In the Comilla program in Bangladesh, Akhter Hameed Khan emphasized the role of training for small farmers and landless laborers; indeed, the whole enterprise revolved around training centers at the *thana* (subdistrict) level. So long as the training, with attendant supervision of operations, was sustained, the LOs performed reasonably well and equitably. When this was relaxed, standards of probity and member control rapidly diminished, indicating that the effect of training without supervision and follow-up is limited (Blair, 1974).

Subordination can also occur as a result of efforts by well-intentioned outside agencies, sometimes with the aim of displacing or preventing government domination of local communities. The Thai Khadi Research Institute in Thailand in its experimental programs deliberately restricted the role of its staff to guidance and resource procurement in order to reduce any "dependency" that could unintentionally result from more active staff roles (Rabibhadana, 1983). The Sarvodaya Shramadana movement in Sri Lanka discovered that with success, rapid expansion, and foreign assistance, it became increasingly centralized and control-minded. Deliberate steps had to be taken to reverse the process (Ratnapala, 1980).

The most important factor in preventing LO subordination—whether by the government, local elites, outside agencies, or its own leaders—is the level of activity and concern by members. Where we find LOs functioning without excessive interference or control, there is usually a mobilized membership. The reasons why an LO is important enough to command the sustained efforts of its members will vary. In an irrigation association, members depend on the water it distributes to produce their crops and feed themselves. Sometimes needed public facilities or services are at stake—

[7] Charlick (1984) found consistent association of most LO performance measures with training (technical plus communications skills) but, curiously, not with his specific measure of increases in technical efficiency. The correlations went *up* if more than the leaders were trained.

such as bridges in the hills of Nepal or water in the plains of Malawi. Often the material interest, however, is reinforced by a common group identity such as ethnicity or caste. Alternatively, avoiding excessive subordination may be a result of tactful government action or of local leadership that maintains a relatively open and participatory style of organization.

We reviewed the two dozen LOs judged to be most successful to see how they coped with the problems of subordination. Almost half were undergirded by strong ethnic or tribal solidarity, which entailed an ethic of egalitarianism that limited leadership excesses and opposed outside domination. Both leaders and outsiders were kept in line by well-conceived practices and group norms in a number of irrigation associations—the Subak in Bali, and the Laur Irrigators Service Association and the Ilocos Norte *zanjera* organizations in the Philippines. On the other hand, in several projects there was a definite governmental role that sought to limit the organization's subordination to local elites.[8] Dedicated local leadership was present in the Baglung bridge committees in Nepal and the Oryu Li mothers' club in Korea. Finally, two organizations in the "most successful" category exhibited a fairly high degree of control from above but nevertheless a rather good spread of benefits among members—perhaps coincidentally one marketing tea (KTDA) and the other coffee (the Portland–Blue Mountain Coffee Cooperative in Jamaica).[9]

Michels argued that organizations of the poor would be more vulnerable to oligarchic tendencies than those of the rich.[10] But he did not consider situations where ethnic or other social factors would lend some force to the egalitarian ethic, or where the government might wish to keep LOs from becoming dominated by elites. We see Michels's prophesy fulfilled quite often. But oligarchic governance can be a matter of degree, and there are bound to be fluctuations over time in the responsiveness of leaders to their members, or the submissiveness of members to their leaders. With supportive social norms, with responsibility for a valued resource (like water) that

[8]These included the Saemaul Undong in Korea, the Farmers' and Irrigation Associations in Taiwan, the SFDP groups in Nepal and Bangladesh (particularly Ballovpur No. 1), and the peasant federation (FCV) in Venezuela.

[9]If we go to the next category of LOs—those judged "very good" overall—the importance of ethnic or tribal solidarity continued to be evident: 8 of the 35 exhibited it as an important factor (four were concerned with irrigation). Only the Puebla and Plan Maize programs, small farmer organizations in Mexico, and the plantation workers union in Zambia had a major government role. Two were founded and maintained with *noblesse oblige:* Lirhembe in Kenya and Heenpitagedere in Sri Lanka. We find these 35 cases on the whole not quite as free from "oligarchy" as the first group reviewed, though still not as subordinated as LOs in lower performance categories.

[10]Michels's "sample" universe was the trade unions and socialist or social-democratic parties of western Europe around the turn of the century. He compared their relations between leaders and followers with the internal democracy to be found in some of the "aristocratic" parties of Europe and found that, paradoxically, the organizations most "progressive" politically were the least "democratic" internally.

is necessary for survival, or with government involvement to neutralize local elites and promote some degree of "accountability" by leaders, we find that subordination need not be the fate of every local organization.

Coping with Internal Divisions

Although resistance and subordination are usually external in origin, the extent of their effect depends partly on internal factors. Conversely, while factionalism, partisanship, and other kinds of cleavage are definitely internal, the attitude of outsiders (officials, party cadres, ethnic leaders, and others) affects their severity. These divisions have to be resolved within LOs, but certain structural or procedural features can help.

Governments and donor agencies need to be tolerant of divisions, knowing that these often exist within their own corridors. To some extent, conflict and competition are signs of vitality, or even spurs to performance. The LO with no such signs may indeed be inert, as suggested by Whyte and Alberti (1976) following their study of Peruvian rural communities. Interestingly, in our work on water management organizations in Sri Lanka, we found that farmers in communities with higher reported levels of conflict, even over water, preferred by a significant margin to have water management handled by a farmers' committee rather than by a government official or an appointed irrigation headman (Uphoff et al., 1981).

At the same time, we would not minimize the damaging potential of internal divisions. In Chapter 5 we described some specific debilitating effects of partisanship, and in Chapter 6 we cited a number of cases where these had contributed to the demise of particular LOs. The qualifications above are intended to prevent readers from assuming that all divisions are a sign of incipient failure. Indeed, outside efforts to eliminate such conflict may well be counterproductive, if the perceptions or inequities giving rise to tensions are not dealt with. In India, when the government decided that factionalism within village *panchayats* had reached too high a level, it introduced a rule that funds would be withheld from those that did not elect officials unanimously. The consequence was that majorities bludgeoned minorities into casting unwilling votes so that the village would not be denied government aid. But this did not make the *panchayats* more effective as local organizations (Nicholson, 1973).

The most obvious antidote to internal division is *group solidarity,* which we saw was an important bulwark against elite resistance and bureaucratic subordination. Offering this observation as advice would be, of course, gratuitous; if there is solidarity, by definition there are no divisions. But actually, there are likely to be some cleavages even in socially homogeneous groups. Some of the measures discussed below may be relevant even for such groups at certain times, and the aim of promoting solidarity should probably be fostered at all times, recognizing that some conflicts are bound to arise.

The most basic measure to minimize divisions seems to be keeping local organizations small and, by implication, relatively homogeneous (Johnston and Clark, 1982:178), though we should note that in our statistical analysis we found no correlation between size or heterogeneity and LO effectiveness in conflict management. The case of the Carrizo Valley *ejido* in Mexico offers an example of how problems of internal division may be resolved by reducing size. Although all the members were landless peasant households with no class or ethnic differences, they came originally from different villages. The stress of starting a new agricultural enterprise under difficult conditions, and in a collective mode of production, caused many conflicts. When the *ejido* was divided organizationally into four units according to previous communities of residence, conflict subsided and production increased—partly because this permitted more individualization of plots, but also because the collective aspects of production became more manageable (Winder, 1979).[11] The Small Farmer Development Program in Nepal found that small size helped avoid the class and caste divisions that beset rural society, though as the program has developed, small farmers themselves are forming *socially* more heterogeneous groups, while preserving essential *economic* homogeneity (B. Pradhan, 1980). This is consistent with our finding in Chapter 5.

One interesting case in this regard involves the village-level irrigation groups in Senegal, where factionalism is considered a serious problem. Two measures have been taken to minimize it. First, the organizations were set up expressly to maximize the homogeneity of membership; second, the government agency that initiated the project deliberately left the responsibility for organizing all collective work to the farmers' representatives, who are selected by the groups themselves. In this way, the agency tried to stay aloof from internal social conflicts, leaving it to the officers and members of the groups to solve the conflicts in accordance with existing social conventions. The officers themselves introduced a system of fines to ensure the fulfillment of collective work obligations, and substantial results have been forthcoming (Fresson, 1979). This is similar to the case of the water users' association in a faction-ridden Indian village, reported by Wade (1982a) and noted in Chapter 5.

Another major method of mitigating divisions is to make *equity and fairness* manifest features of LO operation. The *mahabar* associations among

[11]The greater incentive effect for agricultural production when the units of production are smaller was definitely experienced in China, when it moved from larger to smaller communes, but particularly when it made the production team (20–50 families) the accounting unit for organization of work and distribution of returns (Stavis, 1974b).

the Sidamo in Ethiopia experienced some division between members who came from different villages, but the *mahabar* overcame this problem by establishing their legitimacy and fairness (Hamer, 1980). The potential for divisive rivalry exists within the Local Development Associations in Yemen, since they usually incorporate a number of villages, and there is keen competition for LDA investment in public services. In the Khamer LDA, this problem was resolved by using objective criteria for the allocation of resources. For example, when sites for new wells were being chosen, each settlement was assessed in terms of how far its residents had to walk to obtain water; those who had to go the farthest received the first grants for well construction (Green, 1975).

A number of methods have been developed over the years by the *zanjera* irrigation associations in the Philippines to play down possible disagreement within and between groups by following manifestly fair practices in allocating work assignments and financial contributions (Siy, 1982). The bridge construction committees in the Baglung district of Nepal—a set of organizations as "recent" as the *zanjera* are "old"—dealt similarly with a potentially divisive issue, the sequence in which bridges will be built, by making a plan that was carefully deliberated and widely publicized to achieve consensus. All communities in the district were invited to submit requests for bridges across the ravines and chasms that crisscross Baglung in the middle Himalayas. About 120 requests were submitted, each with justifications in terms of potential traffic, savings in time compared with existing means of crossing, the amount of labor and material that would be contributed by the community, and potential safety improvements (lives can be lost at dangerous fords). A multiyear plan was worked out, giving priority to 62 of them. Various trade-offs among considerations of cost, use, ease of construction, and safety were made by discussion, compromise, and probably some log-rolling—leading to a consensus sequence and a plan that all could accept (P. Pradhan, 1980).[12]

The common image of local communities making extravagant and excessive demands for public services is contradicted by such a case. The requests were reasonable, partly because each community had to contribute a major share of the costs (in voluntary labor and the collection of stones and other building material); it was not competing for free goods handed out by a paternalistic government or a donor agency. In addition, the open process of decision-making contributed to avoiding the kind of conflict between communities, often reported in the literature, that can occur when resource allocation is at stake. One would have expected, from common conceptions, that the peasant confederation in Venezuela (FCV) would have been riven from top to bottom with partisan divisions when three different political

[12]Apparently after this number of "priority" bridges were constructed, the overall committee organization was disbanded (P. Pradhan, personal communication).

parties were vying for representation and influence. The fact that meetings were conducted quite openly, with all members free to take an active part, helped to make that organization viable (Powell, 1971).

The *structure of decision-making* can also help reduce internal divisions, if all relevant groups are represented and can expect that their views and interests will be considered. Even if the resulting decision goes against them, to the extent that they have received a fair hearing, conflicts are less likely. Or if existent factions are not a problem but the danger of internal divisions is to be avoided, a broadly participatory structure of decision-making is advisable. The Mraru women's association presents an interesting innovation in this regard. Policy decisions are made by a full vote of the membership (by secret ballot) after an elected executive committee of nine has considered the issues and presented its recommendations to the group (Kneerim, 1980). The same structure is reported for La Libertad, a women's cooperative in Bolivia. All the decisions of its executive committee, which meets weekly, are reviewed at semiannual assemblies that are usually attended by more than 300 of the 1,200 members (Wasserstrom, 1982).

In general, as discussed in Chapter 5, we found that informal methods of operation are desirable, and this seems to be true with regard to internal divisions. Formal procedures, particularly those that grant the appearance of power to certain roles and give majorities clear rights over minorities, offer opportunities for internal factional domination that can be avoided by informal approaches. Victories by formal rules in LOs usually turn out to be quite hollow, but nevertheless may tempt one group to seek advantage over another. There are, to be sure, good reasons for specifying the rights and responsibilities of members. Bottrall (1980:8) suggests the value of detailed internal procedures to safeguard the interests of weaker members. It was found in the Mmankgodi Farmers' Association that establishing a formal constitution, outlining membership rights and responsibilities as well as delineating the powers and authority of the executive committee and certain procedural rules, helped avoid internal conflict (Kloppenburg, 1983). It should be emphasized that this was a constitution devised by the members themselves, not a boilerplate document handed down by officials. That may help to explain its value, compared with the little value we observed in the majority of cases where such formal instruments were introduced.

Effective performance, in the face of whatever factionalism surfaces, is likely to be one of the best ways to reduce divisions, since it gives all groups a stake in maintaining the organization. We cited previously the example of the Farmech project in Lesotho, where conflict between adherents of the government party and followers of the opposition party immobilized agricultural improvement activities. What finally broke the stalemate was the appointment of an effective manager, who was able to induce enough cooperation and show enough positive results from new practices to undercut previous tensions. Part of the transformation involved making the manager clearly responsible to a control board composed of representatives of the district council (politicians), chiefs, and central government. A staff board was also created to give employees a voice in the operation of the project and to introduce self-evaluation. Such a set of organizational initiatives coupled with administrative leadership put the development effort and the LO back on track (Wallman, 1969).

This specific example is fairly elaborate. The general point, however, should not be lost: one way to reduce internal cleavage is to get beneficial results flowing from the LO. Divisions may, of course, impede good results, but trying to end them as a precondition for LO success may be futile, since there is little reason to change attitudes and social relations in the absence of some change in environment or opportunity (Festinger, 1957). This means that reduction in cleavage and increase in LO effectiveness, discussed next, often need to be managed jointly. The other elements we have identified for reducing the undesirable effects of internal divisions—working with small, relatively homogeneous groups; emphasizing equity and fairness in performance; establishing open, representative decision-making processes—should reinforce one another when there is a legacy of ethnic, religious, party, or kinship differences. As noted in Chapter 5, we found a number of LOs operating quite successfully with a considerable degree of social and even economic heterogeneity. Thus, while we would not want to underestimate the destructive potential of internal divisions, and we recognize that they are common in local organizations, they need not be regarded as fixed obstacles to success.

Reducing Ineffectiveness

Since the large majority of LOs are to some extent ineffective, we want to know how to contribute to their improvement. Environmental factors can affect LO performance but do not do so uniformly. As suggested in Chapter 4, what LOs make of their environment is usually more determinative than the influences the environment has on them. We thus regard the environment as a factor, but not as something that must be changed in order for LOs to operate better. In this connection, while government and outside agencies are part of an LO's immediate environment, they are not the same sort of givens as natural resource endowment, social stratification, or prevailing norms. Their actions and influence can be made more supportive of effective LO performance (see Chapter 9).

One obvious explanation for LO effectiveness or ineffectiveness is the quality of *leadership*. This variable is not readily amenable to policy prescription, however. Good leadership is hard to guarantee by design and, indeed, often hard to recognize in advance. There is a real danger of circularity in dealing with it as a factor, since the discussion readily becomes a tautology: success is due to good leadership and failure to bad leadership; good leadership is indicated by success and bad leadership by failure. Yet however elusive the phenomenon is when we try to analyze it, it is real and important. We address it in Chapter 8. Here we focus on how LOs are structured and operated, indicating what appear to have been helpful features for improving effectiveness in specific instances and amplifying the overall statistical analysis in previous chapters.

One suggestion often made for remedying ineffectiveness is that the organization should become larger, in order to be more efficient or more powerful (depending on whether the focus is on economic or political weaknesses). But this represents a reversal of cause and effect: LOs are more often large because they are effective than effective because they are large. Thus one should not prescribe the creation of larger LOs as a means for promoting rural development.

In a number of cases, practitioners or analysts have suggested that *small-scale organization* is more likely to be effective for the rural majority, at least initially. This was observed for example with the Tambon Yokkrabat organization in Thailand, the Matam region irrigation groups in Senegal, and the Carrizo Valley *ejidos* in Mexico—all previously discussed. In none of these cases was it suggested that larger size would make the LOs more effective. Operations are simpler and more manageable in a smaller group; membership cohesion is likely to be greater; communication is facilitated, keeping officers more accountable and members more active. Small scale may be a starting point, however, rather than an end state.

There is a predictable observed relationship between size and the number of functions performed; the larger LOs tend to perform more activities. The qualities of being small, simple, and single-function tend to go together. But successful LOs get drawn into more activities; case after case shows that once an organization is able to meet one need—improving water supply, for example—it moves on to such activities as offering adult literacy classes, constructing grain storage facilities, seeking fairer prices from middlemen, or starting a cattle-fattening project. Where the additional activities are determined by the LO and are undertaken at a pace it sets, they are usually successful. We did not find many LOs becoming overextended in the number of activities they attempted if they were proceeding under their own direction. The implication is that an LO should begin with an undertaking that has a high priority to its members and master it first before tackling others.

Similar to the proposal that increasing LO size will enhance LO effectiveness is the proposal that making them multifunctional will allow them to provide "integrated" services. The two suggestions are similarly misleading:

like increasing size, evolving multiple functions is desirable but is not to be regarded as a cause of LO improvement. We have observed that sequential rather than simultaneous integration of functions is more likely to pay off for LOs and their members. The successful discharge of multiple functions is more often a consequence than a cause of effectiveness.

A good example can be found in the case of a government-sponsored project at Zacapoaxtla in Mexico, where agricultural officials sought to improve maize production. When they found that indigenous varieties were productive enough that the new varieties available offered little advantage, they turned their attention instead to organization and devised a strategy of starting with consumer cooperatives to build support and capacity for improving agricultural production. For example, farmers were paying exorbitant prices for sugar in private stores. So project leaders purchased sugar in bulk from a state organization at a fraction of the market price and sold it through consumer co-ops, together with other staple goods and canned produce, at attractive prices intended to assist low-income households. The co-ops were thus able to expand their membership rapidly. Project officials then worked with villagers to plan and organize their agricultural production so as to be able to bargain for better prices when selling their crops. This added strength to the movement. Although the first ten co-ops had to be organized largely by project staff, thereafter the idea spread, and other cooperatives formed with little encouragement from the staff. By 1980 there were more than 40 cooperatives involved in group buying of fertilizer and advising members on its best use. By purchasing in large volume, the co-ops were able to specify the exact composition they found most effective for each area and to negotiate the date of delivery for each village. With the organizations providing economies of scale, the farmers in these communities were rapidly increasing their yield and income (Whyte and Boynton, 1983:205–206).

To the extent that an LO is effective, one can expect it to grow in membership as well as in number of functions. The alternatives are to expand existing LOs or to set up similar new ones. Rapid growth of membership is one of the best indicators of success. The 15 market women's cooperatives in Nicaragua grew from 210 members in 1972 to 5,530 in 1979, the average size of each co-op increasing from 15 to 385 (Bruce, 1980). The Sukuma cotton co-ops in Tanzania went from 32 small co-ops in 1952 to 13 consolidated unions in 1955 and finally a federation (the Victoria Federation of Cooperative Unions) in 1959 with 377 local societies, owning and operating their own ginnery, having doubled their cotton production in this period (Lang et al., 1969). The growth of the AMUL dairy cooperative in India to more than 300,000 members is also exemplary. In Nicaragua, the base units expanded; in the other two situations, the number of units increased rather than their individual size. As a rule, we think the latter course is more likely to be successful.

With such growth and proliferation, it is advantageous to increase *vertical and horizontal linkage* to make the base units more effective. To some extent the significant correlations we observed between horizontal or vertical linkage and overall performance may have reflected the basic effectiveness of the LOs involved. The correlation of overall performance with horizontal and vertical linkage represents two-way effects. LOs that are basically effective are more likely to increase in number as well as size and to enter into linkages with other organizations vertically and horizontally. Such linkages reinforce and support the base units' effectiveness but cannot be said to have caused it. That is probably why the statistical relationship observed is not higher, but also why one cannot increase LO effectiveness by simply instituting linkages from outside. Enhancing such effectiveness will require a combination of efforts, most of them internal to the organization.

Efforts to *mobilize resources internally* will be conditioned by how effective the LO is already; people are more likely to contribute to an activity that has some demonstrated payoff. But getting such resource contributions, so that the LO is not wholly dependent on outside sources and so that members regard it as "theirs," seems a critical step.[13] The self-reliance of the Bhoomi Sena organization in India, the Tiv associations in Nigeria, and Ayni Ruway in Bolivia contributed greatly to the dynamic of success that they were able to initiate. Each focused upon utilizing its own resources, though each had some useful outside assistance along the way from political activists, agricultural extension agents, teachers, or other professionals. Even in cases where the government role was large, as with the Taiwan Farmers Associations and the Malawi self-help water committees, it was expected that the members would contribute substantial resources of their own and that, once the activities were capitalized with outside support, they would be largely self-financing.

Various techniques seem to contribute to internal resource mobilization. First, a survey of members, house to house if necessary, has been helpful in a number of cases. We discussed this earlier with regard to the Banki water supply project in India, where such an approach began to break down misunderstanding and opposition to the proposed water system. Elsewhere it has been used to identify both local needs and local resources, helping leaders and organizers match up what they will suggest doing with what people are most willing and able to support. In an experiment in Thailand, such a survey "discovered" three potential leaders from among the poor rice farmers in Tambon Yokkrabat. These were persons who had innovative ideas about how to solve village problems and who were sought out by others for advice. Their base of support was popularity and community-mindedness rather than wealth or formal position. The local organization

[13]We would note that the DAI study of small farmer strategies found "resource contribution" one of the factors most strongly associated with project success (Morss et al., 1976), though Young (1980) has to some extent qualified that finding.

that emerged was headed by these persons, who could easily have been bypassed if the organizers had not known of their presence and capabilities (Rabibhadana, 1983).

Another method is to spread leadership responsibilities more broadly. It is a general though mistaken conception that there is a lack of leadership capacity in small communities, whether because of inhibitions against persons not "born to leadership" or because of exodus to urban areas by the most energetic members. A number of cases indicated that organizational effectiveness resulted from dispersing rather than concentrating responsibility. In the Small Farmer Development Program in Nepal, it was found that most farmers were reluctant to accept leadership positions because these demanded so much time, and they needed to use almost all their time for family survival. The solution was to create committees even within the small group of 12 to 15 members—one for the agricultural improvement activities such as buffalo raising, one for supplementary income from such activities as beekeeping, one for adult literacy and social programs, one for handling group credit and repayments, and so on (Clark et al., 1979). The rather successful SFDP group at Ballovpur in Bangladesh followed a similar course, selecting a two-member executive committee and delegating other responsibilities to members—for animal care, social work, family planning, and health care (Abedin, 1979).[14] We have already discussed how the women's cooperatives in Nicaragua spread responsibilities to guard against elite domination, at the same time enhancing the effectiveness of their operations.

Holding *regular meetings* is another method suggested in some cases. When things are not going well for an organization, there is a temptation not to meet. But the SFDP programs in Nepal and Bangladesh have found regular meetings an essential element of LO development strategy. To be sure, it is often necessary to provide training on how to conduct meetings, so that they are efficient and satisfying. The peasant confederation in Venezuela (FCV) found meetings to be an essential part of its operation at the grassroots level. The interesting situation with the Lirhembe multiservice cooperative in Kenya is that they held only one formal meeting every two months but in fact had informal meetings almost weekly at the social center built in the community as one of the first undertakings of the organization (Morss et al., 1976; II:D20–D30).[15]

[14]Starting in 1976 with a subsidized loan for a cattle-fattening project, the group's activities expanded to milk cow rearing, paddy processing, and paddy cultivation. The members also engaged in latrine digging and road construction, and started a religious school for children with other (nonmember) households—all within three years. The group received loans of 35,000 *taka* for completed activities and another 31,900 for activities in progress, with no overdue payments in three years of operation. In this time, the average household income doubled.

[15]"The project shows the importance of local involvement from the outset; one strength of

Finally we would underscore the contribution of *elections,* meaning the opportunity to replace ineffective leaders. This was graphically shown in the Mmangkodi Farmers Association in Botswana (already cited), where after one year seven of the eight officers were replaced by election (Kloppenburg, 1983). In the Saemaul Undong movement in Korea, elections—with the expectation of eventual rotation in office—have introduced an element of competition among leadership cadres which is positive for LO performance, since good leadership can be rewarded by reelection. There is no direct financial reward, but substantial status benefits and some perquisites help to attract energetic leadership (Goldsmith, 1981).

One important observation in a good number of cases is that illiteracy on the part of leaders is not necessarily a barrier to effectiveness. For some kinds of organizations, or some tasks within them, literacy—and numeracy—are virtually a requirement in leaders. But managing the Subak irrigation associations in Bali has been done quite well by illiterate farmers who enjoy the members' confidence; when necessary, they used the services of a professional scribe for recording procedures or financial management (Birkelbach, 1973). Elsewhere in Indonesia it has been found desirable to remove literacy as a requirement for office in the state-sponsored irrigation associations, the Dharma Tirta (Duewel, 1984). The Nso women's cooperative in Cameroon also found that leaders could function quite well without being literate. The leaders chosen were usually related to the traditional village chiefs or elders, and that status, which enabled them to mobilize and direct activities, was more important than literacy (Gow et al., 1979, II:43–52).

That last observation is the operative one; if the leaders have the personality, motivation, and confidence (in themselves and of the members) to be effective, ways can be found to compensate for limitations like illiteracy. For dealing with the "outside" world, literacy is usually important, if only because of the status associated with it or the stigma attached by the educated to illiteracy.[16] Internal coherence and strength, however, are usually more critical for an LO's effectiveness than are its external linkages. External effectiveness cannot substitute for internal strength.

Regardless of the literacy of leaders—or members—it is desirable to sim-

---

the project was the communication channels, formal and informal, between the people and leaders (and government officials). Of particular importance was that the process included both consultation with farmers and information dissemination at each phase of project development. The project also shows how a social center can assist in bringing about community endeavors, especially in areas where individuals are inclined to pursue their private interests in isolation. It is unlikely that the agricultural program would have gotten off the ground without the help of the social center program" (Morss et al., 1976:D29). The DAI case study details the agricultural improvements in yields of corn and other crops and also in dairy production.

[16]It will be recalled that the only task significantly correlated with literacy was control of bureaucracy (see Chapter 4).

plify regulations, applications, and reporting requirements for local organizations. In the SFDP in Bangladesh, the group organizer at Boyra, who helped members prepare loan applications for the bank, discovered ways to simplify the *pro formas* while still satisfying bank requirements. Consequently, members have been able to handle their own applications and have set up their own savings fund. All the loans have been used for profitable projects and have been repaid, and more are in process (Islam, 1979). While the group has had excellent leadership, has enforced considerable internal discipline (compulsory savings, fines for not attending the monthly meetings), and has diversified into family planning efforts, one of the elements contributing to its success has been the increased accessibility of credit on terms that members could understand and manage themselves. We found a similar situation with the Barpali leather workers' cooperative, which started without government support. When it finally got the attention it needed, one of the things that helped it was the application of simplified regulations for cooperative management (F. Thomas, 1968). There is considerable scope for boosting LO effectiveness in making rules and procedures from the government side less complex and more accessible.[17]

From the LO side, there may need to be some formal *regulations*. We noted in Chapter 5 that the more informal an organization's practices, the better its performance. But we find that several LOs have, on their own, introduced a degree of specificity into role definitions and adopted procedures that are quasi-formal. The value of a written constitution for the Mmangkodi Farmers Association in Botswana has been cited already. The Subak irrigation organizations in Indonesia have codified their own rules for regulating admission, taxation, the powers of members, and the duties of officers; the responsibilities and rights of all are clearly delineated, keeping control and decision-making democratic (Birkelbach, 1973). The successful rickshaw pullers' group at Digharkanda in Bangladesh also developed its own rules; more important than the rules themselves, however, was the fact that these were the members' rules. For an outside agency to have prescribed even the same regulations would probably not have been as effective (Raha, 1979). One way to assist LOs that are having difficulty in operating coherently may be to encourage them to develop rules, but not to

[17]The government in Nigeria required cooperatives to maintain 14 sets of books, according to King (1975:205), and finally had to take over cooperative accounting because the LOs could not handle their own records. King reports that simplified accounting schemes have been developed (for example, for groundnut co-ops in Niger) that allow illiterates to manage co-ops and retain control over this key function. One of the most interesting innovations for administering a rural savings program based on local organizations is the stamp system for savings clubs in Zimbabwe, Lesotho, Zambia, and Malawi (see Howse, 1974; also Smith and Dock, 1981). This technique has expanded local financial resource mobilization and stimulated demand for agricultural inputs at low administrative cost, with minimal malpractices. What spurred the system's formulation and spread was the intention behind it to have something that local organizations could themselves manage.

impose them. Providing guidelines or model rules may be useful but must be done carefully, since anything emanating from the government or from more educated sources may be taken as mandatory. If an agency is giving assistance in this regard, it should probably offer several models or alternatives, conveying the sense that the group must itself make decisions and that there is not one best or required way for all groups to operate.

One of the most commonly proposed solutions to LO ineffectiveness is *training,* and indeed we have found many cases where it was important for success: for example, the Saemaul Undong movement in Korea, the Small Farmer Development Programs in several countries, the Tiv groups in Nigeria, and the peasant federation (FCV) in Venezuela. The SFDP in Nepal has been particularly adept in integrating training into the monthly progress evaluations that are carried on within the small farmer groups, the quarterly and semiannual evaluations that involve both farmers and project staff, and finally the annual evaluation that includes higher government officials (Joshi, 1980). Too often, training is a top-down operation, assuming that the trainers know exactly what the trainees need to know. This one-way transfer of knowledge is seldom effective for LO development.

Some of the most unsuccessful LO training efforts have been found among the cooperatives, where there has often been a great deal of emphasis on the philosophy and goals of cooperation, but not enough on the mechanics of making cooperatives operate effectively (Münkner, 1979:27–28). The emphasis may derive from the fact that it is easier to impart abstract ideas than to explain organizational dynamics, financial management, or legal issues, where more complicated training materials—case studies or balance sheets—are needed. Some such practical training must be available to lower levels of the organization, however, so that there will be enough knowledge of proper practices throughout the organization to check malpractices and to keep the gap between members and leaders (or staff) from growing.

There is some debate as to whether training should focus on effective organizational operation or on the technical skills that members need as farmers, rickshaw pullers, or handicraft makers. As far as we can determine from the case studies, no priority can be generally assigned; both are important. One of the benefits that members may derive from an organization is technical training. At the same time, there are many things about running an organization that leaders and members should know. If these can be presented in a less didactic way than is found in most training programs—if the participants' interest can be engaged by relating the subject matter to their own situation and needs—training will generally be well, even avidly, received. Training sessions can also have a social quality, which makes them enjoyable occasions for imparting and receiving knowledge. And especially among uneducated persons, acquiring knowledge and skill enhances a sense of worth and efficacy.

We would emphasize that training alone cannot ensure effective or honest LO performance, though it can contribute to improved resource management. As part of a larger strategy—including technical advice, the support and supervision of an external agency, a degree of autonomy and self-reliance, and the encouragement of membership participation—training is valuable for making the other elements operative.

One related strategy often proposed for helping LOs to be effective is "consciousness-raising," particularly among the poorer sectors of society. As we did not find this attempted very often, we cannot generalize about it. Certainly if consciousness-raising produces the consequences claimed for it, it would be useful in many rural development situations. Charlick (1984), in his study of *animation rurale* experience in Francophone West Africa and Haiti, found a positive correlation between this approach and LO results. The Bhoomi Sena movement in India is probably the best example of a deliberate effort with some beneficial outcomes. On the other hand, forming "study circles" among the women at Dhulia seems not be have been effective, perhaps because the approach was stereotyped. They often talked about unnecessary things and were perhaps too self-critical, lacking directed conversation and problem-solving orientation (Kanhare, 1980). Efforts at consciousness-raising by the Institute of Cultural Affairs (ICA) at Maliwada, Maharashtra, also in India, was evaluated negatively by De (1979).[18] It is an empirical question whether the "consciousness" of a group needs raising: that is, whether they are unaware of their disadvantaged situation and consequent exploitation. More often it appears that what members or potential members lack is confidence that their collective efforts can improve their situation. What discussions initiated by an outsider can do is to build consensus among the group about tactics, develop mutual confidence, identify obstacles clearly, and get agreement on how each is to be dealt with—all with a view to overcoming the fear or pessimism that often inhibits group efforts among disadvantaged people.

We have noted many ways in which government initiative and intervention can be negative for LO effectiveness, but we have also tried to clarify the methods by which government linkage can be positive and even needed. Several of the studies of more successful cases point to the value of state support. An account of the *mahabar* associations in Ethiopia (Hamer, 1976) concludes that many associations there lost effectiveness for lack of such support. They were started by local initiative, but part of their success

[18] "The organising process has not led to a movement in terms of focusing on some of the strategic issues such as non-availability of drinking water and poor housing conditions for the landless Harijans. . . . The village has been taken as a homogeneous entity which it is not. Leadership is still not shared and the decision-making actors are still those who have more economic power and social status." (De, 1979:11) From our discussions with ICA representatives, we think there has been learning from the Maliwada experience.

hinged on assistance from community development workers of the government who gave technical advice, help in processing loans, and access to contracts whereby the LOs could earn money. The Barpali leather workers in India lacked the technical skills to get into tanning and more advanced marketing so that they could earn more from their labor by making shoes. At first the Department of Cooperatives would give no more than superficial help. The Barpali Village Service sponsored by the American Friends Service Committee had to support the co-op initially. After a few years the department took a more realistic interest; it gave technical advice, provided training, and—as noted already—simplified the applicable regulations, at which point the cooperative became considerably more effective (F. Thomas, 1968). A necessary state role was also seen in Nepal, where committees were set up by *panchayats* to build a road from Illam to Charali. A remarkable mobilization of labor was possible through their decentralized LO system, but the alignment of the road was not good, and there was need for heavy equipment, which only the government could provide. The road was completed in permanent form only when the committees persuaded the government and a foreign donor to become involved. Still, local efforts saved 30 percent of the total cost (P. Pradhan, 1980).

Our point is that government assistance is often necessary, but that it can be counterproductive if given unwisely. In the case of the coffee, cocoa, and other marketing associations in Jamaica, the government's efforts to enhance the LOs' coverage by making membership compulsory led to ineffective organizations. One might say that the government was not so much concerned about the associations as it was about getting more revenue by establishing a monopoly in marketing. This goal was defeated, however, because when the government made the LOs into puppets that could not speak or act on behalf of farmer producers, the result was declining production by smallholders. Thus both farmer and government interests suffered from such domination (Goldsmith and Blustain, 1980, ch. 3).

Creating and sustaining effective LOs, therefore, involves a careful balancing of efforts, internal and external. There is an important role for outside actors in most situations, as discussed in the following chapters. We find that the essential condition is internal leadership and membership participation, which can be reinforced by careful suggestions, supervision, and material support from outsiders. This view of LOs is important. There is a temptation to view them in mechanical terms, assuming that if constructed according to a sound blueprint, they will yield predictable outputs from a given set of inputs. We find this metaphor misleading, since LOs resemble organic rather than mechanical entities. We would underscore the importance of problem-solving experiences, which can be improved by more participative procedures for decision-making and by practical skills acquired through training efforts.

The critical element we see is confidence among members that the LO can help them deal with problems more reliably and more easily than they could handle them individually or by relying on patron-client relationships. Competence contributes to confidence, as does past success. Group solidarity and commitment are often necessary for good performance, but this is a two-way relationship. Unless solidarity is intrinsic because of ethnic, religious, or other ascriptive bonds, commitment must be built through performance rather than vice versa.

Confronting Malpractices

Among the failings of local organizations that most often undermine this confidence and this solidarity are malpractices—misuse or misappropriation of funds, or nepotism and other kinds of favoritism in allocating jobs, services, and other resources. These can be destructive to the functioning and longevity of local organizations. Some of the remedies proposed for other deficiencies can also contribute to reducing malpractices, according to the experience of a variety of organizations. We shall note some of these and then look at more specific remedies.

Smaller groups can most easily discourage abuses. Since all members can know the details of group finances, there is more pressure from the group to handle funds honestly. Where there are intimate connections among the members, there is a sense of group responsibility enforced by social control. In the Njangi rotating credit societies in Cameroon, if a member defaults, members of his or her family made good the obligation (Delancey and Delancey, 1982). This protects the group, but there is also social pressure on an individual to meet obligations, since those who would suffer by his or her misconduct are close relatives. The larger the organization, the easier it is for members to rationalize malpractices as harming only some abstraction or individuals to whom they feel no personal obligation.

As we have said before, there are LO purposes for which a larger organization is helpful, but for these the solution seems to be a multiple-tier structure, in which the members' affiliation is with a primary group at the base. The ability of KTDA and AMUL to maintain high standards of quality in tea and milk procurement rests in part on the small producer committees through which buying is done. Individual temptation to make short-run financial gain by careless tea plucking or unhygienic milking, or by adding branches, stones, or water to increase weight or volume, is checked by the injury it would cause other members if the whole truckload were refused at the processing factory. If 500 or 1,000 rather than 50 or 100 members were delivering to a specific collection point, the interpersonal controls and discipline would be much weaker. Since quality is so important, gains on this front more than offset whatever economies of scale could be gained from larger, more impersonal operations at the base. The principle of group responsibility to reduce malpractices thus applies not only to credit, which is now widely seen as better administered through LO channels than bureaucratically, but also to activities like marketing and irrigation.[19]

The principle of division of responsibility, delegating authority to numerous members in some official capacity, can not only offset the tendency to oligarchy but also serve as a check of malpractices. With financial matters handled by a committee, irregularities are harder to cover up than if one person has full control. One of the most interesting and effective examples of delegation of responsibility to check malpractices was the Quality and Quantity Committee set up by the Irrigators' Service Association at Laur in the Phillipines. In this case, the malpractices were on the side of technical staff who were constructing a dam. Farmers were contributing labor and materials to reduce the total cost, since they were obligated to repay the government's investment in the dam eventually. The Quality and Quantity Committee monitored the materials delivered to the construction site, returning inferior materials and equipment for replacement. It also kept tight control over the use of equipment and fuel, since farmers had to pay the costs of any waste themselves; one of its most decisive measures was to stop engineers from using project vehicles and fuel for personal matters, such as driving into town for lunch while the farmers brought their own lunches to the work site. The committee appointed its own warehouseman and monitored prices to assure that bids were awarded to the lowest-cost suppliers. The resources at stake were, after all, those of the farmers (Isles and Collado, 1979; Coward, 1979a; D. Korten, 1980:492–94).[20]

Simplification of procedures is likely also to help control malpractices, since more members are then likely to be able to understand the financial situation of the organization. Complicated provisions may give government auditors more control over accounts but will forfeit the assistance of members and LO officers who, if they understand the financial dealings, can be more effective than outside auditors in checking on misbehavior.

[19]A very impressive account is given by Anderson (1968) of how the introduction of group credit, where responsibility for making and collecting loans was turned over to local organizations, raised repayment rates from 12 to 75 percent, to 97 to 100 percent in a program including some 42,000 families in Niger.

[20]This case also illustrates the point that local residents may have useful contributions to make to *technical* decisions. The farmers at Laur insisted that the engineers' design for the dam was insufficient and that it would not withstand the force of flood stage during the monsoon. The engineers insisted that their calculations proved the design's adequacy, but the farmers' fears were proved correct when the dam washed out a few months after completion (F. Korten, 1982:14). Cernea (1983:52) reports that Mexican peasants correctly identified the reasons why livestock experts were wrong to propose the introduction of Swiss cattle. The foraging and water constraints in that environment were such that only Zebu cattle, which the peasants proposed instead, could manage.

This relates to the advantage of having regular open meetings. The SFDP has found it important to keep all members informed of their organizations' financial status. The Subak irrigation associations prevent misappropriation by the *klian* (water headman) of the water fees collected from members by assessing all the fees at a public meeting. That way, everyone knows how much everyone else is supposed to pay, and how much should be collected and accounted for in total (Birkelbach, 1973). There is a question as to whether large or small meetings are more desirable. We found cases offering evidence on both sides of the debate, but would cite particularly the experience of one SFDP group in Bangladesh (Ballovpur No. 2), an unsuccessful LO where the leaders had exploited the group's resources and good will. It was said by members who finally denounced and removed the corrupt leader that they found courage to stand up against him only in a large group meeting (Abedin, 1979). Such an occurrence speaks in favor of regular open meetings, which provide occasions for exposing malpractices even if the opportunity is not always used. When meetings stop, one check on leaders—public exposure—is lost.

This points to the utility of what an International Labour Organization report on Philippine development called the "goldfish bowl" approach, in which public exposure through open information channels can be used to deter diversion of public resources (ILO, 1974:68–69). This has been operationalized in Indonesia, where village-level community development committees (VCDCs) were organized in Aceh province by the Save the Children Federation. This followed from an "open management" approach that SCF itself adopted and spread to the VCDCs. "This approach means, in effect, that all expenditures, income receipts, and accounts are routinely published, posted, and made available to everyone. The assumption was accepted that the [higher-level] CDC and VCDCs are not closed groups but are to act on behalf of the community. Therefore, the community has a right to know what is happening. Committee meetings are open. Anyone can come and express their views" (Van Sant and Weisel, 1979:18–19). The organizers found many villagers participating in meetings, aided by training that enabled them to understand the proceedings and records of the committee. Training in basic bookkeeping was emphasized, with the aim of having at least one volunteer bookkeeper in every village. Doing so served not only the project but the local monitoring of funds cycled upward to the VCDC.[21]

[21] Van Sant and Weisel (1979:18) report: "A major result of this openness was the willingness of the community to isolate and even remove corrupt leaders. The availability of information made clear what was not clear before—that the community was being victimized by some of its leaders and representatives." That an accountant, one village chief, and several wayward committee members were dismissed was a significant departure from previous practices, where the poor majority accepted and suffered whatever their "betters" did. "The SCF approach," say Van Sant and Weisel, "is an attempt to manage and structure the acquisition of both skill and information by poor villagers in the context of a development program that creates oppor-

A study of an experiment in decentralization and open decision-making for allocating rural electrification to villages in the Indian state of Rajasthan (only 2 percent a year could be electrified, given resource constraints) found that the "goldfish-bowl" approach increased also the technical efficiency of decisions. Corrupt influence in determining which villages would be hooked up first to the state power grid was reduced when district *panchayats* were given guidelines for determining which villages would make most efficient use of the connection (ground-water supply available for tube wells to increase agricultural production) and which could be connected most cheaply (villages a minimum distance from grid transmission lines). The government's own criteria for proper allocation of hookups were better met by the *panchayats* than by the previous process of administrative decision-making, subject as the latter was to corruption and higher-level political influence. Politics and favoritism could not be eliminated with the LO approach, but each village through its spokesmen could contend publicly against any other that was less qualified by objective criteria (Hadden, 1974).

Decentralization will not eliminate malpractices, but we think there is sufficient reason to regard it as a means to counter them if other measures discussed here are also pursued. There is a common view in the literature that participatory, decentralized development compounds corruption because it creates more opportunities for misuse of funds (Blair, 1982). Decentralization, in the absence of restraining measures, can certainly produce abuses, but the alternative of centralization is not without similar potential for abuse, and on an even grander scale. When program management is decentralized, there is a role for LOs in surveillance, identifying malpractices that occur at field level, beyond the effective scrutiny of central authorities.[22]

Some specific measures are useful for controlling financial and other misconduct within LOs. The most obvious is government audits of LO accounts. These were helpful in the case of the Nso women's cooperatives in Cameroon, though it is probably more important that the 62 member co-

---

tunity for decision-making on choices that directly affect the poor. That there has been a significant effect on the attitudes and behavior of the poor is widely acknowledged by both participants in and observers of the CBIRD process in Aceh" (1979:19).

[22]This was suggested by the chief engineer for construction in the Gal Oya project in Sri Lanka. He was concerned that if the building of irrigation structures was contracted out to private firms, substandard construction—which is a common means of seeking higher profit—would result. Even if his own lower-level staff did the work, this could happen. He suggested that farmers' committees oversee and inspect the construction work. Since they would be served by the structures, they would have incentive to report malpractices. Alternatively, he suggested that they might bid on the contracts to do the work themselves and would probably do it as well or better, since they would have more incentive to dig deep enough footings, use enough cement with the sand, etc. (notes by Uphoff from project meeting in Ampare, June 17, 1981). See also Cernea (1983:66–72) on Mexican experience in this regard.

ops had a rather high level of member involvement in all aspects of management (Gow et al., 1979, II:43–52). The women's rice noodle cooperative in Malaysia also benefited from the accounting and auditing services of the government-operated co-op with which it was affiliated (ESCAP/FAO, 1979). Such measures can deter malfeasance among LO officers and among officials in agencies working with LOs but cannot in themselves contain such behavior, since ways can be found to circumvent almost any formal checks unless someone else is prepared to make a challenge and point out where the records may have been falsified. Government audits can produce accountability of officers *upward,* to complement accountability *downward* to members, but (as sometimes occurs) the government auditors may themselves be corrupt.

One measure that helps to produce the intended effect is to provide outside support in kind rather than in cash. The Baglung bridge program in Nepal had a high ethic of leadership integrity and of popular participation, but the relatively unblemished performance of the committees can be attributed partly to the fact that most of the resources they handled were labor contributions or materials donated from the government, particularly steel cable (P. Pradhan, 1980). Cash grants from the district *panchayat* were a small share of total resources and were used mostly to pay the skilled workers.[23] In the case of the Manyu oil palm cooperative, too, it was found that financial assistance given in kind was not so easily subject to abuse (Gow et al., 1979, II:53–66). The same observation was made at a workshop held at the Bangladesh Academy for Rural Development (Comilla) in May 1978, for the Small Farmers and Landless Labourers Project: provision of loans in cash encouraged dishonesty and corruption of officers, so SFDP assistance would best be given in kind.[24] Wherever there are material goods of some obvious value, that is, for which demand exceeds supply, however, even assistance given in kind can be abused. This can be seen from a study done by the American Public Health Association (1981) of LO activities to improve health in a number of developing countries by operating village pharmacies or simply medicine supply chests. What stands out from this study is that considerable resources have been and can be mobilized by cost-

[23]Interestingly, these were mostly low-caste villagers—blacksmiths and masons. Higher-caste persons in this instance were "unskilled" and thus provided unpaid labor.

[24]We think there is some validity to this, though giving in kind is not easy, as the Saemaul Undong movement in Korea learned. In its first year, it gave each rural village 335 sacks of cement to improve its environment—to upgrade roads, bridges, meeting halls, to construct small irrigation systems, and so on. The intention in the next year was to give more cement and materials to the villages that had made best use of the initial contribution. In fact, because of little advance warning or planning and because all villages received the same materials and amounts, considerable waste and inertia were observed. By the third year, villages were classified into three categories: basic (needing much assistance), self-reliant (needing some help), and independent (not needing aid). SU contributions were then varied according to needs and capabilities—with much more success (Lee, 1979; Goldsmith, 1981).

sharing provisions if there are appropriate safeguards against malpractices.

This brings us to a general observation: that malpractices are likely to increase as a larger proportion of LO resources comes from outside sources. There can be thefts of members' own resources by their leaders, but this is less common. Members are not likely to keep contributing resources that are diverted to improper uses, but it can be presumed that "outsiders" will not know about such diversions or will be forgiving, since they have abundant resources. One of the experiences of the Comilla experiment in Bangladesh was that standards of probity became much more difficult to maintain once the government began pouring money into the cooperatives for agricultural loans (Blair, 1974, 1982). Inasmuch as outside agencies may validly seek to increase the resources available to LOs for development work, this poses a dilemma. There will always be some risk of corruption whenever funds are given for desirable purposes, whether through bureaucratic or local organization channels. One solution is cost-sharing; increasingly practiced in public finance, it should apply as well to the "third sector"—local and voluntary organizations that are neither public nor private. The "matching" or "mixing" of resources—as in the Laur irrigation scheme described above, the Malawi self-help water supply program, and DESEC in Bolivia resulted in little misuse of funds.

Where there are malpractices, it is important to take proper and effective legal action against offenders. Local organizations may take the initiative, but if crimes or misdemeanors have been committed, the regular processes of the law should go into action. In the case of the Puebla cooperative in Mexico, membership almost doubled after the LO took persons who defaulted on group loans to court. Lamb (1974:115), in his case study of the Murang'a coffee cooperative in Kenya, recounts some crude frauds but reports that discipline greatly improved once a few offenders were convicted and there was a general tightening up of supervision.

Lamb argues that if supervision is undertaken carefully and consistently, it can make a significant difference in LO standards of performance. This was one of Akhter Hameed Khan's major concerns at Comilla, and it is evidently one of the key elements in the success of the Farmers Associations and Irrigation Associations in Taiwan (Stavis, 1974a). But government intervention can have the opposite effect if done in a heavy-handed manner. In Northern Tetu, Kenya, the government decided to amalgamate cooperative societies into multipurpose organizations, in part to reduce opportunities for abuse. But because the decision was made without any input from members, it did not lead to a revitalization of the LOs (Karanja, 1972).

The analysis of experience with cooperatives in Thailand by the Thai Khadi Research Institute (1980) provides another trenchant case study. Abuses in these LOs were rampant. The committee members and managers were local political leaders and wealthy farmers who retained their positions year after year. Even honest managers were under strong temptation to accept kickbacks because corruption was so pervasive. While it is true that local leaders lacked formal qualifications, a program to upgrade the quality of management backfired when better-trained managers resorted to more subtle but equally corrupt practices. Stricter controls from above are not the answer, the TKRI concludes. They either stifle the growth of self-help organizations or encourage collusion between management and government officers. The best control on corruption, according to the TKRI, is effective membership participation, though instigating it in this case was made difficult by the legacy of abuses in which government personnel had participated.

The mobilization of members to confront malpractices is difficult but not impossible. In the San Jose credit cooperative in the Philippines, local officials were hostile to the idea of exposing the corruption of co-op officers, but they were eventually routed by the insistence of members. In the process, the LO itself was transformed. In a new election by secret ballot, "the winners were those men and women who had emerged as effective leaders in the previous months of countless meetings and mobilizations. They showed markedly different characteristics from the old officers. With the exception of one teacher, a former president, all lacked a completed high school education" (Hollnsteiner, 1979). The LO emerged much stronger from the struggle with a membership representing 60 percent of all households in the community, many of them among the poorest. On the other hand, there was Huizer's unhappy case (1963) in El Salvador, where LO members were most reluctant to press complaints against an official who was in effect "stealing" a road from them.

The government's attitude toward malpractices is an important variable affecting the way local leaders and organization members will view the LO. High standards of conduct by government staff, setting a positive example and taking quick action to punish offenders, will encourage LO members to expect and demand proper performance. Nothing is so likely to discourage them as the expectation that government will do nothing about their complaints, even if they bring evidence—for which indeed they may be victimized locally unless protected by officials.

As suggested at the beginning of this chapter, none of the measures we have observed is foolproof, and the problems they are meant to curb, as discussed in Chapter 6, are common. We have reviewed abundant evidence that well-conceived support by governments and external agencies can be repaid by LO vitality, creativity, and responsibility. This leads us to focus more specifically on elements of strategy. In the following two chapters we look at the roles that may be played in LO development by local residents and community leaders on one hand, and by government and external agencies on the other.

CHAPTER 8

Strategies for Strengthening Organizations at the Local Level

The main locus of effort for establishing or strengthening rural organizations is at the local level, though as elaborated in Chapter 9, initiatives and policies from the national government and external agencies have a major influence on the performance of LOs. Here we outline the main elements of strategy that emerge from the experience reviewed. In addition to the cases examined from the literature, we draw on the experience of the Rural Development Participation Project with rural local organization in countries as diverse as Botswana, Costa Rica, Jamaica, Yemen, and Sri Lanka, supplemented by what we have learned through colleagues who have worked in other countries.[1] The formulation represents a general approach, which may seem a contradiction of our previous emphasis on the diversity of circumstances and tactics for LO development. In fact, we find that some elements are generally valid, though they need specific adaptation to the situations and problems being confronted. We suggest principles of action, with the recognition that specific variations in application will be required.

Assessing Organizational Channels: New or Existing Groups

As noted in Chapter 3, one of the leading controversies in the study of rural local organization is the advisability of working with existing LOs or establishing new ones (e.g., Gow et al., 1979, I:19–21; Buijs, 1982c; Galjart and Buijs, 1982, passim). One of the conclusions of Curtis (1980), based on his work with village-level institutions to supply water in Lesotho, was that new institutions should be organized rather than existing ones utilized. His arguments are strong but derive from the specific weaknesses of govern-

[1]These include Bangladesh (Harry Blair), Bolivia (John Hatch), Egypt (Iliya Harik), Guatemala and Honduras (William F. Whyte and Lynn Gostyla), India (K. C. Alexander and K. K. Singh), Nepal (Prachanda Pradhan), the Philippines (David Korten and David Rosenberg), Indonesia (DAI staff), Niger (Robert Charlick), and Tanzania (Louise Fortmann).

ment-established village development committees, which catered to the chiefs' interests and overrode those of the villagers. In a concurrent publication of the Overseas Development Institute, Bortei-Doku of the University of Ghana took a contrary view, based on Ghanaian experience:

> Traditional institutions when fully understood can play a very useful role in rural development. . . . [They] are traditional human resources with defined leadership, and with proper planning can make an excellent contribution towards supplying some of the manpower needs and skills needed for rural development. It is regretted that studies in this field are limited and we have not made adequate use of the institutions in the rural areas for promoting rural development.[2]

Certainly there will be circumstances where new organizations offer the better prospect of furthering rural development objectives, but on balance we find that the burden of proof is on those who would set up new groups. We say this despite our own encouraging experience with introducing new farmers' groups for water management in the Gal Oya project of Sri Lanka, a settlement scheme that had no long-standing social organization. Also, some of the most effective outside interventions to improve the productivity and well-being of rural households have been made through the FAO-supported Small Farmer Development Program (SFDP), particularly in Nepal, where groups were established *de novo* (Ghai and Rahman, 1979 and 1981). Nevertheless, the most fruitful approach is to begin by identifying and assessing whatever organizations already exist.

We would characterize such LOs as "existing" rather than "traditional" for several reasons. They could include long-established but "modern" organizations like the Jamaica Agricultural Society, started in 1895, which has branches throughout most of the country (Goldsmith and Blustain, 1980). Even "indigenous" local organizations like the burial societies in Botswana can be quite "modern" in their features: deliberate (and recent) creation, well-defined roles and rules, monetized transactions, functional specificity, highly instrumental tasks (C. Brown, 1982). Further, it should not be assumed that existing LOs are static. Okonjo (1979) has shown how women's rotating savings and credit groups in Nigeria have become much more sophisticated financially and functionally than similar groups were just 10 to 15 years ago. Existing LOs thus may have either remote or relatively recent origins, and may stem from previous local or outside initiative. They vary in their degree of formalization but have a history of membership involvement with performance of certain tasks and some demonstrated utility.

[2]*Agricultural Administration Network Newsletter*, no. 3, July 1980, 4–6. For data on the use of traditional labor-exchange groups for development tasks in Ghana, to which Bortei-Doku was referring, see Seibel (1981).

Existing organizations are too readily ignored in most project planning, usually on the presumption that since the existing situation is deficient, it must be totally changed. Too often, little is known in detail about the prevailing situation. This applies particularly to local groups; if recognized at all, they are likely to be viewed in stereotyped terms by outsiders.

A large-scale irrigation project undertaken by the Japanese government in the northern Philippines illustrates the common neglect of existing LOs. In a project area of some 8,000 acres, nearly 100 *zanjera* irrigation groups were already functioning. As is documented by Coward (1979a) and Siy (1982), these are very effective organizations for managing water distribution efficiently and equitably, for handling system improvement and maintenance, and for resolving disputes. But the engineers designing the new project proceeded as if these LOs, with their intricate method for apportioning responsibilities according to landholding, did not exist. The new channels would have eliminated well-functioning systems of channel maintenance by scrambling landholding patterns and sizes, thus "disinvesting" in the social infrastructure built up by Philippine farmers over generations.

In an AID-assisted project for integrated rural development in Jamaica, it was understood that some LO capabilities were important for implementation of the project, and the Cornell Rural Development Committee was requested to assess local organizational alternatives. The analysis concluded that it would be reasonable for the local branches of the Jamaica Agricultural Society to participate in local-level planning and implementation (Blustain and Goldsmith, 1979). The study pointed out certain limitations in JAS performance, but it was the most widely recognized LO in the project area. It has broad-based membership in terms of size of landholding, and an unusually high proportion of women members and officers.[3] Nevertheless, the project management decided to establish its own development committees, which Blustain observed over the next two years. The outcome was that the new committees became de facto branches of the JAS. Because the JAS was the accepted and legitimate channel for calling meetings of farmers, the persons elected as committee officers were often JAS officers, and eventually the project began referring to its LO effort as "revitalization" of the JAS in the project area.[4]

A study of 164 World Bank agricultural projects determined that more than 40 percent of them used existing or new organizations. "The main reason for the presence of such project provisions, in the absence of a formal

[3]Goldsmith and Blustain found that the percentages of JAS farmers according to farm size corresponded quite closely to the distribution of holdings by size in the project area (1980: 78–79). Moreover, women's participation and office-holding was reasonably proportional (1980:83–85).

[4]See *IRDP Annual Report, for Year Ending April 30, 1981,* 22–23. These issues are analyzed in Blustain (1982a).

policy, was the intuitive perception or the empirical conviction of the individual staff involved in the design or appraisal of projects that the development process needs to rely upon and promote the structured self-organization of the small producers for their own interests." However, it was also evident that there was an "absence of a practical methodology . . . for identifying and effectively supporting existing traditional organizations" (Cernea, 1982:132–33).

The Jamaica analysis described above was done in six months' time and represents one such methodology, but it built on a base of existing knowledge about the rural sector (some of it misleading, to be sure). Similar tasks undertaken in Yemen and Botswana presented more of a challenge. The AID-assisted project in North Yemen to strengthen existing but often recently established Local Development Associations, linked loosely within the Confederation of Yemeni Development Associations, was operating with so little systematic knowledge of the LDAs that they represented *terra incognita* (Cohen et al., 1981). Usually formed around traditional leaders (sheiks and family or clan heads), they were undertaking modern tasks of road, school, water supply, and other construction, using mostly funds generated by migrant employment in the Persian Gulf states. Given the need to tailor training, technical, and financial assistance to the LDAs in ways that would augment (not simply substitute) resources and would improve the technical quality of investments, some rather basic kinds of field studies were needed. Two anthropologists with prior experience in Yemen conducted a simple descriptive survey of 25 *uzlas* (subgovernorates) in the highland and coastal areas to determine the range of socioeconomic conditions within which the LDAs operated. This was to be followed by more detailed studies of seven or eight *uzlas* that represented modal types within the range, and two in-depth studies: one in a reasonably representative highland community and one in a parallel coastal community. The strengths and weaknesses of the existing LDAs could be determined by this methodology and related to ecological, social, and economic variations in the country.[5]

The task in Botswana was to discover why government-sponsored groups set up to operate and manage the small catchment dams being constructed by the Ministry of Agriculture were not functioning better by the ministry's criteria. The groups did not meet regularly, did not do regular maintenance of the dam bunds, did not regularly use the dams, and did not collect the agreed fees from users. Assessing this situation turned out to be difficult, not because the rural area was unknown—as in Yemen—but because of

[5]These studies have been synthesized into a single volume (Swanson and Hebert, 1982). Unfortunately, because of logistical and agency difficulties, the middle level of analysis had to be omitted, but its desirability as a methodology is evident.

inadequate understanding by government officials of the highly varied and changing ecological relationships governing the use of water and rangeland (Fortmann and Roe, 1981). The dam groups, it turned out, functioned reasonably well during the few months in a year when the dams were important to herders as part of their "fall-back" strategy of following the cyclical pattern of water and pasture availability. Herds would move out into the veldt after the rains to take advantage of ephemeral water sources and grazing, and move back eventually to the base villages with deep boreholes to survive through the driest months. It was best for a number of reasons that the small dams *not* be used year round. From the users' viewpoint, therefore, it was not necessary that they be managed continually, or even that the LOs have any continuous activity. The water-use levy set by the government was not collected by the LOs because it was much too high and because it was unnecessary—the maintenance could be done by group labor or ad hoc contributions.

Understanding the working of these organizations required an extended analysis of interactions, from the perspectives of household and of government, in the ecological, socioeconomic, and administrative policy realms. Also, examining the existing organizations without evaluating the physical and technical aspects of their task environment would not have been useful. Our researchers had to devise methodologies for dealing with those aspects in tandem with the variables more commonly dealt with by social scientists.[6] Appropriate methodology needs to go beyond identifying and assessing existing groups; it must also include an understanding of their task environment.

The choice between working with existing organizations—which can include some modification or expansion—and establishing new ones has to be context- and task-specific. The working assumption should be that existing LOs represent valuable *social capital* that should be discarded or bypassed only when they are clearly unsuitable. One should start by presuming that externally introduced LOs will not be initially understood by members or potential members, and will not engender their commitment. Understanding and support can be built up, but the performance that would elicit support is likely to be dependent on having members who are understanding and supportive. This creates a chicken-egg situation where progress must proceed iteratively. If the new LO meets some clearly recognized need and if the activities involved are reasonably familiar or simple, progress can be accelerated, as we found with the water management groups in Sri Lanka.

[6]Such an analytical integration for range and water management in Botswana is presented in Roe and Fortmann (1982). A similar synthesis is offered in Blustain (1982a) for agricultural development and soil conservation in Jamaica.

If the LO provides valued goods or services that are channeled to it from government or other donor sources, this may elicit quick responses to establish organizational structures. But according to our reading of dozens of case experiences, such "pump-priming" of LOs that are based primarily on outside resources seldom results in effective, sustainable organizations. The leadership that emerges is likely to be less well-motivated and may indeed be inclined to self-enrichment when resources flow freely. Members will take what is free but will not develop a sense of responsibility for the LO. The twin evils of corrupted leadership and a psychology of dependency among members are the most probable outcomes of introducing new LOs through substantial resource transfers from the outset.

These dangers associated with externally introduced organizations are familiar to observers of LOs. We would note an additional reason for caution. When LOs originate from outside the community, the roles specified (president, chairman, secretary-treasurer) appear to members and officers alike to have outside authority behind them; therefore, the offices are likely to be exercised with less restraint and consideration than if they were created—not just filled—by the members. "Traditional" LOs differ from "modern" ones not simply because they are more informal but because they are more consensual. Majority voting is less common because it is understood that agreement needs to be fashioned by discussion and persuasion, if implementation is to follow. Local leaders can be more effective functioning as intermediaries than as "authorities" (Duewel, 1984). "Modern" officer roles may be exercised more peremptorily because there appears to be influential backing for the positions and for the decisions taken. If the accountability of officers to members can be maintained in new organizations, however, they may differ little in effectiveness from existing groups.

One of the principal reasons often given for avoiding existing LOs is that their leadership, drawn from elite strata, may be unrepresentative, domineering, and exploitative. This question is taken up below, but we would note here that the same potential exists in new organizations (the *ejidos* set up in the Mexican land reform, for example). Also, over time, the oligarchic grip on rural communities may be loosened: such changes are coming now in India after some years of parliamentary and local *panchayat* elections.[7] This said, it is certainly true that existing LOs may be inappropriate for certain development tasks. Buijs's argument (1982c:80)—"The new functions and objectives usually require different ideas on financial management, administration, affinity and efficiency from those held in existing

[7]On this, see Blair (1982) and compare with (1978). Volken et al. (1982:6–9) have observed this while organizing "the poorest of the poor" in India. See also Manor (1977, 1982) and M. Robinson (1983), who has traced changes over the past 25 years in a subdistrict in Andhra Pradesh, India. Her earlier book (1975) described the process of loosening the grip of traditional elites at village level in Sri Lanka.

traditional organizations"—is correct.[8] But it equates all existing organizations with "traditional" ones and regards all new organizations as "modern." Buijs notes both that "there do exist new organizations which have taken on traditional elements" and that some "traditional" organizations—such as rotating credit societies—have "spontaneously" taken on new functions, as have the Bamileke in Cameroon (Boumann, 1977). This confirms our view that each case and each situation needs to be assessed for the potentials and obstacles involved in working with existing organizations.

Understandably, one would like to have the best of both worlds. This may mean grafting new functions onto existing organizations, or patterning new LOs after familiar modes of operation. We do not accept the categorical conclusion of the University of Leiden group that "participatory projects require a modern organization" (Buijs, 1982c:82).[9] Our view is that one should look first for and at existing LOs, recognizing that there are limitations on the new tasks they can or will engage in.[10] An LO may be unsuitable for development activity because its structure is not appropriate, its mode of operation would bias the results, or its norms of cooperation would not support the new effort. But if an existing LO's lack of interest inhibits its engagement in a project, one may need to question the value of the project to its members and consider adjustments in objectives and activities, rather than dismiss the local organization.

It may be possible to work with a combination of new and existing LOs, as Charlick (1984) discovered in 56 percent of the *animation rurale* projects he analyzed. When he compared the respective contributions of new and existing organizations, he found no significant difference in technical performance criteria. Interestingly, more equitable outcomes were associated with working through existing organizations than through new ones, though the latter were better at achieving participation (empowerment) beyond the village, as they acted more like "constituency" groups (Esman, 1978b). Overall, Charlick found that capability for local action was most enhanced where both new *and* existing LOs were associated with a project (1984:103).

[8]She cites a U.N. report: "Most cooperative tasks are different from traditional collective tasks, which are generally intermittent in nature and aimed at meeting collective needs rather than directly raising individual income levels."

[9]Boer (1982:159) says: "Customary cooperative institutions generally do not seem to form a firm basis for participatory projects." Buijs writes, however, that especially in the "first phase," traditional ideas on rights and duties and on division of labor will have important influence, so it will be desirable, as well as possible, to incorporate elements of traditional cooperation into modern organizations. So this project, which has paralleled our own, with field studies and literature review, comes out not as differently on this issue as it might at first appear.

[10]March and Taqqu (1982) argue that attempting to transform a traditional, informal organization to perform new tasks may destroy its capacity to accomplish even its original limited objectives.

Another possibility is to structure the organizations themselves as hybrids. Haeri and Farvar (1980) describe efforts to introduce cooperatives in Iran: initially, the organizations were accepted and successful because they were compatible with existing traditional forms of social organization, and LO officers were not regarded as government officials. Transforming these LOs into governmental organizations in 1963 proved fatal to them, as officials became more distant and corrupt. Obviously, not all governments are able or willing to make the accommodations necessary to sustain "mixed" organizations.

Where existing organizations are not suitable, or there are none, the introduction of new LOs should build as much as possible on accepted roles, principles of obligation, group sanctions, criteria of status and success, and common work ways. In any case, we would underscore the importance of proceeding in a "learning process" mode and avoiding a "blueprint" approach (D. Korten, 1980). It may be good to work with more than one organizational model, and the best structure and practices may change even within the life of a project, a process that can be observed if participation and results are carefully monitored. Smith, Lethem, and Thoolen, in their analysis of organizational strategies for the World Bank, state the issue well: "The key failure is to attempt to move too rapidly to the choice of a single organization. Not enough time is spent in exploring and weighing alternatives and having those who will have a role in implementing the final choice themselves suggest alternatives" (1980:38).

Local Leadership and Membership Participation

Leadership and participation are two sides of the same coin, though leaders and members may be juxtaposed analytically and may have different interests and needs. One can say that without leadership there is no sustained or sustainable participation, but the reverse is equally true. Unfortunately, the study of leadership is one of the most neglected areas of political science (Almond, 1973:17). This is partly because it is the most complex and contingent of subjects, as transitory and variable in its manifestations as it is central and determinant in shaping outcomes. Leadership attracts much attention but in itself remains difficult to account for, perhaps because, as Rustow observes: "In a variety of ways the leader may be said to be more dependent on the supporting movement than the movement on any particular leader. There can be no leadership without followship, and since the followers may be many, they are in a sense less accidental than the individual leader" (1967:160).

We cannot deal here with the broader theoretical arguments surrounding the analysis of leadership. These might be summarized in the question of whether leaders are "made" or "born." We would begin by observing that leadership is important for its role in identifying and articulating problems to be acted upon, in formulating plans of action, in mobilizing resources and managing them, in resolving conflicts—in short, in the various LO tasks analyzed in Chapter 3.[11] What concerns us here is the operational choice between working with existing leaders in a community and trying to elicit new leadership. We need to address also the issue of accountability, as it links the factors of leadership and participation.

The importance of leadership is not in dispute. Analytically, there is the danger of tautology, so that good LO performance is attributed to good leadership, and failures to ineffective leadership. But even such circularity cannot negate the evident necessity of some adequate performance of this function, which aims to guarantee the success of the enterprise. A typical comment is that of Hafid and Hayami, assessing local irrigation projects in Indonesia: "the main determinant of success in effective organization of community projects is adequate leadership, which differs in form for villages with different socio-economic environments" (1976:8). Similarly, Sharpe, in assessing a cooperative in the Dominican Republic, concludes: "The most important factor in overcoming the barriers of distrust and peasant lack of knowledge and self-confidence appeared to be the quality of the *local* leadership . . . and not of the outsiders like El Padre and Fidel" (1977:221; emphasis added).

Certainly one should not generalize about all existing or new leaders at the community level. Even in the literature on a single country, Senegal, we find very different views on whether "traditional" leaders at the village level will strengthen development efforts (Fresson, 1979) or impair them (Bergmann, 1974). We identify three criteria for "developmental" local leadership, and the first question is how well the existing leaders meet them: (1) ability to mobilize local resources for development efforts, (2) capacity to acquire outside resources to complement local resources, and (3) willingness to use these resources for broad-based benefit in the community. No leadership is likely to be fully or equally successful in all three dimensions, so the assessment is a matter of judgment. We assume that there is potential leadership talent in all communities, unless a high out-migration draws off the most energetic and inventive people. But the conditions attracting such talent into positions of responsibility may be highly variable, and communities can be saddled with leaders who lack capacity for developmental performance or are disinclined to act in a public-regarding way.

Rather than pass general judgment on all existing leadership, it makes more sense to try to seek conditions of leadership selection and retention

[11]One task more specific to leadership than to LOs generally—finding ways to minimize "free rider" problems so that members have incentives to contribute to LO enterprises—is well analyzed by Popkin (1979:252–66).

that permit members to assess the performance of their present leaders and the potential of possible alternatives. For many development projects, one might not be satisfied with the existing leadership group; it may be ineffective, corrupt, or biased toward the more privileged classes. In particular, if conventional selection processes are followed, the same leaders will be put forward as before—those with political connections, accumulated wealth, or traditional status. There is always the danger, as Chambers (1975:58) says, "of creating a local dictator who will exploit his power to his own benefit and that of his friends." However, among existing leaders there can be persons with experience and dedication, and with sympathy toward and from the membership. Clearly, existing leadership should be neither automatically condemned nor automatically approved.

Leadership selection is a delicate process, one that outsiders must approach with circumspection.[12] We have seen in the case studies and our own experience that it is the *overall* process of leader-member interaction, perhaps best characterized in terms of effectiveness and accountability, which is crucial. One basic suggestion is that leadership selection, even with continuing LOs, be preceded by as much discussion as possible about the purpose of the LO and about what its members want and need in their leaders. This communicates expectations to whomever is chosen, but it also gives cues for the possible selection of new leaders who meet the criteria better than existing ones. In Gal Oya, Sri Lanka, the Agrarian Research and Training Institute (ARTI) promoted farmers' groups for water management by using "institutional organizers" as catalysts (Uphoff, 1982b). The boundaries of the functional group were determined hydrologically: all those farmers drawing irrigation water from a single field channel, perhaps having a history of some cooperation. By starting the process of group formation from the bottom up, talking with farmers first individually and then in small groups before calling the whole set of 10 to 25 (or more) farmers on the field channel together, the organizers communicated to all potential members (and leaders) in advance the expectation that this new group would operate in an egalitarian, participatory manner.

The result was an unexpectedly high level of activity, including unprecedented cooperation and sharing of water, even in two water-short years. Once selected, the farmer-representatives (unpaid) worked more capably

[12]It is possible that communities will put forward surrogate leaders in positions of formal responsibility vis-à-vis state agencies, not the "real" ones to whom they look for ideas, judgments, and advice. A study by Ganewatte (1971–72) of land settlement schemes in Sri Lanka compared the pattern of formal organizational leadership with patterns of influence and respect measured by sociograms. These showed that there were a number of "real" leaders in the communities who had no formal office, though their opinions and example determined collective attitudes and activity more than what the "open" leaders said and did. Uphoff's study (1979) of 16 communities in two other districts of Sri Lanka, on the other hand, using similar methodology, did not find such a two-tiered pattern, perhaps because these were longer-established communities or because the formal offices were more meaningful to villagers than in the settlement scheme. Ganewatte's finding should be kept in mind.

and conscientiously than expected and made a positive impact on lower-level government staff and district-level officials, gaining their cooperation and support as never before. Most important for our consideration here is the fact that even two of the established "leaders" in the 4,000-acre pilot area, who initially opposed the organizations, after some months came to support the group effort actively and assume constructive leadership roles (Uphoff, 1981, and 1982a). Our best explanation for the change is that the groups started from below were willing to bypass the old leaders if they did not cooperate. The leaders seem to have joined on the members' terms. The organizations did not stop at this level but continued upward through their representatives to distributary, branch canal, and district levels.

Our concern is with the principle that the group creates the leaders. In order to achieve ostensible economies of scale and speed when starting local organizations, outside agencies frequently choose to call communities together to "elect" leaders. The most prominent members are invariably selected and then given training and control over resources for the community, without any detailed and extended communication with the other members about objectives, rights, or duties. Creating the groups through these leaders, in effect, establishes a power relationship that is open to abuse. The agency has little or no communication with the community except through these leaders. The more training and resources they are given, the more distance is created between leaders and members. The short cut of trying to mobilize rural people from outside through leaders, rather than taking the time to gain direct understanding and support from members, is likely to be unproductive or even counterproductive, entrenching a privileged minority and discrediting the idea of group action for self-improvement.[13]

There is no question that certain leadership roles with delegated authority and responsibility are needed. All LOs of any size and significance need some specialization and division of labor, though authority and responsibility may be more or less broadly shared, and incumbents may have limited or unlimited terms of office. Michels (1915) correctly saw that such delegation creates the potential, even the tendency, for leadership to rise above the members and pursue its own interests, against or simply in addition to the interests of the membership. This is why we stress the importance of formal mechanisms of accountability, as discussed in Chapter 7, and a broad-based

[13]One problem reported from several countries in South Asia with regard to the "training and visit" (T & V) system of agricultural extension is that the "contact farmers" selected to work with the extension agents do not relate intimately and equitably with the farmers to whom they are supposed to report. The transmission of information downward may reach the contact farmer, but he often lacks the means or incentive for further transmission if he is not really accountable to the other farmers in his community (see Chapter 2, footnote 10). Extension agents in the Gal Oya area asked to work with and through the farmer-representatives chosen by the field channel groups, since such linkage was more vital than that provided by the "contact farmer" system (Uphoff, 1982a).

process of consensus building about goals, powers, and obligations. Leadership training certainly can be useful for equipping LO spokespersons to deal with their internal and external tasks, but the members, too, should gain knowledge and experience, in part because they are the pool from which additional leaders must be drawn to achieve continued effectiveness. As reported in the previous chapter, Charlick (1984) found better performance where the rank-and-file as well as leaders received training.

A promising approach that combines the informal training of leaders with membership involvement is the periodic evaluation meetings in the SFDP of Nepal. The groups of 10 to 15 members, including their leaders (who, as we have reported, constitute up to half the membership through various committee assignments), evaluate their own progress, usually every month. Then a number of leaders chosen by the members meet on a quarterly basis to discuss both their problems and their achievements. These discussions are reported back to members. Such training keeps leaders linked more closely to members. Some variation on this combination of participatory evaluation and training through group problem-solving could probably be worked out for most LO programs.

One of the themes running through the case studies is that the performance and accountability of leaders depend heavily on active participation by members. That this sounds circular perhaps helps to explain why very active and effective LOs are the exception rather than the rule, since some combination of leadership and participation is needed, and each requires the other to be sustained. Leaders accomplish group goals through the work of members, and having more persons willing and able to take initiative and responsibility multiplies group capacity. A variety of mechanisms for holding leaders accountable—regular elections, secret ballot, frequent meetings, rotation of officers, and management or vigilance committees—can provide the means to constrain poor performance, deter malpractices, or at least facilitate criticism of faults and replacement of leaders. The membership needs to know how to use such mechanisms and be motivated to do so when necessary; these are often culture specific.[14]

The aims of membership participation need to be realistic. As we have suggested elsewhere, there are many different kinds of participation, not all of them relevant or effective in all situations for all tasks (Cohen and Uphoff, 1977; Uphoff, Cohen, and Goldsmith, 1979). It makes no sense to think

[14]The checks on malpractices may not be very formal. The Thai Khadi manual for working with farmers (Heim et al., 1983:48) discretely advises group organizers simply to stop working with a group if corrupt practices appear. "Leaders are not changed automatically in the Thai society. The Thai way of showing objection is to abstain and keep away and not to revolt against a person in public." Members' efforts to curb corrupt practices will depend very much on whether they value the actual or potential performance of the LO. This underscores the extent to which LO success or failure has an inbuilt cycle of reinforcement. It is important to provide incentives for both members and leaders to be active and responsive, not only economic gains but also status and roles in self-governance.

in terms of achieving *maximum* participation, since participating in making or implementing decisions, for example, entails costs as well as benefits to individuals. It is more reasonable to seek some *optimum* participation, which justifies to members their expenditure of resources or time because the benefits are greater, more prompt, more appropriate, better distributed. Some delegation of decision-making and some specialization within the group can contribute to better performance so long as inefficiencies equivalent to "monopoly" do not result and informal social sanctions operate to discipline members as well as leaders. Michels saw delegation and specialization as twin threats to democracy within organizations. Sadly, the poorer and less educated an organization's members, the more likely it is that leaders will become self-serving—even though it is precisely such members who most need the benefits of organizational initiative on their behalf.

In practical terms, there may be no alternative but to work with existing local leaders. Bypassing the "natural" leadership found in communities is difficult and always puts a strain on any program attempting to do so (Ralston et al., 1981:18–21). The DAI study of local organizations, found that in 40 of the 41 cases, "the local power structure either actively participated in the new development initiative or formally approved of it. The one exception was Argub [in Yemen], where the opposition of the traditional leader led to armed conflict with the more progressive elements in the area. In virtually all cases, village leaders were also leaders of the local organizations involved" (Gow et al., 1979, I:146). This can produce good or bad results, but on the whole these leaders provided appropriate impetus and guidance for local action. Even if traditional leaders are not dynamic and supportive, their position is usually so dominant that it cannot be simply ignored.[15] Many national and international advisors probably have an aversion—in the name of democratic and egalitarian values—to working with existing leadership, which is usually already advantaged within the local context; but as Johnston and Clark have suggested, bypassing such leaders may not only be impossible but—instead of helping the less advantaged—may result in an ineffective LO helping hardly anybody.[16]

[15]In his study of 50 village cooperatives in Senegal, Bergmann (1974) found that elected presidents of co-ops, if not themselves the village chiefs, intentionally or unintentionally oriented their behavior toward the latter. Thus, proposals to try to keep "traditional" and "modern" roles separate by barring chiefs from holding office in the cooperatives would likely be ineffective, Bergmann concluded. It would be better to try to organize the collaboration of the two subsystems, providing traditional village leaders with a defined function in the cooperatives.

[16]In a communication to Guy Hunter, Johnston and Clark comment on a disposition among many outsiders to advocate "superficial democracy and egalitarian forms of organization [which often amounts to] hostility towards emergence of forceful, innovative leadership—emphasizing instead a vague blend of local (leaderless) autonomy plus 'professional' (read 'central' or 'imported') program management" (quoted in Hunter, 1980b:23). Tendler (1983) gives specific examples where less than "ideal" leaders in fact contributed to commendable LO results.

One should not assume that traditional and more advantaged leaders will benefit only themselves. In a study of LO performance in the Philippines, Montiel (1980:189) found that the larger the socioeconomic differences between leaders and members, the more the organization was able to produce service outputs. Their distribution was not precisely ascertained, but poorer members were getting access to services not otherwise available to them. Self-help efforts in Kenya, entailing some community organization, involvement, and initiative plus a community contribution of finance or labor (or both), have persisted in the post-independence period, as they did not in Tanzania to the same extent, because of the leadership of the rural elite (Holmquist, 1979:130–32). While the bureaucracy in both countries tried to muzzle—if not curtail—self-help activities, these were promoted by the rural elite in Kenya as a means of improving its political and economic position. The power of the rural elite has not destroyed the self-help process, which continues to provide benefits to rural communities (B. Thomas, 1980; Barkan, 1981).

The extent to which local elites can be controlled by the community depends on three factors, according to M. Moore (1979:247–48): the level of social interaction, the degree to which elites are subject to the moral pressures of neighbors, and whether they live in close proximity. If leaders are themselves involved in agriculture, they will perceive their self-interest as dependent on improvement of local production and are likely to be willing to assume LO responsibilities with little direct reward. Tendler (1983) gives examples of such incentives for local leaders in Bolivia. Problems arose for the Cultivation Committees in Sri Lanka when politicians, in lieu of holding elections for CC officers, appointed as leaders persons who had their economic base in trading or landownership and who were outside the moral influence of the community.[17]

Under various circumstances, it is more or less likely that existing leaders will be responsive to the needs of the poor majority in their communities. Leonard (1982a:15–20) has analyzed them in terms of three factors. First, if leaders share the interests of the poor (for example, when they have need of the same facilities, such as drinking water or transportation), many improvements can be attempted through LOs to provide fairly wide benefits. Second, if there is competition for leadership among elites at the national and/or local level, some elements are likely to appeal for support to the poor

[17]A. T. Ariyaratne, founder and president of the Sarvodaya Shramadana organization in Sri Lanka, tells the story of a rich man who was so proud and supercilious that he never attended the funeral of anyone in his village who was not of equal or greater status; instead of going himself, he would send a servant, carrying the master's walking stick, in his place. When this man died, however, his neighbors had their revenge. Nobody came to his funeral, and his courtyard was filled with others' walking sticks. In rural communities where status and honor are greatly prized, such social inhibitions weigh on the elite as long as they wish to remain part of the community. These considerations will not negate all selfish economic interests, but they are reckoned especially by persons who desire prominence. The Cultivation Committee experience is analyzed in Uphoff and Wanigaratne (1982) as well as M. Moore (1979).

majority by making commitments that confer benefits in new and needed ways. This was seen at the national level in Mexico in the 1930s (Cornelius, 1973) and at the local level in Egypt some 20 years later (Harik, 1974). In practice, when such competition arises and the most privileged persons are excluded from leadership roles, their place is likely to be taken by persons from the next most advantaged level. Harik documents this sequence in Egypt during land reform, and Leonard points out that it happened even in China after the revolution. The terms on which the new leadership holds its position are altered by the new circumstances, however; even if the poor have not taken over authority roles themselves, those roles can be made more productive for them—especially if Leonard's third condition is met: that the poorer sectors have organizations which can speak on their behalf and bargain with others by giving or withholding support.

This brings us full circle, in that local organization—particularly for the disadvantaged—becomes a means of extracting benefits from and through the leaders, while at the same time leadership is needed for effective organization. Although some may be concerned with the analytical problem of determining "which comes first," it is more important to address the practical problems of getting both started. Leadership that arises from within the poor strata does not of itself ensure benefits for them, for social origins are an imperfect predictor of performance on behalf of the disadvantaged, as Sharpe (1977:220) discovered in working with cooperatives in the Dominican Republic. Leaders from impoverished backgrounds may seek to enrich themselves once invested with whatever power comes from holding office, while those from more privileged circumstances may be motivated to assist the poor—out of a sense of *noblesse oblige,* because it is expected of them by superiors, or because it gains them popularity, security, or satisfaction. When it comes to bringing forward either old or new leadership and, particularly, encouraging leaders to operate in a new, more developmental mode, the literature and our own experience offer good reason to work through persons acting in special catalytic roles.

The Paradoxical Role of "Promoters"

It is often necessary for some outside person to initiate or encourage LOs. Since a good number of the successful groups in our sample of 150 stemmed from local initiative, we would not go as far as the Leiden group in attributing virtually all effective participatory local efforts to outside initiative (Buijs, 1982a; Grijpstra, 1982a).[18] But in the absence of local initiative, there must be some outside impetus if the status quo is to change. A government

[18]One of the most detailed case studies of an exception to their generalization, and an exceedingly significant case in itself, is by Womack (1968), esp. ch. 1.

or private agency could simply communicate a willingness to work with any LOs a community might set up, perhaps offering inducements in the form of building materials, subsidized credit, or technical assistance; but a more active role can lead to more rapid changes, subject to the cautions we have expressed already about not imposing a blueprint. It is paradoxical but true that top-down efforts are often needed to promote bottom-up development (Stiller and Yadav, 1979).

The best channel for such efforts appears to be a person engaged by an agency, private or public, specifically to promote and assist local organizational development. As noted in Chapter 5, we are referring to a *style* of interaction with rural communities. It is most likely to be consistent with the initiatives for rural development coming from, say, a private voluntary organization or foundation, such as those Fremerey (1981) describes for programs in Thailand, Indonesia, and Sri Lanka—the German Friedrich-Naumann Foundation, the Indonesian Council of Churches, and the locally founded Sarvodaya movement in Sri Lanka (Ratnapala, 1980).

But government staff may also play such a role, as did the "promoters" described by Sharpe (1977:212) who assisted rural cooperatives in the Dominican Republic:

> They helped educate and organize the leaders to take action; they explained contracts, fought through bureaucratic red tape, arranged for exporting licenses, took care of the complex shipping procedures, and arranged for the buying and installation of processing machinery; they helped the leaders struggle with officials and understand the hidden traps in contracts placed by incompetent, unconcerned or deceiving officials; and ultimately their access to traditional sources of power gave the cooperative access to the power it needed to gain at least some economic independence.

The encouragement for co-op leaders came partly from the Catholic Church but mostly from government employees, some of them concerned and dedicated, and a few even antigovernment. This case, however, illustrates the limits on officials who act as catalysts: the government agency assisting the cooperative had enormous control over the promoters by virtue of paying their salaries; it could rein them in and put pressure on the cooperative whenever the LO was seen as too assertive or independent. As it happened, however, there was an unusual staff member who sometimes defied his superiors' pressure and risked his job to defend peasant interests: "Without Fidel, the cooperative might not have survived and certainly would not have gained the independence it did." Since his most important supportive actions were taken in spite of or in opposition to the agency that employed him, it was clear that how the government agency chose to use its power was critical for LO success. When governments introduce a catalyst approach, they need to be prepared for, and even to welcome, enhanced local capacity to meet people's needs, directly through self-help but also through lobbying and other tactics that may create stress for bureaucrats and policy-makers.

Many names have been used for such a role. In an earlier mode, more directive than the participatory mode intended now, there were "village-level workers" promoting community development, with or without institutionalized LOs, and "*animateurs*" fomenting *animation rurale*. The term "promoter" has been used in many projects and by the Leiden group. The approach of "consciousness-raising" advocated by Paulo Freire (1970) called for work by "facilitators," whose efforts were expected to culminate in a local organization acting on behalf of the poor. In an earlier analysis of cases for the Rural Development Committee, Lassen (1980) used the word "catalyst," which suggests that the person induces a reaction but is not part of it. In extension work, the conventional term has been "change agent," which unfortunately implies a one-way effect and focuses on individuals. The "change agents" currently trained and deployed in Sri Lanka by the Rural Development Department to stimulate LOs among the poor, however, appear to perform quite an appropriate role (Tilakaratna, 1982; Talagune, 1982). The most complex terminology is that used in FAO's Small Farmer Development Program: "group organizer/action research fellow" or GO/ARF, for short.[19] In the Philippines, the role introduced by the National Irrigation Administration with Ford Foundation assistance was called "community organizer" (D. Korten, 1980), while a similar role in Sri Lanka—established by ARTI with Cornell involvement to help form water users' associations—was called "institutional organizer."[20] In any case, each situation needs a uniquely fitted designation.

In general, it seems that promoters should be selected from outside the community, individuals more educated and less vulnerable to attacks on their persons and reputations (Grijpstra, 1982a:200). They are not tied up with the local structure or involved in feuds and factions, and their indepen-

[19]The latter part of the term refers to the expectation that such persons would perform a monitoring and evaluation role as well as an activist one. The Philippine and Sri Lankan experiments cited next assigned such responsibilities to a "process documenter" (described below), who was separate in the former case but in the latter had some combined responsibilities.

[20]This was the English term; the Sinhala term used, *govi sangvidayake*, translates as "farmer organizer." The word "organizer" by itself had been used in connection with party-political initiatives so farmers and officials would probably have misunderstood the role if so designated. Much time was spent considering alternative terms. "Community organizer" was thought to connote too multifunctional an organization, and "irrigation organizer" would identify the person too narrowly—and too closely with the Irrigation Department, given the farmers' negative view of the department at the outset of the program (their attitude has since become more favorable, thanks to the LOs' operations). The AID-assisted water management project sponsoring the work had proposed creating a new Division of Institutional Development within the department to assist and maintain liaison with farmers' groups; that tipped the choice toward the term "institutional organizer."

dence can make it easier for them to work with less advantaged groups (Devitt, 1977), though Volken et al. (1982) have had success with persons chosen from untouchable or tribal communities in India. Outsiders may have more standing and credibility with a funding agency than would insiders. A compromise approach is found useful by the National Community Development Service in Bolivia (Savino, 1984), where local residents are recruited competitively from the target communities but usually assigned to communities other than their own. Those who had shown leadership in their own organizations were usually effective in stimulating such development elsewhere. These communities, it should be said, are relatively homogeneous ethnically and relatively equal socioeconomically. In more stratified circumstances, local people even a few villages removed might have more difficulty in such roles, suggesting the need then for outsiders as catalysts.[21]

The approach is one of sending the promoter into the community (or set of communities) to become acquainted with the people and their problems, and to initiate discussion of these problems and of the people's capabilities and expectations. Together with the prospective members, the promoter will identify measures to be taken by the group, on its own or with outside help, to deal with the problems of highest priority or those most amenable to solution. In the process, the basis for self-sustaining organized activity should be laid by demonstrating local capabilities, bringing forth local leadership, and motivating members to become and remain involved.

This is called a "lodging strategy" in the Leiden project's analysis; the promoter "approaches the village where he wants to work and lodges with someone, or moves into an empty house" (Buijs, 1982a:53–54). The first step is usually a village survey (the Banki water supply project—Misra, 1975) or profile (the NIA program in the Philippines—F. Korten, 1982:17–22). A methodology for surveys of community health or water programs has been described by Isely et al. (1979) and Isely and Hafner (1982). A less formal method was used in Tambon Yokkrabat, Thailand (Rabibhadana, 1983). In the process, the promoter can become personally acquainted with all the households that might join the LO, and its purposes can be explained directly, without intermediation, distortion, or suppression by local elites. The latter may wish to disrupt or prevent the survey or profile, but with perseverance and tact on the part of the promoter, plus support from some quarters of the community combined with some official sponsorship, the effort can continue. Personal rapport is the key. As Buijs puts it: "By this

[21]An innovative rural education program in Mexico (CONAFE) started with instructors assigned to teach in the communities they had grown up in. "Soon this was found to be unsuitable and therefore certain modifications were introduced. Community response seemed to be better when instructors came from other places. But in all cases, instructors had a rural background and were familiar with the local customs and mores" (Paul 1982:94). This program expanded rapidly between 1973 and 1979, to involve 19,000 instructors in as many rural communities.

willingness to live in the village and to take part in village life, discussing and advising often till late in the evening, the field worker builds up a trust relation. Through this trust, people are willing to accept his ideas about a project" (1982a:54). This result can be seen in the SFDP projects in Nepal, Bangladesh, and the Philippines; the DESEC program in Bolivia; and the Gal Oya water management project in Sri Lanka.

What is sometimes called "consciousness-raising" in such a process is probably more appropriately understood as trust-building. The ideas that are communicated are important, but the nonverbal communication of attitude and values through such signs as dress, posture, tone of voice, and eating practices may be more important in creating a positive (or negative) response among rural residents. The qualities for which promoters can be selected are so varied that no certain criteria appear, beyond commitment to program goals and willingness to live in the village. As Oakley (1980) observes, no candidate will have all the ideal qualities, and we find that persons with quite different personalities can achieve similar results—good or bad.[22] Even verbal facility, which is often seen as a requisite for organizing, does not appear to be universally needed. Some organizers who are reserved and quiet can be quite effective, particularly among farmers who may be as impressed by what is communicated nonverbally as by what is said. Promoters must learn the simple truth that *listening is a communication skill.*

According to Buijs's review of field studies (1982a), the first phase of getting mutually acquainted appears to require a minimum of three months and an average of six (though in the Gal Oya project six weeks was sufficient under crisis conditions). Where there are difficulties, more time may be needed. In some circumstances, it may even be better not to force the issue but to move on, in the hope that observable results in a nearby community will provide a positive demonstration effect.[23]

The length of time the promoter should stay in the community is highly variable. In the Nepal SFDP, it was estimated that six or seven years would be needed before the small farmer groups would be completely self-sustaining (Ghai and Rahman, 1979:22). In Gal Oya, farmers and organizers think one to two years will be enough to get viable organizations launched, at which time the organizers can move on to other communities, although a

[22]In the water management programs in the Philippines and Sri Lanka, women have proved to be at least as effective, possibly more so, than men in doing organizational work among farmers. They are shown more respect, and they do not get involved in drinking, which—as farmers in both countries report—interferes with an organizer's effectiveness.

[23]A promoter should not abandon an effort in a way that jeopardizes or leaves vulnerable to reprisal those who have worked with him or her in good faith. This presents a dilemma where it arises. In such circumstances, the organizational effort should probably have been broken off earlier, hoping to build up pressures for group work by showing real benefits from LO activity in a nearby area. Some continuing involvement with a community, even if full-fledged organization does not emerge as intended, may be appropriate where active organizing efforts are withdrawn.

few will remain in the area as ombudsmen to assist a larger number of groups. Since some continuing connection with the sponsoring program is likely to be useful for LOs, even if it is much less intensive, a complete divorce may not be effected unless the LOs desire it. For the program to be cost-effective, some thinning out of involvement is necessary, but only at such a rate that the LOs do not collapse. If at some point it appears that an LO can never become self-sustaining, a decision to disengage should be made, since one of the goals of the effort would be to avoid creating "dependency" relationships that are the antithesis of development.

The promoters must be carefully coached and supervised so that they help the LO gain experience and capacity for self-management; they are not to become permanent fixtures in the community. This means they must assiduously avoid assuming leadership roles themselves. Some pump-priming may be appropriate in terms of planning activities with more interested members and setting up initial meetings, but always in the effort to bring forth community leadership. Local-level workers must guard against what Toth and Cotter (1978:28) call "the Lawrence of Arabia syndrome," even though it may be hard to get promoters to take more satisfaction in the emerging capacity of a group than in its visible accomplishments.[24]

Some LOs have started and sustained themselves without outside promoters or catalysts, though most of the successful ones subsequently established linkages to sympathetic and supportive public or private agencies. But the promoter role is coming into prominence, as a number of government-aided projects and independent LO experiences are suggesting its utility if viable LOs have not emerged from purely local initiative. A bonus value is that the LOs stimulated in this manner seem better able to induce or compel local-level government staff to perform their duties more regularly and conscientiously, because there is some initiative, capacity, and even pressure from below. The cost of introducing catalysts, therefore, should be compared not only with the direct benefits produced by LOs but also with the added benefits of better administrative performance.

The Paradox of Assisted Self-Reliance

The strategy of development outlined here points to a three-tiered effort. It may differ according to local circumstances and the complexity of the task at hand. But basically, accepting the paradox discussed above of using top-down initiative to get bottom-up development, we see local self-reliance as the keystone for effective interaction and exchange between local organizations and government or private agencies.

[24]As the Thai Khadi manual says: "If you are asked by the group to do this or to do that for the group, don't do it. Learn to say no in a diplomatic way as the members have to learn to do everything by themselves" (Heim et al., 1983:49).

The expectations and behavior of rural people are crucially affected, for better or worse, by the attitudes and performance of officials as well as other outsiders. One can expect that rural people will not change what they do unless others who are taking advantage of their weakness or ignorance, or who are simply indifferent, change *their* practices. At the same time, mobilization of whatever talent and resources local residents can be induced to contribute provides some incentive or leverage to get others to make such changes. Possibly the new efforts will be resisted and repressed, but to the extent they are clearly constructive, protective backing can often be gained from bureaucratic, political, or private voluntary sources.

As the first tier of effort, the LO—whether a cooperative, an LDA, or a women's association—should identify and diagnose local problems, determine what solutions lie within local means, and carry out such measures. Doing so not only provides direct benefit (or relief) but also demonstrates a serious desire to develop—not merely to attract handouts from the government. It runs counter to the images that officials and other educated outsiders often have of rural people as passive, fatalistic, mendicant, not competent to help themselves. Much of the "dependency" observed in rural communities has actually been induced by possibly well-meaning politicians, planners, or bureaucrats, or by persons intending to establish relations of subordination. That pattern needs to be reversed.

Depending on the stage of development and the actual resources, human and material, in the community, there will be more or less scope for self-help (B. Thomas, 1980). But the diagnosis and planning process at the community level should include an assessment of priority needs for which local-level officials can be of assistance. Indeed, if relations have become good enough, such officials may participate in the discussion and decision-making. Otherwise, delegations of local residents may meet with them, or they may be asked to come and meet with the full LO membership. The purpose is to get local officials to do as much as is within their scope, given their mandate and resources. Usually they can accomplish much more than they have been doing, especially if there are local contributions. There is evidence that field-level officials, when confronted with an active LO, begin performing their duties better, keeping office hours more regularly, making requested visits to the field, coming up with suggestions of possible higher-level assistance. The second tier of activity, then—building on the first—is joint action with government or other agencies at the local level to supplement what community members are already doing for themselves.[25]

[25]De Silva (1981), as deputy director of the Sri Lanka Irrigation Department, experimented in one scheme involving 15,000 acres by introducing farmers' organizations similar to those assisted by ARTI at Gal Oya. He had the same experience of getting more energetic and constructive performance from local officials of several ministries, once there were elected farmer representatives to deal with. Rabibhadana (1983) takes an opposite view, based on experience in rural Thailand, of the prospects for cooperation or support from local officials.

The step beyond this is for the LO, aided by documentation and specifications from local staff and supported by the federated structure to which it is linked, to seek direct investments or policy actions by the government at the district level or higher. Having already utilized local resources, LOs are in a stronger position to make claims and to get responses. Government support in this context should encourage members to continue a degree of self-help, which at least some leaders in virtually all governments would like to promote, if only to alleviate pressures on the central budget. There is some assurance to decision makers that when requests come through such organized channels, meeting them will satisfy a broader constituency than when dealing with individual demands. Arrangements for maintenance or conditions governing utilization of resources can be made when bargaining with organized claimants, and some regional or national policy considerations, or the need to distribute facilities spatially, can be introduced. Not all demands may be fulfillable, but the planning and implementation process in this mode incorporates a substantial element of initiative from below, meshing national with local priorities and perceptions of need.

The balance of local, joint, and government action will vary by sector and certainly over time, and it will depend on the strength of local capabilities. The advantage of the framework described here, synthesized from various LO experiences, is that the accomplishments at each level support and stimulate action at the other two. We would expect that eventually the activities that can be performed purely by local effort would be completed, and the relationship would evolve into a more complex one; multipurpose local government plus larger, commercially successful cooperatives and various special interest groups would emerge as centers of development initiative and performance, working closely with government ministries as they have in Japan (Aqua, 1974).

We have already seen in Chapter 5 that because of the many intervening variables, small size and homogeneity of the base group are not in themselves determinants of success. Nevertheless, most of the literature on this question (reviewed in Buijs, 1982a) suggests that preference should be given to small size and homogeneity in forming groups, provided they are not isolated units but are linked horizontally and vertically, as we concluded in our 1974 study.[26] Membership requirements can exclude certain persons, such as farmers with holdings over a specified size, or the program can be formulated in ways that are not attractive to larger farmers, such as offering

[26]This is the premise of the Small Farmer Development Programs in Asia, which have often been cited. See also the Thai Khadi manual, which suggests homogeneous group membership of 20 to 40, and not to exceed 60 (Heim et al., 1983:45). Volken et al. (1982) started with such smaller groups but found it useful at some point to expand them—when problem-solving discussions identified constraints arising from small size.

services not of interest to them (hand pumps instead of motorized ones) or setting conditions that are unattractive (such as limits on the size of loans available).[27]

The strategy of having fairly small, homogeneous base organizations linked to higher levels of organization permits agencies assisting LOs to pursue the advantages of one of our most interesting empirical findings: that informal LO roles and procedures seem advantageous. It appears that when there are several tiers, the lowest one can operate quite informally so long as there is some clear and acceptable designation as to who will represent the group at the next higher level. This higher-level organization can then operate more formally, with such trappings of modernity as a constitution, bank account, and minutes of meetings. Indeed, legal recognition may adhere to the higher body, which rests on the solidarity of constituent small-scale organizations.[28]

A similar evolution may be appropriate with regard to the number of functions the LO undertakes. LOs do best to start with a single valued function, taking up additional activities as members and leaders are prepared to do so. We would suggest leaving open—and up to the LOs themselves—the extent, pace, direction, and methods of multifunctional evolution. Once an LO is clear about what new activities it would like to undertake, it can determine in what ways, if any, government or private agencies can assist. If agencies start the process by offering to provide certain services or resources, this may elicit affirmative responses which are not based on LO understanding or commitment. That the activities of LOs should spring from needs recognized by their members is crucial to the strategy of assisted self-reliance that we have outlined.[29]

[27]For an analysis of the kinds of assistance programs that are more or less vulnerable to capture of benefits by more advantaged members of the community, see Leonard (1982a:8–15).

[28]Duewel (1984) has observed that outside agencies in Indonesia preferred to introduce more formal, "modern" roles in water users' associations. But informal organizations at the base, encompassing 10–50 members cultivating 10–15 hectares, worked best. Offices at this level were often not specified, and matters affecting the village as a whole were referred to the village assembly. We found in Gal Oya that when it was left to farmers to determine the size of their irrigation groups, they generally chose to have one representative for every 12 to 15 farmers. Only two of 71 groups fell outside the range of 8 to 25 members. Although the organizations started in Mexico under the CONAFE program were not aggregated into higher-level bodies, the base-level approach was similar to ours in Gal Oya, Sri Lanka. Paul reports: "The transactions with the community were conducted informally to begin with. As experience accumulated, communities were encouraged to elect small committees which negotiated with the instructor and CONAFE on the terms and arrangements for setting up the school" (1982:94).

[29]This approach is similar to the earlier community development method of identifying and responding to "felt needs" (Holdcroft, 1978; Blair, 1982). But it stresses having organized local capacity to sort out and reconcile contending needs within the community, and mobilize resources to meet such needs locally as much as possible. This creates some bargaining power with outside agencies whose assistance is needed.

Inductive Planning and Implementation

Approaches at the local level are to be formulated both in terms of what local residents can do and in terms of cooperation between outside agencies and the community. The "inductive" approach formulated here represents a synthesis of what members and leaders and assisting outside personnel can do to support the development of LO capacity. This is consistent with the "learning process" strategy in which D. Korten (1980) emphasizes the need to proceed flexibly and with periodic assessment and reassessment.

This approach enjoins planners and implementers to formulate in advance the action hypotheses they think are reasonable to guide decisions, and then to monitor and reconsider them continually. An example from our work in the Gal Oya project is the proposition that LOs in an irrigation system undergoing rehabilitation could be best promoted if initiated in conjunction with the planning and physical reconstruction of the system, rather than independently of such work; otherwise, the LOs that were expected to take over the management of water once the system was redesigned and rehabilitated would have no prior involvement in and sense of responsibility for it. This seems a well-founded proposition based on what is known from other experience; but it might turn out, in a particular instance, that the unpredictability and urgency of rehabilitation efforts impeded organizational development. No proper "experiment" is possible in that there is no intention of carrying a particular premise through to the end of the project to see whether or not it was "correct." It is appropriate to make action hypotheses explicit among those working in the project and then to take stock periodically to assess how the work is proceeding. If contradictory evidence emerges, the approach should be modified as soon as is practical.

Flexibility of this sort implies the need to take advantage of unexpected opportunities. Solomon (1972) has commented on the importance of using crises or sudden external problems to consolidate support for LO development. In the Gal Oya project of Sri Lanka, the fact that water levels in the reservoir at the time the organizational effort began were unusually low, even for a dry season, caused ARTI to accelerate the effort, cutting short the time planned for training organizers by having them construct "profiles" of each field channel area and the farmers cultivating there. Going "prematurely" to informal organization for the special tasks of water rotation and water saving demonstrated to engineers the farmers' organizational competence in water management. If the original organizing timetable had been maintained, an opportunity to help farmers and simultaneously strengthen the program would have been passed up.

The acceleration ran the risk of discrediting the organizational campaign, which would have been a handy scapegoat if the crop failed in spite of the water management efforts. Farmers were not pressured or forced to cooperate, and they adopted the special measures surprisingly quickly and freely. ARTI impressed upon the organizers that the actual decisions and plans for water rotation and saving had to be left to the farmers, with organizers acting only as facilitators. In this case, the rapport among farmers, organizers, and ARTI led to a good decision, aided by some unexpected rains toward the end of the season.

In support of such a flexible approach, it is useful to have what David and Frances Korten and their associates in the Philippines have called "process documentation." This involves specially designated persons with some training and reporting skills who will regularly write down and transmit observations on how the LO formation and functioning are progressing. Documenters speak at length and privately with members (and non-members) to get their views of the process, and with officials interacting with the organizations. Further, meetings of the LO and of its members with officials are observed by the process documenter, who must play a less active role than any person acting as a promoter.

The information thus compiled helps persons having responsibility for the program to assess what is going on in the field and to make judgments about the validity of the hypotheses on which the program rests. The documentation helps them determine and calibrate rates of change, isolate patterns of causation, pinpoint weaknesses in the program, pick out the most effective promoters and local leaders, and develop materials for training subsequent promoters. If the program is proceeding with multiple organizational models, it can show which are probably better and provide evidence for preserving several different LO forms.[30]

As their title indicates, the Group Organizers/Applied Research Fellows of the FAO-sponsored Small Farmer Development Program were expected to do something like process documentation in the field. The National Irrigation Administration program in the Philippines, on the other hand, recruited independent process documenters from a national university to observe and report the work of its community organizers. The ARTI approach in Gal Oya was intermediate. About 20 percent of the institutional organizers (IOs) were assigned to spend the majority of their time in process documentation and the rest in organization work, giving them firsthand experience of the subjects they were reporting. The SFDP approach has produced some reports, but the organizational tasks have readily overwhelmed the documentation effort, and it may be difficult for the GO/ARFs to be entirely objective. The NIA's autonomous process documenters generate much more documentation and more objectivity; reports are fed

[30]Seibel and Massing (1974) inform us that in Liberia alone, residents had developed 16 different forms of rotating credit and savings associations to meet various needs and conditions. It is likely no single form of organization will be best for all tasks or circumstances.

into a high-level policy body for NIA review on a monthly basis. Timely information is thus available for making corrections and building commitment to the program. The "feedback loop" is fairly long, however. The ARTI compromise has produced less documentation but more response in the field; each team of organizers is to go over the fortnightly report and discusses the issues and problems identified to see whether any immediate improvements can be made. Their supervisor can subsequently make suggestions based on the reports, as can the ARTI professionals involved intermittently with the program.[31]

With process documentation, the learning process can become more institutionalized. To the extent that the work is guided by explicit hypotheses about farmer and official behavior and about group performance, these can be checked through a process of "inductive" planning and implementation. This stands in contrast to a "deductive" approach, which assumes that the necessary knowledge is in hand and needs only to be applied (Uphoff, 1982b). With the inductive approach, one draws on varied experiences and general principles, but for specific situations one can only hypothesize cause-and-effect relations. These need to be checked continually and appropriate adjustments made in the program if accumulating evidence so indicates. The effort is experimental, but not in the conventional sense of setting up a test of ideas to be evaluated when the effort is finished some years later.

Local organizational development does not lend itself to planning of the sort most project designers and administrators are accustomed to. Instead of "building" LOs, which suggests a mechanical process, we need to think in terms of "nurturing" them, which implies the more organic processes of evolution and growth. Because a learning-process approach to rural development places such a premium on innovative implementation, it requires even more skill and creativity than the planning.[32] Therefore, the prevailing

[31]Assigning the IOs in teams to large areas and having them meet regularly, with one of their number serving as coordinator, has proved useful, making the program itself more bottom-up than originally designed; supervision has been less extensive than planned, and the organizers' self-criticism and self-correction has been greater than expected. Had they been assigned individually or in pairs, there would have been less synergy of effort. Also, if one organizer gets sick or has to take leave, several others are sufficiently familiar with the area's farmers and situation to cover for the absentee. Close supervision is almost impossible in this kind of program, as CONAFE in Mexico discovered: "Once in the field, very little direct supervision of instructors could be organized from headquarters. Communities were often difficult to reach and means of communication were extremely limited. A system of 'assessors' was established to visit the community schools periodically to assist the instructors in solving their problems and answering their questions, if any" (Paul, 1982:95–96). Since monitoring was difficult, CONAFE determined that the community itself would have to exercise effective control. We have experienced similar difficulties of supervision in Sri Lanka, for which the team approach of deployment and operation has compensated. So far devolution of responsibility for supervision to the communities themselves has not been attempted.

[32]On the contrast between "process" and "blueprint" approaches to rural development, see also Sweet and Weisel (1979). This is amplified in D. Korten (1980).

belief that design and planning need to be done by the most intelligent and educated persons, but can be implemented by others less educated and imaginative, needs drastic revision. Initial plans and formulations demand insight and systematic analysis, but carrying them out with suitable modifications and reformulations based on experience requires even more talent and commitment.

Especially when it comes to creating local capacities for rural development, we need to break with the notion that planning is for the queen bees and implementation for the drones. Indeed, the dichotomy between planning and implementation that generally distorts development efforts (Rondinelli, 1982) is probably nowhere more debilitating than in this area of LO development. The basic feature of an effective LO program is to link planning and implementation in a way that engages the membership and leaders to sustain a continuous problem-solving effort in conjunction with appropriate, supportive outside agencies. Approaches at the local level need to have what Meehan (1978:63–64) calls "strong local anchor points," where bottom-up development is based on LO efforts at self-help. As we know, however, anchors are of little value without lines connecting them to something else. It is with this understanding that roles for governmental, international, and private agencies take on more significance and potential.

CHAPTER 9

Contributions of Governments and External Assistance Agencies

We are confronted with evidence indicating that local organizations are most likely to be effective for development when they originate from the initiatives of local people, yet to expand the range of their services, they are likely to require assistance from governments and even international agencies. Further, when they do not already exist, we find that LOs can emerge in communities if appropriately conceived and stimulated by catalyst agents, so that top-down initiative can yield bottom-up impetus for development. Such paradoxes represent a special challenge to students and supporters of local organizations. In the preceding chapter, we considered means by which local efforts can be focused with greater probability of success. Here we examine complementary activities by governmental and nonofficial agencies—means by which they can enhance programmatic efficiency, promote social equity, and assist in the empowerment of rural publics through their local organizations.

What Can Governments Do?

During centuries of minimal government, rural societies functioned for the most part with minor contact and little concern for the state. For most rural people, life was primarily a struggle to maintain subsistence. Commercial exchanges, limited as they were, took place with little interference from the state. Rural areas were commonly isolated by poor transportation. Central authority was fragmented territorially and exercised by revenue officers or military agents. The state encroached on rural areas to levy taxes and requisition foodstuffs, but for the most part rural societies were left alone to be governed or exploited by landowners or officials whose power base was local.

During the past half-century, this situation has changed fundamentally.

The end of European colonialism and the drive of successor states for national integration and economic development have set in motion efforts to penetrate and incorporate rural areas. Whether local associations develop spontaneously or are established by voluntary agencies or by government, much that they do involves governmentally sponsored policies and programs.

Some writers on rural development cling to a lingering nostalgia for the simpler days when rural life was untouched by government and bureaucracy. Others retain a deep suspicion of the state, considering it at best incompetent or irrelevant to the needs of the productive members of rural society, at worst as the corrupt and inveterate exploiter of the rural poor, usually in league with local elites (De, 1979). From different points of departure, both schools of thought fear and distrust the state. Even when they do not entirely reject it, they emphasize the need of rural local associations for autonomy and self-reliance.

We believe that neither of these perspectives can provide an adequate orientation for the development of local organizations in most rural areas. Some governments may indeed be so inept or corrupt that small-scale, informal organizations, aided by private voluntary agencies, are better off avoiding official bureaucracies altogether.[1] For better or worse, however, government influence is here to stay; whether leftist, rightist, reformist, or technocratic, government activities in rural areas are likely to increase, not diminish. While governments can be incompetent or brutal and exploitative, they can also be helpful to and supportive of the poor and disadvantaged. Indeed, most successful local associations enjoy the support or at least the acquiescence of government and are linked to services or resources that originate in the state.[2] Even local organizations that adopt a confrontational posture toward local elites or toward government itself must establish some relationships with officialdom if they are not to be systematically bypassed and eventually repressed.

Whether and under what circumstances a particular local organization should relate to government or avoid government is a major strategic choice. Our study of local organizations persuades us that general prescriptions for absolute autonomy, for implacable confrontation with the state, or for limitation to single, time-bound concrete tasks are not suited to the trends that are evident in rural life. Maintaining distance from government

[1]This was the explicit premise of the Thai Khadi Research Institute (1980) in its study of self-help organizations in Thailand.

[2]The success of local organizations enjoying government support may not be caused by that support. The Zanjera Padong Irrigation Association illustrates how well some indigenous organizations have been structured and operated without government support. Once the government does enter the scene, the organization may need some government cooperation to continue to function successfully, since the government's presence changes the environment (Coward, 1979b:28–36).

may be a rational strategy where the state is incompetent or hostile. But in the great majority of cases, the ability of local organizations to help their members depends on linkages worked out with institutions that control and allocate various services and resources.[3] The parochialism of rural life is passing everywhere, and local organizations must cope with this reality.

We identified in Chapter 1 a number of considerations that can motivate political leaders in developing countries, in their own interest, to tolerate and even to support or encourage rural local organizations. If they wish to maintain pressure on sluggish bureaucracies to implement rural development activities as the government intends; if they want rural development programs to be responsive to the needs, priorities, capabilities, and convenience of diverse rural publics; if they wish to induce counterpart resource contributions from rural publics, while expanding their own political support base and reducing the risks of social protest and violence; or if they are interested in projecting a favorable image with development assistance agencies, they have reason to work with and through rural membership organizations, including associations of depressed and disadvantaged groups. These are reasons to be taken seriously by regime politicians. They are not derived primarily from considerations of justice or ethical motivations, though the influence of these should not be entirely discounted.

Governments that are interested in promoting rural development thus have reason to establish effective communication with and enlist the energies and support of rural producers through local membership organizations. How, then, can interested governments help to foster and sustain such organizations? We identify five areas in which governments can work out productive relations with local groupings of the rural majority: financial allocations, direct assistance to local organizations, special organizations for disadvantaged publics, reform in the provision of public services, and reinforced accountability within local organizations.

Financial Allocations. Urban bias is a well-recognized feature of public expenditure and investment in most developing countries (Lipton, 1977); weaker and disadvantaged groups have suffered especially from government neglect of rural interests. There is growing evidence, however, that lagging agricultural productivity, the urgent need to increase food production, and the chronic poverty affecting large and growing numbers of rural residents are leading governments to shift their expenditure priorities to discriminate

[3]The Tiv in Nigeria refused to accept government credit, because of the conditions imposed by the state, and set up their own credit associations (*bams*), which were quite successful. In order to achieve more comprehensive development, however, they found that they would have to work with government. The farmers associations that they subsequently formed were closely linked to government extension services, which had in the meantime become more responsive to Tiv rural interests (Morss et al., 1976, II:231–41).

less against rural areas. Favorable price incentives to stimulate agricultural production and relieve pressure on the balance of payments for food imports are now more common in developing countries. Encouraged by a growing consensus in international circles on the importance of rural development, this desirable shift in budgetary priorities is likely to continue.

Increased investment in rural programs is limited, however, by the chronic resource shortages facing most governments in less developed countries. This makes it all the more important that any increments in public expenditures and any new and expanded programs oriented to rural development be buttressed by policies and institutional infrastructure that enhance their effectiveness. The effectiveness of economic incentives and technological innovations intended by governments to increase the productivity of small and marginal cultivators will depend in large measure on responsive behavior by farmers and on complementary activities by their local organizations.

In order that scarce resources not be wasted on well-intentioned activities which are nevertheless unproductive because they fail to respond to the needs, priorities, capabilities, and convenience of marginal farmers and other disadvantaged groups, it is important that government agencies know what these needs and preferences are.[4] This is not likely to be possible in the absence of authentic spokespersons for these usually silent constituencies; these spokespersons must be rooted in organizations that genuinely represent the rural poor.

Educated technicians and senior officials are often ignorant of the real conditions, needs, and preferences of small-scale cultivators; their backgrounds and experience provide little reliable insight into these needs.[5] In the absence of reliable information both about these needs and about the impact of government-sponsored programs on intended beneficiaries,

[4]The government-sponsored attempt to provide market women with credit in Nicaragua failed the first time around, not because the resources were not needed or because the terms were onerous but because the *method* of providing the credit was too unpleasant for the women. Rather than approach the unfamiliar, impersonal, air-conditioned loan office ("after a long day in the market a woman might catch cold here"), the women continued to deal with the traditional moneylender, whose charges were ten times higher. His style was familiar and intimate, and he was conveniently located in the market, available day or night to make an emergency loan. The government's second effort was based on a community-level cooperative. "The project has been successful because the cooperatives have built upon existing market women's culture, utilizing all the subtle and complex interpersonal relationships established over the years" (Bruce, 1980:1).

[5]Chambers (1980) discusses a series of biases that tend to keep staff members uninformed of real conditions, regardless of their level of personal motivation. Perhaps the most powerful of these is a professional one: technicians' specialization and training usually focuses their observations on those situations relevant to their area of expertise. They do not see the holism of poverty. Another factor is the common civil service practice of purposely posting people to regions other than their home areas to avoid entanglements with family, friends, and so on. Similarly, the practice of rotating officers in their posts every two or three years works against technicians' ability to build up effective local working knowledge and personal relationships.

scarce financial, technical, and administrative resources are often dissipated. If services are to be relevant to specific needs as the beneficiaries themselves experience them, the rural poor must be consulted, must have a voice in the choice of services and how they are provided. This voice can be provided by membership organizations.

It is a great temptation for governments eager to demonstrate their solicitude to provide services in paternalistic ways that foster dependency rather than self-reliance. When governments are constrained by very limited resources, this can lead to perverse outcomes.[6] The indiscriminate use of Food for Work programs, in which rural residents are compensated for community work, has undermined local self-help traditions in a number of areas. In many of the Village Development Committees in Botswana, for example, the food given for work became the only incentive for contributing labor to build roads, schools, and hospitals (Vengroff, 1974:303–9). In the absence of a requirement that beneficiaries also contribute within their capabilities, local publics are likely to sit back and wait for government to deal with their problems. In the San Antonio irrigation scheme in the Philippines, the National Irrigation Administration built the system with the agreement that the costs would be repaid by its users. But villagers had little say in construction and no control over the cost, and they now claim that the system is poorly constructed. In addition to refusing to repay construction costs, they refuse to maintain the system, saying that is the responsibility of the government. Since it is not *theirs,* they feel no obligation to maintain it (de los Reyes, 1980).

Where governments neither consult local publics nor ask for counterpart contributions to development projects, local organizations limit themselves to petitioning and otherwise making claims on already overburdened governments, rather than engaging in collective self-help, mobilizing and contributing resources, and participating in the management of programs and facilities. Government help should encourage and where possible require local publics to mobilize resources that complement those provided by the state, and to reduce the costs of administration by assuming part of the administrative burden. This amounts to assisted self-reliance as proposed in Chapter 8.

For this to be successful, governments must provide assistance in ways that are compatible with the convenience and work schedules of local people and must be reliable in delivering the resources or services that they

[6]Writing about the government's rural works program in Pakistan, which channeled the bulk of funds through elected local governments, Garzon commented: "Paradoxically, the sudden flow of funds to the Unions, which overnight doubled their budgets, made them more, not less, dependent upon the central government and its district administrators. Often [local governments] responded to the windfall by lowering taxes or becoming lax in collection [to incur] the favor of their constituents" (1981:16).

promise as counterparts to local self-help. Frequent failure to do so results in discrediting local self-help and weakening local organizations. Governments should be able to reward local organizations that contribute resources and accept responsibility for program management so as both to increase the efficiency of resource use and avoid the dependency syndrome. By doing so, governments can encourage effective resource mobilization and management, which we found in Chapter 3 to have the highest correlation with overall LO performance.

Direct Assistance to Local Organizations. While local associations of rural publics can greatly facilitate program management, they will run into difficulty unless supported or at least tolerated by government. A hostile or indifferent government encourages local elites and locally based civil servants to undermine or bypass local associations, relegating them at best to small-scale activities and often contributing to their extinction. The risk to government in encouraging local organizations is that they will make excessive claims which cannot be accommodated, or that they will be taken over by opponents of the regime. The main political benefit to government is the expanded support base that local associations can bring to a regime that uses them to extend services and maintain communication with rural constituencies. LOs can also facilitate local initiative and thus divert demands from the central government and its agencies; this was found to be the case with the Village Development Committees in Botswana (Vengroff, 1974), the Farmers' Associations in Taiwan (Stavis, 1974a), and the Saemaul Undong movement in Korea (Goldsmith, 1981). In the absence of sponsorship and encouragement of local organizations, governments risk the alienation of rural publics and sacrifice the contributions that these organizations can make to the effectiveness of government programs intended to improve the quality of rural life.

Governments can support local organizations by direct sponsorship, by targeting services to disadvantaged publics through the membership associations, by counteracting the opposition of local elites, by requiring government agencies and staff to protect such organizations and the rights of members to participate in them, and by providing technical assistance and specialized training to leaders, members, and staff. As the counterpart for such support and assistance, governments can require local organizations to contribute a share of jointly provided services and to participate in their management and maintenance.

There are very practical measures by which governments can lend assistance and support directly to local organizations. One of the most useful is to provide sound technical assistance to projects undertaken by LOs, thus compensating for the limited skills of members; this kind of service is well illustrated by the road-building project in Nepal cited in Chapter 7

(P. Pradhan, 1980). Another is to train leaders, both in technical subjects and in the skills necessary to maintain and manage organizations. Learning how to operate a meeting, keep books, and let contracts, as well as how to benefit from the services offered by government departments and voluntary agencies, are skills inculcated by the more successful training programs. It is useful to reinforce formal training with periodic meetings and newsletters or other communication media. Different sorts of training should be available for persons with different needs; the unpaid chairman of a small local unit needs skills and information different from those of the full-time accountant of a federated center. Training blends into technical assistance as government staff, or PVO (private voluntary organization) personnel working with government, visit local organizations to learn specifically about their needs, to impart information, and to mediate between a local organization and a government agency where linkages are not working effectively.

Special Organizations for Disadvantaged Publics. Government-supplied resources intended for disadvantaged publics may be vulnerable to preemption by local elites, often with the connivance of opportunistic, corrupt, or unsympathetic officials. The diversion of resources is especially likely when such services are provided through bureaucratic agencies whose primary clientele are the nonpoor. If resources and services are to reach specific disadvantaged publics, they should be distributed by agencies specially committed to the task. The Small Farmer Development Programs operating in several Asian countries use membership rules to keep wealthier farmers out of the organization, and then funnel resources through the organization only to members. Even with such precautions, some resources may still be diverted by landowners or merchants. This abuse can be diminished, if not entirely prevented, by providing services that are of little interest to more advantaged groups (such as primary education), cannot be hoarded and resold (such as inoculations), or represent no financial subsidy (Uphoff, 1980:43–48).[7] Intensive management can oversee and reward staff to ensure that services reach their intended beneficiaries.

Since government agencies seldom have sufficient personnel to provide services to the rural poor on a one-to-one basis, they have begun to experiment with unconventional methods, including indigenous paraprofessionals, to reach all those who are eligible for services and to bridge the gap

[7]Leonard (1982a:8–15) suggests the following design conditions for reducing the vulnerability of services to capture by local elites: (1) benefits are indivisible and widespread (e.g., village sanitation); (2) they are linked to use of a resource the poor have in abundance (e.g., labor-intensive production); (3) services deal with problems or opportunities more common to the poor (e.g., diseases that afflict the poor, or crops they depend upon, such as cassava or sorghum); (4) supply exceeds demand (market is flooded) or demand is limited (e.g., smallpox vaccinations); (5) units of service exceed the demands of local elites (e.g., primary school rooms).

between government technicians and LOs (Esman et al., 1980). Agencies providing services to disadvantaged rural publics are strengthened if these publics are also brought together in parallel membership organizations that assume some of the burdens of program management. The special bureaucratic agency and its intended beneficiaries are thus linked in a mutually supportive network. Special organizations for the disadvantaged may be initiated by PVOs, which take the initial risk. When they prove successful, governments may begin to provide assistance through these channels.[8]

Administrative Reform. For the provision of services to rural publics, governments rely primarily on bureaucratic agencies, but in most developing countries those are ill prepared to serve the rural poor directly or to interact with local organizations. Staff members in the field are often too few, insufficiently trained, poorly motivated, and inadequately equipped; structures are too centralized and procedures too rigid to permit timely service in response to needs that vary greatly with distinctive local conditions. Given the importance of linkage, governments cannot seriously consider helping local organizations without improving the capacity, motivation, and skills of the bureaucratic agencies and their staff members who are in direct contact with rural publics.

Staff members of the increasing number of agencies that work in rural areas may regard local membership organizations as nuisances that limit their power and unnecessarily complicate and politicize their technical work. High-handed attitudes and sometimes corrupt methods of relating to rural publics must be changed if officials are to work compatibly with, rather than bypass or undermine, local organizations. A service ethic must replace the conventional command ethic. New communication, negotiation, and learning skills should complement purely technical skills, and bureaucratic systems should reward service and assistance to rural publics rather than encourage rigid adherence to formal rules, technical standards, or quantitative targets. Specifically, government staff must have incentives and be rewarded for success in working with and through local organizations.

No strategy to strengthen local organization in support of rural develop-

[8]This occurred with the Barpali leatherworkers' cooperative in India, started with the help of the American Friends Service Committee and given government assistance once it was established (F. Thomas, 1968:364–87). The Social Work and Research Centre in India was also started by a private voluntary organization, with its benefits targeted to the underprivileged in the rural Ajmar District of Rajasthan. That different government agencies joined the activities of the SWRC was a major factor in the success of the organization (Kale and Coombs, 1980). In Thailand, the Lampang Health Development Project was initiated in 1971 by USAID with the help of the University of Hawaii, and was later supported by the government (Coombs, 1980). Another good example is the International Planned Parenthood Federation's youth project in Indonesia. After it was started by an international PVO, government family planning services began to cooperate with it (Soenarjono, 1980:635–98).

ment can avoid or bypass reform of the instruments of public administration through which the state relates to local publics (Esman and Montgomery, 1980; D. Korten and Uphoff, 1981). A number of measures can be combined as elements of a strategy to reorient public administration to the needs of rural publics.

(1) *Alternatives to Bureaucracy.* The first involves the search for alternative means of reaching the rural poor, relieving the strain on bureaucratic structures and drawing on other capabilities that may be better suited to serving rural publics or may complement the activities of bureaucratic agencies. Some activities, especially those involving the supply of production inputs and marketing, can frequently best be handled through commercial channels. As indicated in our introductory analysis, the private sector offers one of the three major channels for linking rural people to national centers of service and opportunity. The main limitation of this approach is that small farmers, tenants, and landless workers are seldom a profitable market for legitimate business enterprises. Because of their weak bargaining position, the rural poor can be victimized by unscrupulous merchants and moneylenders. When the state relies on market processes to provide commodities and services, measures may be required to protect disadvantaged rural publics from abuses, especially those arising from monopoly, monopsony, or debt bondage.

For this purpose, rather than undertaking the administratively more costly process of providing services directly, governments can resort to regulation, which is less expensive in scarce administrative resources.

Local organizations can increase the effectiveness of such regulation, as well as the bargaining power of rural publics, in their market activities. One example is the Puebla project in Mexico, where the desire to escape the inordinately high prices charged by local fertilizer distributors provided much of the impetus to form local organizations. Although the *campesinos* encountered resistance from local distributors, who were selling fertilizer above the government-set price, they won their case by appealing to the government (CIMMYT, 1974). Effective local organizations can help to curb monopoly practices by businessmen and ensure that regulations designed to protect local publics are enforced. The broader their membership base, the more effective local organizations will be when bargaining with rural elites.[9]

Private voluntary organizations (PVOs) can often provide services that help the rural poor in ways that are relatively free from the rigidities of bureaucratic operations (Hyden, 1983:119–27). Though they are not immune from elitism and paternalism, PVOs can attract staff members, both domestic and foreign, who are more willing to live in rural areas and serve

[9]An example is the Khet Mazdoor Union of landless workers in India, cited in Chapter 5.

the poor, more committed to developing their capabilities and institutions, and more prepared to experiment with unorthodox ideas and practices. With some notable exceptions, PVOs usually operate on a small scale.[10] While some governments are suspicious of private voluntary organizations, especially those of foreign origin, others are willing to take advantage of the flexibility, skills, and commitment that they can bring to their work on behalf of the disadvantaged.

Thus for many activities, especially those requiring experimentation and the patient development of local leadership and local participatory institutions, PVOs are a useful alternative to governmental bureaucratic agencies and provide a supportive linkage for associations of the rural poor. Some of the more successful small-scale cooperative movements have been initiated through the efforts of private voluntary groups.[11] An example of beneficial interaction between a private entity and the government is the Banki water project in India. Initiated through a local leader who requested assistance from a nongovernmental Indian institute, it eventually tied together funds and technical assistance from the institute, the World Health Organization, and the Indian government. The project successfully supplied a target area with piped water, and the system eventually was operated entirely by a local organization (Misra, 1975).

To complement and exercise surveillance over bureaucratic agencies, some governments may rely on political means. Party organizations can become both sources of information to the political leadership about how government-sponsored programs are working, and channels for the ventilation of grievances by organized but unsatisfied publics.[12] To many civil servants, political "interference" may be an irritating and chronic source of corruption and distortion of orderly and efficient administration. It is true that politicization can undermine administrative discipline, but it can also provide a safety valve for public grievances, a means of mitigating the rigors of impersonal rules, and a way of identifying and reporting inadequacies and breakdowns in bureaucratic performance. Political channels do not *substitute* for bureaucratic operations, because they are not designed to deliver services or resources on a regular basis, but they can serve as a means of pressure to monitor and improve the quality and the responsiveness of bureaucratic services.

[10]Among the well-known exceptions are BRAC in Bangladesh and the Sarvodaya movement in Sri Lanka (Coombs, 1980).

[11]For examples, see the World Neighbors newsletter, *World Neighbors in Action* 1:1, 1:2, 3:2E, 7:5E, 10:4E. In the Manyu Oil Palm Cooperative in Cameroon, assisted by the Catholic church (Gow et al., 1979, II:53–66), project staff members were able to devote the time necessary to increase the awareness of members and cooperative officials alike of the goals and objectives of a cooperative and the means of achieving them.

[12]Two examples of such organizations are the Federacion Campesina de Venezuela (FCV) and the Confederacion Nacional Campesina (CNC) in Mexico.

Governments must therefore confront directly the structural and behavioral deficiencies of their bureaucratic agencies, which remain their most important instrument for linking with organized rural publics, for extending information, for providing resources and services, for managing development projects, and for keeping the regime informed of rural needs and preferences. Whether a government relies primarily on the line agencies of its ministries and departments or on such "independent" entities as statutory corporations or autonomous authorities seems to make little difference.[13] The establishment of autonomous agencies frequently reflects the unwillingness or inability of governments to attempt to reform their bureaucratic structures. Once rural development becomes a priority, however, and governments decide to work with organized rural publics, administrative reform becomes a necessity. Two kinds are required: reform in structures and procedures, and reform in the behavior of civil service personnel.

(2) *Structural Reforms: Decentralization and Integration of Services.* Even when a particular public service is provided efficiently by a specialized government bureaucracy, its consumers may not be able to use it, because complementary inputs that they require are not available on time from other agencies. Since local organizations are in a position to pressure the locally posted officers of specialized government agencies to integrate their service delivery, agency staff members need sufficient discretion to adjust their activities and schedules to accommodate local conditions. Governments therefore should relax the prevailing centralization of decision-making, allowing and indeed requiring local staff to interact more responsively with organized publics and with other agencies providing services. Horizontal linkages need to be formed among specialized government agencies in the field as well as among local membership organizations. In the absence of such adjustments, neither local publics nor governments will be able to realize the benefits of constructive linkage on which the efficiency of service delivery and the effectiveness of local organization depend. Service centers located in market towns, where all the specialists can have their offices in proximity to one another and easily accessible to their clientele, can also facilitate field coordination. Locating the offices of local membership organizations in the same complex promotes easy communication with government agencies and horizontal linkages with other LOs with which they share common interests.

[13]Though independent agencies benefit formally from greater flexibility in personnel and financial management, the operating styles of the public bureaucracies soon assert themselves in these nonministerial entities. The rivalry and hostility between independent agencies and established departments, whose personnel consider that the former are receiving favorable treatment at their expense, will often cancel out the limited benefits and flexibility that initially accrue to independent agencies. The ambivalence that development specialists feel toward autonomous agencies can be seen in responses to Guy Hunter's query on "Enlisting the Small Farmer: The Range of Requirements" (1980b:47–48).

What is widely regarded as the most needed structural reform is to *deconcentrate* decision-making and action within specialized bureaucratic agencies.[14] Administrative deconcentration is not an all-or-nothing relationship but a matter of degree; some specific decisions and all revisions in general rules will be reserved to higher authority. Nor can the power to act be deconcentrated overnight without risking administrative chaos. Agency rules, financial management practices, reporting systems, and supervisory methods must be revised, and training upgraded, so that the field staff can be entrusted progressively with increased authority to act in response to local requirements.

A more thoroughgoing form of decentralization is the *devolution* of functions to local government authorities or to local associations, accompanied by access to financial resources that permit local organizations to assume greater responsibility for program management. This can be done by making general or specific conditional grants to LOs to cover the cost of performing services otherwise rendered by the bureaucracy, or by giving them authority to levy taxes or user charges, or some combination of both. Devolution affords rural people the opportunity to function more independently through the "third sector" by exercising direct influence over locally based staff who become employees of their organization, as well as over program activities. The Farmers' Associations and Irrigation Associations in Taiwan, for example, by themselves hiring and paying extension agents and local water managers received more responsive and accountable service. The main risk for national governments seeking to benefit the rural populace in this way is that local elites will assume control over the resources and programs devolved by government and divert the resources to their own benefit at the expense of the intended beneficiaries.[15] This is less a risk when local authorities, particularly local governments, are involved with public goods and social services rather than with economic activities. Elite control can also be mitigated by the reservation of positions on governing councils to members of disadvantaged groups.

(3) *Interdependencies between Agencies and Publics.* One means of encouraging supportive bureaucratic performance is to make the success of technicians and administrators dependent on securing the cooperation of the proposed beneficiaries. In Mexico, for example, in connection with the PIDER rural development project, a special office—*Direccion de Caminos de Mano de Obra*—was set up within the Ministry of Works for labor-intensive construction of rural roads. Since the DCMO had access to very little heavy machinery, and no funds to pay for outside labor, it had to rely on rural communities to achieve its agency's goals. Road committees were therefore

[14]For a good treatment of this issue, see Bryant and White (1982:160–63).

[15]The decentralization versus equity dilemma is discussed, too pessimistically in our opinion, by Lele (1981).

formed in the communities eligible for assistance, and plans for new or improved roads were worked out jointly with those committees. Engineers knew that if the plans were not acceptable to the community, no local labor would be forthcoming. As a result, the rural roads component was the most successful part of the larger project. According to a World Bank report, "the success of the local level committees for the labor-intensive rural road construction program suggests that an effective organizational structure at the local level can be established and can facilitate the execution and operation of the program" (Cernea, 1979:70). Indeed, the same organizational strategy that required the engineers to work with and through local committees was proposed for the water supply sector in the next phase of PIDER.

The importance of this innovation in bureaucratic *structure* should not be overlooked. As a separate unit that did not have the means to achieve its goals *except* through cooperation with the local committees, the technical staff of DCMO had to develop techniques for working with communities if they were to meet their bureaucratic targets and technical performance objectives. In the process, they also developed more respect for what rural people could and would do if approached in a cooperative rather than a paternalistic manner.

A similar built-in pattern of interdependency can be been in the National Irrigation Administration (NIA) in the Philippines. There, the improvement of small-scale "communal" irrigation schemes was entrusted to a number of NIA engineers, and plans had to be worked out jointly with communities of irrigators. The latter were expected to repay the capital cost of the improved diversion dam and canal structure over 30 years and to contribute labor and materials to reduce the initial capital outlay. For each proposed project, a local Irrigators' Service Association (ISA) was set up with the help of community organizers.[16] If the ISA refused to accept the engineer's plans and to sign the contract for a loan to cover the construction's costs, no work was done. After investing much time and energy in designing a new dam, the engineers were under some pressure to be accommodating to the association and to arrive at a mutually agreeable plan of work. This principle, creating "structural interdependence" between administrators or technicians on the one hand and rural local organizations on the other, seems a generalizable one for promoting effective participatory organization. A new reward system can change the terms on which officials relate to rural people, once there is something that the latter must do voluntarily to help the former be successful by bureaucratic criteria.

(4) *Improved Personnel Skills and Reward Systems*. Technical, organizational, and communication skills for government field staff working with

[16]The Laur case included among our 150 case studies was the first instance where NJA introduced this role and activity (Coward, 1979a; also F. Korten, 1982).

local organizations can be cultivated and improved by training. Successful training encourages responsive service to disadvantaged publics and aims to impart a more sympathetic and realistic understanding of the problems encountered by rural people—especially by the more disadvantaged groups, including rural women—in their daily struggle for livelihood. The training program for field organizers developed by the Thai Khadi Research Institute, as a result of the institute's previous experience with field projects, places considerable emphasis on this sort of understanding.[17] Similar methods can be employed with even greater effect by government agencies, but this requires a serious effort to understand and appreciate the dynamics of peasant life in the specific area to which government staff are posted. Gooch (1979:8–9) reports the favorable outcome of a training program for civil servants in Upper Volta, designed to sensitize crop-oriented field personnel to the needs of cattle growers and help them find methods of working with the local cattle owners' union.

Training is likely to be ineffectual, however, unless it is reinforced by subsequent experiences, formal or informal, that refresh the original training messages and help the staff members to upgrade their technical and communications skills. The regular scheduling of training sessions, as in the World Bank–sponsored "training and visit" system, helps to ensure that training will actually take place on a predictable basis (Cernea, 1979), though one disadvantage of this regimented system is its inhibition of a two-way flow of information. The practice of human relations skills can be reinforced also by supervisory methods that emphasize the ethic of service to rural publics and that reward both the substance and the style of performance which is responsive to local needs. Management incentives for more responsive staff performance can be strengthened by the influence and activities of local organizations; government technicians and administrators in the field can be expected to work more efficiently and responsively when they must take account of the demands of rural people through their LOs. Thus LOs can provide a decisive supplement to the government's incentive and reward system.

Governments can also help to ensure better staff performance by recasting incentive and reward systems that have been heavily biased in favor of detailed compliance with rules, procedures, and centrally prescribed targets. It is sometimes difficult to maintain high morale in rural areas, far from the urban amenities and contacts with authority to which educated civil servants and technicians aspire. The sense of abandonment by their headquarters, the feeling that they are in dead-end jobs, frequent and apparently random transfers between jobs and locations, the suspicion that they are subject to

[17]The TKRI manual (Heim et al., 1983) even includes a discussion of the value of a sense of humor in dealing with rural publics.

blame by local people for program failures for which they are not responsible—all these can undermine morale and impair the ability and incentives of staff to work effectively with local publics. Low morale combined with weak and inactive local publics contributes to indifferent performance and to the petty corruption so pervasive in rural public administration and so corrosive to public confidence in government. In many countries these negative practices are so deeply institutionalized that changing them will require a determined and sustained effort by government (Wade, 1982b). As we indicate below, staff training activities and management innovations can be facilitated by development assistance agencies.

Rewards to field staff in the form of promotions, pay increases, and recognition should emphasize successful performance in working compatibly with local publics and in achieving rural development objectives. It is doubtful, however, that training alone, or training plus supervisory and procedural reforms, or training and procedural reforms combined with improved staff incentives and rewards will be sufficient to reorient the behavior of bureaucratic agents in rural areas—necessary as all these measures certainly are. Given the distance between government policy-makers and program designers on the one hand and field personnel on the other, and given the frustrating conditions under which field personnel must often work, improved management methods are likely to be fully effective only if they are buttressed by the continuing activity, influence, and pressure of organized publics.

Reinforced Accountability and Internal Discipline. While it is important to avoid imposing formal procedures on local organizations, government can help to maintain the confidence of members by enforcing standards of probity, by periodically auditing accounts, and by otherwise helping to foster and preserve the accountability of officers to the body of members, once local organizations exceed the size at which informal social controls can perform this function. The accountability of leaders and employees, especially in larger organizations and in federated units that control substantial resources, should be both downward to their members and upward to government. In this way, government can help to protect the interests of members and the integrity of their organization when leaders and employees are tempted to indulge in corrupt practices.

Here government must draw a fine line between too much and too little—there should be sufficient oversight to prevent and correct serious abuses, but not at the price of domination or of preventing members from learning from their mistakes and solving their own problems. It is better that some local organizations should fail than that their survival should depend entirely on the patronage and protection of government. With learning, new and more effective LOs can succeed those that fail. Government assistance should therefore be predicated on practical self-help by local organizations and their members. Governments should maintain businesslike relationships, especially when financial considerations are involved. It is especially important to avoid subsidized credit or production inputs which tend to be preempted by local elites and officials and are associated with high rates of default. Failure by governments to enforce financial obligations against LOs contributes to dependence and eventual failure.

Governments can help to support the internal discipline of local organizations that are threatened by forces beyond their members' control. The police and judicial machinery can be made available to enforce the collection of loans when social control fails. In the case of the Puebla farmers' groups in Mexico, reported at the end of Chapter 6, the ability to invoke state sanctions against loan defaulters gave the organizations a boost in membership support. Governments can make membership a condition for certain benefits, when free ridership would inflict an unfair burden on the organization and its members. Official initiative and support may be especially useful when the local power structure and social norms are unfriendly to local organization among disadvantaged groups. Governments can help local organizations, especially those among socially and economically depressed groups, to vindicate the rights of their members to minimum wages, tenant security, or public services to which they are entitled by law, and to protect the organizations themselves from the pressures of hostile interests when the latter resort to physical or economic intimidation (Espiritu and Dias, 1982; Rahman, 1983). Intervention by the Mexican government to protect the rights of *ejiditarios* to irrigation water is a case in point (Hunt and Hunt, 1974:149–50). Governments can provide a climate of support for LOs to demonstrate their legitimacy, and compel local officials and local elites to recognize and deal with them—so long as they too work within the law.

Where governments are hostile to local organization, because they support local elites or prefer atomized rural communities, none of these prescriptions will be feasible. Such conditions do prevail in some countries. Our study indicates, however, that the image of the monolithic regime conspiring with rural elites in the systematic exploitation and repression of the rural poor hardly describes the orientation of most governments. Among regimes with widely different ideologies and political structures, rural membership organizations are tolerated and often encouraged. In a number of cases, governments do attempt to maintain tight control over such organizations and limit the scope of their activities, but others permit considerable opportunity for self-determination, and some governments positively support and foster LOs, so long as they do not threaten the regime. Our suggestions, based on our examination of the experience of successful local organizations in many countries, are directed to the growing number of governments that recognize the importance of self-help and collective effort to the development of their growing rural societies and to the effectiveness of the public services that they provide.

What Can Development Assistance Agencies Do?

Bilateral and multilateral assistance agencies usually work in developing countries in direct support of government-sponsored projects and programs. Voluntary agencies are sometimes allowed to work outside the framework of government sponsorship, but only with the acquiescence of governments. During the past decade, development assistance agencies have shifted their interest and their priorities away from urban infrastructure and industry in the direction of agricultural and rural development. This trend in donor interest is likely to continue as both governments and donors confront the prospects of chronic food shortages, continuing poverty, and the unremitting growth of the labor force in rural as well as urban areas.

Since local membership organizations are associated with both increased productivity and improved welfare, they are likely to attract greater interest among donors as one component of successful rural development strategy.[18] Donors are inclined to support local organizations not as ends in themselves but as instruments for the success of such development projects as community irrigation, smallholder credit, employment generation for the landless, or improved health facilities.

Donors can use their influence with governments to emphasize the importance of local membership organizations for general rural development and for specific development activities. They can help with the formulation of projects that incorporate local organization and ensure that the designs of specific LOs are consistent with successful experience. Donors can encourage governments to maintain an appropriate time perspective, recognizing that the building of new institutions and of local action capabilities on a sound foundation is a learning and adaptive process for all parties involved and that the expectation of early dramatic results may undermine the process of institutional development (Esman, 1972). Through the monitoring of specific projects, comparative research, and support for institutions that systematically build knowledge, provide specialized training, and make expert consultants available, donors can help to build a body of operational doctrine and practice that will reduce the risk of failure and increase the probability of success for local organizations oriented to rural development (D. Korten, 1980).

Donors can contribute resources to development projects that include a

[18]Cernea (1982) cites a survey of World Bank–sponsored agricultural projects which indicated that 40 percent included rural local organizations, not because of policy but because the World Bank professionals judged that they were essential to project success.

local organization component. One common form of external assistance is training facilities for the leaders, members, and staff of local organizations; the World Bank, for example, has provided support for such facilities for water users' associations in Indonesia and the Philippines. Governments are seldom inclined to be generous with training opportunities even for their own civil servants, much less for nongovernmental personnel. It is not that they oppose training but that they are reluctant to assign scarce resources to this purpose. This is an area where the intervention of international donors can be beneficial at relatively low cost. The development of practical and appropriate training materials and methods in conjunction with a governmental or nongovernmental training and research center can contribute to both leadership and staff development. A good example is the USAID-supported Gal Oya project in Sri Lanka, where training is provided by the local Agrarian Research and Training Institute (Uphoff, 1982b).

There can also be donor assistance for technical support and guidance to local organizations. The tasks of assisting LOs are not yet so documented or predictable that routine efforts are likely to suffice. This means that serious and sustained efforts—involving experimentation, continuous evaluation, and innovation—are needed. Government departments seldom have enough experienced and qualified staff to be able to promote these learning processes without the help of some added personnel who have more than routine skills. A government agency working with LOs should not, however, be displaced by external assistance. Its active involvement with the experimental aspects of the enterprise is vital, so that its staff acquire not only knowledge and skills but also commitment which will carry the effort through. Technical assistance—from the donor agency itself, from PVOs specializing in organizing skills, from universities or research institutes having interdisciplinary expertise, or from experienced individuals available as consultants—should be treated as an investment but certainly not as a substitute for the participation of the relevant line agencies.

The behavior of civil servants who interact with local organizations in the field is an important factor. That behavior must usually be reoriented, as we have indicated, so that public services can be more sensitive to the needs of organized local publics (D. Korten and Uphoff, 1981). Development assistance agencies can help governments to strengthen staff training activities and to design and support changes in administrative structures, rules, and procedures that increase the responsiveness of government staff to organized publics and strengthen the links between them. Such efforts at bureaucratic reorientation cannot be achieved rapidly or on a wholesale basis. For that reason they are best limited initially to agencies responsible for specific donor-assisted programs. When such pioneering efforts encounter obstacles resulting from government-wide policies, traditions, or methods, development assistance agencies can help to negotiate exceptions that permit the introduction and testing of innovations. They can then support the general adoption of changes that demonstrate good results.

Donor assistance to LOs may require fundamental revisions in the procedures and criteria by which assistance agencies design, review, and implement rural development projects. As Rondinelli (1982) has demonstrated, the leading international donors have adopted engineering concepts of project design and management derived from the experience of manufacturing and large construction industries. Projects must be justified by elaborate analytical exercises based on a priori economic calculations even when reliable data are not available; they incorporate rigid quantitative and time-phased financial and output targets; their management methods are highly centralized and control oriented; and the affected publics are seldom if ever consulted. Unfortunately, donor agencies often foster similar rigidities among the planners and project managers of the governments they are assisting. This style of project design and management is manifestly unsuited to rural development projects, which tend to be characterized by pervasive uncertainty, varied conditions, and changing circumstances.

Rural development projects, particularly those that incorporate LOs, should instead be treated as action hypotheses, subject to continuous testing and modification while project managers and the affected publics learn from experience. Such projects are not a matter of transferring known techniques but of attempting to adapt existing, contingent knowledge to social structures and relationships that are only imperfectly understood and are not passive objects for transformation. This calls for managerial flexibility and a spirit of experimentation incompatible with the rigid, control-based methods that the major donor agencies employ to justify and implement projects.[19]

Donor agencies must prudently subject the projects they are asked to support to technical, economic, social, and administrative tests of feasibility in order to reduce the risk of failure and to direct scarce resources to the most promising efforts. But when working with rural development activities and local membership organizations, their criteria must be attuned

[19]Rondinelli puts it this way: "Under these conditions, the most valuable managerial skill is not necessarily the ability to conform to preconceived project plans or to networking charts and project schedules, but the ability to innovate, experiment, modify, improvise, and lead—talents that are often discouraged or suppressed by rigid designs and centrally controlled management procedures. What leads to success is the ability of managers to design and manage simultaneously, and to test new ideas and methods continuously no matter what the circumstances under which they find themselves. This managerial flexibility, however, is often squashed by officials in headquarters of international agencies or national ministries who insist on conformance to detailed plans and rigid management systems. In such situations, the major criterion of success for many project managers turns on their ability to conform to plans or programs designed in aid agency or national ministry headquarters, rather than on their ability to seize local opportunities in order to achieve a project's purposes or to modify goals to reflect changing or unanticipated conditions" (1982:52).

to the uncertainties inherent in promoting institutional development and behavioral change among populations whose needs are complex and whose dynamics are not well understood. When bureaucratic reorientation is also involved, strategies must be aptly and subtly conceived, and results will be represented by leaps and lags rather than schedules of "progress." Donor agencies, by appropriate revisions of their own review criteria, can help governments to recognize the inevitably experimental dimensions of most rural development projects, to encourage and reward managers for timely and flexible adjustments, rather than seek adherence to predetermined methods and targets. The best use of scarce donor resources will usually entail refined responses to local differences and, above all, involve project beneficiaries through their local organizations in continuous information exchange, negotiation over resource mobilization and allocation, and shared responsibility for the management of project activities. This implies, it should be clear, that donor agencies must embark on their own "bureaucratic reorientation."

Institutional Development

During the 1960s and 1970s, the governments of developing countries, often with foreign aid, made large investments in educational, research, and service delivery institutions and in infrastructure related to agriculture and rural development. These investments and the annual expenditures to operate and maintain them reflect the experience of every country that has achieved dynamic agricultural development during this century. Relatively small-scale production units can be highly productive, but they need continuous appropriate technological innovation. Further, to compensate for the natural and economic risks involved in farming, governments find it necessary to provide large and continuing networks of supportive services. If small-scale farming, including farming under less than favorable agronomic conditions, is to be socially and economically viable, governments must orient public policies and services to its distinctive needs, region by region and commodity by commodity. As off-farm employment becomes a priority for rural areas, similar networks of state-sponsored institutions will be required to stimulate and support nonagricultural rural enterprises.

Such investments and expenditures are likely to yield low returns, however, unless their intended beneficiaries are brought together in counterpart organizations that enable them to benefit from favorable public policies and services, contribute to their costs, participate in their management, promote and defend their common interests in relation to other organized groups, and influence the action of government on their behalf (Whyte and Boynton, 1983). Governments, especially those that face severe limitations in funds and in trained manpower, cannot manage rural development without the participation of rural people of all classes. This participation requires social organization. Establishing successful local organizations that can make an effective contribution to broadly based rural development requires a sustained effort; such institutions cannot be imposed or brought forth by standardized, centrally imposed blueprints, but must evolve inductively from the specific experience, needs, and preferences of their members. Moreover, reasonable provisions must be made to accommodate the social learning and adjustments in structure and procedures that are essential to the elaboration of successful institutions.[20] Governments must be prepared to tolerate considerable pluralism and to work with diverse forms of local organization so long as they reflect the genuine needs and preferences of their members.

Governments and international donors interested in promoting rural development cannot confine their efforts to economic policy, infrastructural investments, and technological assistance. Institutional and human resource development is also an essential component. Viable membership institutions must be cultivated—preferably by local initiative, where possible by voluntary agencies, where necessary by government or a combination of all three. Except in conjunction with locally organized publics, the costly networks of agricultural and rural service institutions in which governments and donor agencies have invested during the past two decades will not yield commensurate economic and social returns.

Our 1974 study showed the association at the macro level of local organization with successful rural development. The present study, examining local organization at the micro level, has confirmed and elaborated the conclusions reached in the earlier study. It has disaggregated the concept of local organization and indicated which structural and operational features are associated with various dimensions of successful performance under different environmental conditions. It has indicated that while it is harder to form and sustain viable organizations among poorer and weaker rural publics than among better-endowed groups, such organizations have been successful in a variety of natural, political, and social environments. It has identified policies, methods, and lines of action that appear advisable and others that should generally be avoided by sympathetic governments, voluntary agencies, and development assistance organizations. While these findings should be refined and modified by further research, we are confident that they trace the features of local membership organizations that can

[20]At the project level one can use a monitoring method called "process documentation" to permit timely learning from ongoing experience and consequent adjustments in the structure and procedures of local organizations (see Chapter 8; and D. Korten, 1980, and F. Korten, 1982).

be successful and of strategies for promoting their role in rural development.

The authors are sympathetic to rural local organizations because LOs can help the diverse publics that constitute the rural majority to improve the quality of their lives and gain greater control over their environment. They can help to achieve greater efficiency, equity, and empowerment for rural people at relatively low cost to government. We have, however, carefully avoided the populist fallacy that ascribes only good motives to the poor and malign intent to those in positions of power. These stereotypes distort the reality both of rural publics and of governments and public bureaucracies in most countries. The rural poor and their leaders are no less vulnerable to egoism than individuals and organizations in other classes of society. Governments are not inherently benign but neither are they invariable exploiters of the rural poor. Market mechanisms are not necessarily antithetical to the interests of the rural poor, nor are public bureaucracies irremediably corrupt or incompetent. Philosophies and policies based on such categorical premises confuse and distort the complex realities of rural society and of government operations. They are of little use in shaping viable rural development strategies or in providing practical guidance for local organizations.

Our theoretical orientation, while recognizing the severe institutional handicaps encountered by disadvantaged rural publics, assumes substantial degrees of freedom both for government and for the rural populace. Our statistical analysis and our observations confirm the proposition that neither is fully determined or constrained by the rules of the system; the limitations and opportunities they face vary among and within countries and need to be empirically identified rather than deductively assumed. The challenge for sympathetic governments and donor agencies is to discover what is possible, what is likely to work for specific rural groups in specific situations.

In most developing countries local membership organizations can and do function. If they are to function effectively, reciprocity and exchange must supplant the prevailing authoritarian and command methods that governments and bureaucratic agencies commonly use in their relations with the rural poor (Uphoff, 1983b). A more balanced relationship does not mean that governments should cease to govern or that they should relinquish leadership over public policy in any area, including rural development. On the contrary, an exchange strategy enables governments to increase the efficiency and the beneficial impact of their public services by encouraging and supporting the mobilization of the latent skills, resources, and commitments of rural publics and joining them with the efforts of government. This can be accomplished only by conceding a degree of self-determination to rural organizations that enables them to act in the authentic interest of their members. While the mobilization of new interest groups in rural areas may be perceived as a threat by privileged elements, including some civil servants, and may therefore provoke conflict, a supportive regime stands to benefit greatly by extending its base of political support through organizations that represent the rural majority.

We are persuaded that the recommendations emerging from this study are both politically and administratively feasible. Development assistance agencies and, increasingly, officials in developing countries can profit from strategies that benefit the rural majority through local membership organizations not only because they may be normatively desirable but because they are economically rational and politically wise.

Afterword

We began in our preface with comments by a diverse set of observers on the empirical contributions of local organizations to development. We would add, as a postscript, the prescriptive conclusions of two writers who have studied Latin American and Asian development experience extensively. Keith Griffen and Ajit Kumar Ghose (1979:382–83) advance the following summary views which converge with our own conclusions on the role and possibilities of rural local organizations:

> The centralized planning procedures existing at present are biased in favor of large projects designed to high engineering standards, amenable to sophisticated cost-benefit analysis and perhaps suitable for foreign financing. Anti-poverty focused rural development, however, requires a great many small, labor-intensive investment projects dispersed throughout the countryside. . . . [This] implies that plans for rural development should be made locally, by those who will implement them, benefit from them and bear the major cost of them.
>
> Government, of course, has a major role to play in helping rural people to plan their own activities, to articulate their demands and to become organized. The poor are unlikely to organize themselves spontaneously; some outside stimulus normally will be necessary and government can assist the process by at least being tolerant of rural peasant movements and better still by encouraging and supporting local organizations.
>
> Under favorable conditions a large number of such organizations could emerge. Some may form around a piece of technology, as with irrigation cooperatives; some may be interest groups, including women's organizations; and some may be class-based groups, such as unions of landless laborers or plantation workers, small farmers associations, etc.
>
> Ideally, these organizations should have their own funds and be independent of government yet have ready access to it. The government in turn, should regard local organizations as partners in rural development, sources of ideas, places where priorities can be hammered out, recruitment grounds for "barefoot" doctors and engineers, peasant leaders and local government personnel. In this way it should be possible to approach the ultimate objective in which planning for the poor becomes planning by the poor.

APPENDIX A

Listing of Cases

Name	Country	Type/Score
Agricultural Cooperatives	Thailand	CO 1
Agricultural Cooperatives	Tunisia	CO 1
AMUL Dairy Cooperative	India	CO 5
*Anta Cooperative	Peru	CO 3
ASAR/ARADO Potato Production and Seed Improvement Association	Bolivia	IA 4
Association of Agricultural Credit Users (AUCA)	Paraguay	CO 2
Ayni Ruway Village Committees	Bolivia	LDA 5
*Baglung Bridge Construction Committees	Nepal	IA 5
Bagoy Irrigation Association	Philippines	IA 4
Bakel Farmer Association	Senegal	IA 3
Ballovpur Midpara SFDP Group No. 1	Bangladesh	IA 5
Ballovpur Eastpara SFDP Group No. 2	Bangladesh	IA 1
*Banes Irrigation Association	Philippines	IA 2
Bangladesh Rural Advancement Committee (BRAC)	Bangladesh	LDA 4
*Bani Awwam Local Development Association	Yemen	LDA 3
*Banki Water Supply Project Committees	India	IA 4
Bara Village Development Committee	Malagasy	LDA 1
Barangay General Luna SFDP Group No. 3	Philippines	IA 4
Barangay Liberty SFDP Group No. 1	Philippines	IA 3
Barangay Liberty SFDP Group No. 4	Philippines	IA 4
Barpali Leatherworkers' Cooperative	India	CO 5
Barpali Vegetable Farming Cooperative	India	CO 3
Barpali Weavers' Cooperative	India	CO 1
Bawer Sagidan Irrigation Association	Philippines	IA 3
Benduguba Association of Village Cooperatives	Mali	CO 2
Bhoomi Sena	India	LDA 5
Boiteko Women's Self-Help Groups	Botswana	IA 2
Borehole Syndicates (Kgatleng District)	Botswana	IA 2
Buba Tombali Water Supply Committee	Guinea Bissau	IA 3
Carrizo Valley Ejidos	Mexico	CO 2
Cauca Rural Development Project Committees	Colombia	IA 1
*Centers for Social and Economic Development (DESEC)	Bolivia	LDA 5
Chipko Movement	India	LDA 5

*Cases included in Appendix C.

(*continued*)

Name	Country	Type/Score
Coffee and Cocoa Marketing Associations	Jamaica	CO 1
Community-Based Integrated Rural Development Committees	Korea	LDA 3
Coffee Cooperatives (Murang'a)	Kenya	CO 3
Community Development Societies (Murang'a)	Kenya	LDA 3
Confederacion Nacional de Campesina-Taretan	Mexico	IA 3
Cooperative for Rural Area Planning	Benin	CO 3
Cooperative Movement	Zaire	CO 2
Cultivation Committees	Sri Lanka	IA 3
Dana Sehat Irrigation Committee	Indonesia	IA 4
Daudzai Irrigation Associations	Pakistan	IA 4
Dhulia Women's Movement	India	IA 4
Fakirakanda Fishermen SFDP Group	Bangladesh	IA 1
Fakirakanda Marketing SFDP Group	Bangladesh	IA 3
FAO Fertilizer Use Cooperatives	Ghana	CO 3
Farm Mechanization Groups	Lesotho	IA 2
*Farmers' Associations	Taiwan	IA 5
*Federacion Campesina de Venezuela (FCV)	Venezuela	IA 5
Federation of Free Farmers	Philippines	IA 3
Federation of Regional Agricultural Cooperatives	Guatemala	CO 2
Gojjam Peasant Associations	Ethiopia	LDA 3
Gondo Cotton Marketing Cooperative	Uganda	CO 1
Gwarzo Farmers' Cooperatives	Nigeria	CO 2
Hanover Street Women's Cooperative	Jamaica	CO 2
Health Committees	Guatemala	IA 2
Heenpitagedera Credit Union	Sri Lanka	CO 4
Henna Project Peasant Association	Ethiopia	LDA 4
IBRD Agricultural Development Project Committees	Gambia	CO 3
Illam-Charali Road Committee	Nepal	IA 3
Jamaica Agricultural Society (JAS)	Jamaica	IA 2
Jharkand Mukti Morcha	India	LDA 4
Kagawasan Movement	Philippines	LDA 5
Kenya Tea Development Authority (KTDA) Committees	Kenya	IA 5
Khamer Local Development Association	Yemen	LDA 4
Khari Irrigation Association	India	IA 3
Khet Mazdoor Union	India	IA 3
Kou Valley Cooperative	Upper Volta	CO 2
Krishi SFDP Groups	Bangladesh	IA 4
Kweneng Rural Development Association	Botswana	LDA 2
Lampang Health Committees	Thailand	IA 3
Land and Food for People Project Committees	Jamaica	CO 2
Laur Irrigator Service Association	Philippines	IA 5
Leribe Pilot Agricultural Committees	Lesotho	IA 2
Lirhembe Multi-Service Cooperative	Kenya	CO 4
Los Pinos Settlement Groups	Costa Rica	IA 4
Mabati Self-Help Water Committees	Kenya	IA 3
Maghlaf Local Development Association	Yemen	LDA 2
Mahi-Kadana Irrigation Associations	India	IA 3
Maharashtra Village Development Committees	India	LDA 4
Malaking Arado Irrigation Associations	Philippines	IA 4
Manyu Oil Palm Development Cooperatives	Cameroon	CO 3
Market Women's Cooperatives	Nicaragua	CO 4

*Cases included in Appendix C.

(*continued*)

Name	Country	Type/Score
Mmangkodi Farmers' Association	Botswana	IA 4
*Morningside Delightful Buying Club	Jamaica	CO 1
Mothers' Club of Doo Kok Li	South Korea	IA 3
*Mothers' Club of Oryu Li	South Korea	IA 5
Mothers' Club of Wae Am Li	South Korea	IA 1
Movement for the Development of the Community of Le Borgne (MODECBO)	Haiti	CO 3
Mraru Women's Society	Kenya	IA 4
Multipurpose Cooperative Societies	Sri Lanka	CO 2
National Union of Plantation and Agricultural Workers (NUPAW)	Zambia	IA 4
Ngok Dinka Consumer Cooperatives	Sudan	CO 1
Ngok Dinka Group Farms I	Sudan	CO 2
Ngok Dinka Group Farms II	Sudan	CO 1
Njangi Rotating Credit Society	Cameroon	CO 3
Nogar Associations	Nepal	CO 3
North Clarendon Processing Company	Jamaica	CO 2
Nso Women's Cooperative	Cameroon	CO 4
Paddy Processing SFDP Group	Bangladesh	IA 1
Pelebo Health Committee	Liberia	IA 4
Penny Foundation–Sponsored Committees	Guatemala	CO 3
Plan Maize Farmer Groups	Mexico	IA 4
Portland–Blue Mountain Coffee Cooperative Society	Jamaica	CO 5
*Puebla Farmer Committees	Mexico	IA 4
Regional Union of Agricultural and Marketing Cooperatives (URCOMAYA)	Upper Volta	CO 2
Rice Noodle Cottage Industry Women's Cooperatives	Malaysia	CO 3
Rickshaw Pullers SFDP Group No. 1 (Digharkanda)	Bangladesh	IA 4
Rickshaw Pullers SFDP Group No. 2 (Digharkanda)	Bangladesh	IA 1
Rural Development Societies	Sri Lanka	LDA 3
Saemaul Undong	Korea	LDA 5
San Antonio Irrigation Association I	Philippines	IA 3
San Antonio Irrigation Association II	Philippines	IA 2
*San Jose Credit Cooperative	Philippines	CO 4
*San Luis Reconstruction Committees	El Salvador	LDA 2
Santa Valley Cooperatives	Peru	CO 3
Sarvodaya Shramadana Movement	Sri Lanka	LDA 4
*Self-Help Water Supply Committees	Malawi	IA 5
Shabgram SFDP Group No. 5	Bangladesh	IA 4
Shabgram SFDP Group No. 2	Bangladesh	IA 3
Sashemene Peasant Associations	Ethiopia	LDA 3
*Sidamo Mahabar Associations	Ethiopia	LDA 4
*Small Farmer Cooperatives	Ecuador	CO 5
Small Farmer Development Project Groups	Nepal	IA 5
Social Interest Agricultural Society (SAIS)	Peru	CO 2
*Subak Irrigation Associations	Indonesia	IA 5
Sukhomajri Water Users Association	India	IA 4
Sugar Production Cooperatives	Peru	CO 3
Sukuma Cattle Cooperative	Tanzania	CO 2
*Sukuma Cotton Cooperative	Tanzania	CO 5
Sukuma Credit Cooperative	Tanzania	CO 3
Sukuma Fish Cooperative	Tanzania	CO 1
Swanirvar Movement (Jessore)	Bangladesh	LDA 2

*Cases included in Appendix C.

(*continued*)

Name	Country	Type/Score
Taiwan Irrigation Associations	Taiwan	IA 5
Tambon Yokkrabat	Thailand	IA 4
Thana Irrigation Program	Bangladesh	IA 3
Tiv Bams Saving and Credit Associations	Nigeria	CO 4
*Tiv Farmers' Associations	Nigeria	IA 5
Ujamaa Village Committees	Tanzania	LDA 2
Ulashi-Jadunathpur Self-Help Project Committees	Bangladesh	LDA 3
Union of Bougouriba Village Groups	Upper Volta	LDA 1
United Workers Ltd. of Tucuman	Argentina	CO 3
Vicos Project Committee	Peru	LDA 4
Village Development Committees	Botswana	LDA 3
*Village Health Care Teams	Niger	IA 1
Village Level Irrigation Groups	Senegal	IA 3
Village Patrols	Senegal	IA 1
Women's Vegetable Cooperative	Gambia	CO 3
Zanjera Irrigation Groups (Ilocos Norte)	Philippines	IA 5

*Cases included in Appendix C.

APPENDIX B

Protocol for Coding Cases

Overall Performance = summary assessment of the LO's performance of tasks and its ability to improve the lot of its members, especially of poorer members:

5 = outstanding (multiple successes, no significant failures)
4 = very good (a number of successes or a few solid successes, no significant failures)
3 = good (some successes, few if any failures)
2 = poor (few successes, some obvious failures)
1 = very poor (no real successes, multiple failures)

Tasks

Planning and Goal-Setting: assessments of community or group needs and of various problems, means, and strategies; formulation of plans to deal with needs and problems.

Conflict Management: efforts to resolve conflicts within community or organization, to facilitate production or maintain social harmony.

Resource Mobilization: gathering community resources for development effort, or gaining resources from outside sources through LO effort.

Resource Management: efficiency and correctness in resource use, including financial, organizational, and natural resource management.

Provision of Services: delivery or distribution of services, either those of the LO or from outside sources with LO involvement.

Integration of Services: coordination of services, either LO's or outside sources, so that they most efficiently and effectively meet members' needs.

Control of Bureaucracy: efforts to make government staff work harder, more flexibly, and more cooperatively with and for members to ensure attendance at office, field visits, lack of corruption, etc.

Claim-making: efforts to get government decision-makers to deal with community problems and needs; may include getting rules altered, budget allocations changed, etc.

5 = quite effective
4 = effective
3 = average
2 = ineffective
1 = quite ineffective
0 = not relevant or too little information to score*

Structure

Functions: number of kinds of activities; not the same as operations, since a single function may have several operations.

5 = comprehensive (half a dozen, many sectors)
4 = multiple (more than three, many sectors)
3 = multi (at least three, more than one sector)
2 = dual (two functions, one sector)
1 = single (one function, one sector)

Formalization: characteristics such as written constitution and by-laws, government recognition and regulation, and other characteristics of "modern" institutions; opposite to what is normally referred to as "traditional" organization involving community-established roles and sanctions.

5 = highly formal
4 = formal (may have some informal features)
3 = quasi-formal
2 = informal (routine operation largely traditional)
1 = highly informal

Government Linkage: interaction between government staff and LO in terms of frequency of communication and cooperation, as well as government control over resources LO depends on. Score reflects degree of LO autonomy from government control.

5 = directed (government controls, no reciprocity)
4 = high (high interaction, some reciprocity)
3 = moderate (interaction, though not regularly)
2 = low (little interaction)
1 = autonomous (no interaction)

Vertical Linkage: frequency of interaction between base-level organization and higher tiers, in which the highest level is still a representative organization and not a government agency.

*This option for scoring applied to all variables.

5 = national federation (of LOs)
4 = loose federation
3 = three tiers (usually to district/regional level)
2 = two tiers
1 = no linkage

Horizontal Linkage: interaction between and among base-level organizations.

5 = network (regular cooperation among LOs)
4 = extensive
3 = frequent
2 = sporadic
1 = no linkage

Initiative: who initiated efforts to set up the LO.

5 = outside (international agency/NGO; catalyst mode)
4 = government (bureaucratic mode)
3 = shared (with government or outside agency)
2 = leaders (community leaders)
1 = members (community effort)

Economic Composition: homogeneity or heterogeneity of membership in terms of economic status.

5 = great heterogeneity
4 = relative heterogeneity
3 = moderate heterogeneity
2 = low heterogeneity
1 = relative homogeneity

Social Composition: homogeneity or heterogeneity of membership in terms of social status, ethnicity, religion or other characteristics.

5 = great heterogeneity
4 = relative heterogeneity
3 = moderate heterogeneity
2 = low heterogeneity
1 = relative homogeneity

Sex Composition: gender of members.

5 = all female (over 95%)
4 = mostly female (60–95%)
3 = mixed
2 = mostly male (60–95%)
1 = all male (over 95%)

Incentives: basis for membership, both taking up and maintaining it, including decision to join, contributions and dues, attendance at meetings, etc.

5 = voluntary (completely—no requirements)
4 = quasi-voluntary (entry voluntary—some requirements to maintain membership)
3 = mixed (combination of rewards and sanctions)
2 = quasi-compulsory (some pressure to join—some sanctions)
1 = compulsory (membership required—strong sanctions)

Decision-Making Structure:

5 = assembly
4 = assembly plus committee(s)
3 = committee(s)
2 = executive plus committee(s)
1 = executive

Normative Orientation: climate within LO supporting or not supporting egalitarian/participatory operation.

5 = highly egalitarian/participatory norms
4 = egalitarian/participatory norms
3 = neutral (indifferent)
2 = inegalitarian/participatory norms
1 = highly inegalitarian/participatory norms

Size: membership in the base-level organization (such an organization may be federated at a higher level).

5 = very large (membership over 1,000)
4 = large (500–1,000)
3 = medium (100–500)
2 = small (50–100)
1 = very small (under 50)

Environment

Topography: terrain, rivers, valleys or other natural features which facilitate or hinder LO operation.

5 = very favorable
4 = favorable
3 = neutral
2 = unfavorable
1 = very unfavorable (e.g., very mountainous terrain)

Resource Endowment: soil, climate, and other such features facilitating/hindering agricultural and other production.

5 = very favorable
4 = favorable
3 = neutral
2 = unfavorable
1 = very unfavorable (e.g., arid or semi-arid conditions)

Infrastructure: road network, communication facilities, and other support features.

5 = very favorable
4 = favorable
3 = neutral
2 = unfavorable
1 = very unfavorable (essentially no physical infrastructure)

Income Level: estimates of per capita income, according to LO source material or general reference material.

5 = very high ($1,000+)
4 = high ($400–$1,000)
3 = middle ($200–$400)
2 = low ($100–$200)
1 = very low ($50–$100)

Income Distribution: concentration of income within the community, reflecting in some degree the proportion of landlessness and land distribution in general.

5 = high inequality (small percentage receive large income)
4 = substantial inequality
3 = definite inequality
2 = modest inequality
1 = relative equality (moderate differentials in community)

Diversification: structural differentiation of the economy between agricultural and industrial activities.

5 = under 20% agricultural
4 = 20–40% agricultural
3 = 40–60% agricultural
2 = 60–80% agricultural
1 = 80% agricultural

Settlement Patterns: concentration of people in their habitual living patterns, which may or may not facilitate LO operation.

5 = very favorable (communication easy)
4 = favorable
3 = neutral
2 = unfavorable
1 = very unfavorable

Social Heterogeneity: distribution of common social characteristics such as ethnicity, caste, religion, etc.

5 = heterogeneous (strong or diverse cleavages)
4 = mostly heterogeneous
3 = moderate heterogeneous
2 = mostly homogeneous
1 = homogeneous (few or no cleavages or diversity)

Social Stratification: social mobility and social distance between upper and lower classes, largely based on subjective assessments of the situation.

5 = extreme (status pyramid steep and rigid)
4 = relative (some movement)
3 = average
2 = little
1 = very little

Sex Discrimination: against women; *Social Discrimination:* against social groups, such as castes or races.

5 = very high
4 = high
3 = moderate
2 = low
1 = very low

Literacy: from general reference sources.

5 = over 80%
4 = 60–80%
3 = 40–60%
2 = 20–40%
1 = under 20%

Partisanship: history of competition along party or other factional lines.

5 = very high (history of conflict)
4 = high
3 = moderate

2 = low
1 = very low

Group Patterns: most common basis for grouping people to accomplish work or make claims (there may actually be several in some situations).

5 = associational (voluntary)
4 = patron-client (following)
3 = familism
2 = kinship (clan, tribe)
1 = caste (ascriptive)

Community Norms: prevailing acceptance within the community of egalitarian outcomes and people's participation in decision-making, implementation, etc.

5 = highly participatory (acceptance of wide participation and egalitarian outcomes)
4 = participatory (acceptance of participatory opportunities and egalitarian outcomes)
3 = neutral
2 = nonparticipatory (little acceptance of participation and egalitarian outcomes)
1 = highly nonparticipatory (rejection of participatory goals and action)

Societal Norms: prevailing acceptance within society as a whole of egalitarian outcomes and people's participation in decision-making, implementation, etc. Scored similarly to Community Norms.

Political Support: degree of support or opposition for the LO in this case, indicated by policies and resources provided.

5 = active support
4 = support
3 = indifference
2 = non-support
1 = active opposition

Administrative Support: inclination and ability of the administrative staff to assist the LO. Scored similarly to Political Support.

Performance

Agricultural Productivity and Income: increases in (a) *yield per acre* attributable to LO activity; (b) *area planted* due to LO activity; (c) *diversification* of production; (d) *income gains* through marketing, infrastructure improvement, etc.; or some combination of these.

Nonagricultural Productivity and Income: increases in (a) *output* of nonagricultural commodities/products; (b) *diversification* of production; (c) *employment and income increases,* or some combination of these.
Education: educational programs or services, with outcomes measured in terms of enrollment or literacy.
Health: improved access to health services, improvements in public sanitation, and/or reductions in mortality or morbidity attributable to LO.
Nutrition: improved food intake and nutritional status.
Water Supply: provision of wells or piped supply.
Transportation: provision of roads, bridges, bus service, and other facilities.
Public Facilities: facilities such as community centers, meeting halls, etc.

5 = significant gain
4 = identifiable gain
3 = some gain
2 = no improvement
1 = decline

Income Distribution: relative increases in income for disadvantaged groups in the community.
Asset Distribution: relative gains in control over assets, usually land.
Access to Public Services: relative gains in access to schooling, health care, water supply, etc.

5 = absolute gains (poor gain while rich do not)
4 = relative gains (poor gain)
3 = some gains (for poor but gap may not narrow)
2 = few gains (for poor)
1 = worsening (situation for poor)

Sex Discrimination: reduction in sex prejudice and new opportunities for women in education, employment, etc.
Social Discrimination: reduction in racial or other prejudice and new opportunities for disadvantaged social groups in education, employment, etc.

5 = significant reduction
4 = definite reduction
3 = some reduction
2 = no improvement
1 = worsening

Government Participation: increased role and effectiveness of members in decision-making and resource allocation beyond the community.
Community Participation: increased role and effectiveness of members in decision-making and resource allocation within the community.

5 = significant increase
4 = definite increase
3 = some increase
2 = no improvement
1 = worsening

LO Problems (not scored but entered descriptively).

Resistance: opposition by government or national elites; by groups within the community; or passive or active resistance by the majority of the community.
Subordination: manipulation by the government to serve its interests; suffocation by political or administrative paternalism; bureaucratic competition subordinating LO to an agency; overburdening of LO with too many responsibilities; capture by local elites; or hegemony of continuing leadership.
Internal Divisions: partisanship, factionalism, or between-village competition which is obstructive to the operation of the LO.
Malpractices: corruption, favoritism, or use of the LO for personal advancement.
Ineffectiveness: internal LO problems such as a lack of skills; lack of realistic goals; instability of leadership.
Differential Participation: lack of participation by targeted groups, such as the poor, women, minorities, geographically remote groups.

LO Roles (entered descriptively).

Membership Roles: who is a member of the LO; their initiative in setting up and directing it; their resource contributions; and control over leadership.
Leadership Roles: who are the leaders of the LO; their initiative in setting up and directing the LO; their duties and activities; their favorable and unfavorable characteristics; and maintenance of accountability to members.
Government Staff Roles: which staff are involved with the LO; any initiative in setting up and directing the LO; technical backup; enforcing rules and guidelines; and leadership recruitment and support.
Outside Agency Roles: activities of nongovernmental organizations and private voluntary organizations; initiative in setting up and directing the LO; their resource contributions; and their interaction with the community and with government.

APPENDIX C

Case Summaries

Anta Cooperative, Peru.
Type: Cooperative.
Performance: 3—Good.

Background

The Agrarian Reform Law of 1969 set up a cooperative model for the peasants. The cooperative movement failed, but the land seizures were successfully implemented. Peasants mobilized to claim their land and distributed it to all, whether they were active in the movement or not, whether they were rich or poor. Part of the peasant mobilization was directed to disbanding the government-imposed cooperatives.

Source

Santiago Roca, *Participatory Processes and Action of the Rural Poor in Anta, Peru.* Geneva: ILO, 1980. See also *Development: Seeds of Change.* Rome: Society for International Development, 1981, 1:12–15.

Tasks

Planning/goal-setting	4—Effective
Land-seizing required planning and division of tasks among villages	
Conflict management	4—Effective
Recaptured land divided among all peasants without discrimination	
Control of bureaucracy	3—Average
Claim-making	3—Average

Structure

Functions	1—Single
Land acquisition	
Formalization	4—Formal
Started informally, but required to have General Assembly and Vigilance Committee	
Government linkage	3—Moderate
Vertical linkage	2—Two tiers
Horizontal linkage	2—Sporadic

Initiative	2—Leaders
Economic composition	3—Moderate
Social composition	1—Homogeneous
Sex composition	2—Mostly male
Incentives	5—Voluntary
Decision-making structure	4—Assembly plus committee
Normative orientation	4—Egalitarian
Size	3—Medium
Environment	
Topography	2—Unfavorable
Resource endowment	2—Unfavorable
Infrastructure	2—Unfavorable
Income level	1—Very low
Income distribution	4—Substantial inequality
Diversification	1—Over 80% agricultural
Settlement patterns	3—Neutral
Social heterogeneity	5—Heterogeneous
Indian community versus Ladinos	
Social stratification	5—Extreme
Sex discrimination	4—High
Social discrimination	5—Very high
Literacy	1—Under 20% literacy
Group patterns	4—Patron-client
Community norms	4—Participatory
Societal norms	4—Participatory
Officially government, highly participatory	
Government orientation	5—Active support
Bureaucratic capacity	4—Support
Performance	
Agricultural production/income	3—Some gain
Asset distribution	4—Relative gains
Land distribution among all peasants	
Participation in government decision-making	1—Worsening
Participation in community decision-making	3—Some increase

Problems

Resistance: active resistance from large land owners and their political allies

Subordination: government established unpopular cooperatives, which peasants disbanded

Differential participation: poor generally active in LOs

Roles

Membership: need for subsistence land leads to active participation; richer peasants take advantage of benefits but do not participate in land seizures

Leadership: peasant leaders are those with "higher level of consciousness and experience"; a dynamic group usually takes the initiative in each community, centering on two or three leaders who mobilize the peasants

Government staff: ambiguous position; sometimes they try to fragment peasant organizations, other times legally sanction them

Baglung Bridge Construction Committees, Nepal.
Type: Interest Association.
Performance: 5—Outstanding.

Background

In 1974 a member of the national *panchayat* began a campaign to build bridges, using local technology and voluntary labor. There was initial resistance by government officials who felt the bridges would not be safe. Eventually, the Local Development Department contributed materials (plus a little money) to support the project. A Baglung district committee was formed, outside regular bureaucratic channels, to plan and oversee the project. Local committees did implementation. Sites for bridge construction chosen on the basis of need and practicality, though all sites given consideration. Labor contributed, except for some paid skilled labor (by low caste). Local people also contributed money. The district now approaches complete satisfaction of bridge needs.

Source

Prachanda P. Pradhan, *Local Institutions and People's Participation in Rural Public Works in Nepal.* Ithaca, N.Y.: Rural Development Committee, Cornell University, 1980.

Tasks

Planning/goal-setting	5—Quite effective
District-wide plan, 118 sites considered	
Conflict management	4—Effective
Resource mobilization	5—Quite effective
Large-scale local input of labor and materials	
Resource management	5—Quite effective
Cost-effective use of resources, including indigenous technology	
Provision of services	5—Quite effective
Claim-making	4—Effective

Structure

Functions	1—Single
Formalization	3—Quasi-formal
Formal at top, informal at bottom, outside regular bureaucracies	
Goverment linkage	4—High
Linked with Local Development Department of government	
Vertical linkage	3—Three tiers
Horizontal linkage	1—No linkage
Initiative	2—Leaders
Economic composition	5—Heterogeneous
Social composition	4—Some heterogeneity
Sex composition	3—Mixed
Incentives	4—Quasi-voluntary

Fines for not contributing labor; tradition of public works and labor exchange	
Decision-making structure	3—Executive committee
Normative orientation	4—Egalitarian
Size	3—Medium
Environment	
Topography	1—Very unfavorable
Resource endowment	2—Unfavorable
Infrastructure	1—Very unfavorable
Income level	1—Very low
Income distribution	3—Definite inequality
Diversification	1—Over 80% agricultural
Settlement patterns	2—Unfavorable
Social heterogeneity	4—Mostly heterogeneous
Social stratification	4—Relative
Sex discrimination	4—High
Social discrimination	4—High
Literacy	1—Under 20% literacy
Partisanship	2—Low
Group patterns	1—Caste
Community norms	3—Neutral
Societal norms	4—Participation
Government orientation	4—Support
Bureaucratic capacity	4—Support
Performance	
Transportation	5—Significant gain
62 bridges at 1/4 to 1/8 normal government cost	
Access to public services	3—Some gains
Sex discrimination	2—No improvement
Social discrimination	3—Some reduction
Wages go to low caste craftsmen, other labor is unpaid	
Participation in government decision-making	3—Some increase
Participation in community decision-making	4—Definite increase

Problems

Resistance: government initially did not believe in value of local technology; pilot project proved the worth of this approach

Roles

Membership: local committees mobilize local labor, materials, and techniques

Leadership: M.P. initiated project, became District Committee leader; outside *panchayat* system, but works with *panchayats*

Government staff: Local Development Department contributed materials, assisted the committee; final responsibility for implementation rested with committee

Banes Irrigation Association, Philippines.
Type: Interest Association.
Performance: 2—Poor.

Background

In 1971, through a former barrio captain, the farmers organized an irrigation association, partly in order to receive financial assistance from the National Irrigation Administration. The members did not realize they would need to repay NIA. In addition, its leaders were told that the Farm Systems Development Corporation would organize the farmers. While the association operated, water use was organized; maintenance was the responsibility of those in an area of damage. There were fines on those not participating in association activities. Management of the irrigation system eventually reverted to the council of elders and village government officials.

Source

Romana P. de los Reyes and Ma. Francisca P. Viado, "Profiles of Two Communal Gravity Systems," *Philippine Agricultural Engineering Journal*, 10:14–18.

Tasks

Planning/goal-setting Unable to improve system	2—Ineffective
Conflict management Water stealing only slightly hindered by guards	2—Ineffective
Resource mobilization Operated traditionl system	3—Average
Resource management Imposed some sanctions	3—Average
Provision of services Unable to serve entire area during dry season	2—Ineffective
Claim-making Got NIA loan, though had to turn it down	3—Average

Structure

Functions	1—Single
Formalization	3—Quasi-formal
Government linkage	2—Low
Vertical linkage	1—None
Horizontal linkage	1—None
Initiative	1—Members
Economic composition	3—Moderate
Social composition	2—Low heterogeneity
Sex composition	2—Mostly male
Incentives	4—Quasi-voluntary
Decision-making structure	2—Executive plus committee
Normative orientation	3—Neutral
Size 55 members	2—Small

Environment

Income level	2—Low
Income distribution	3—Definite inequality
Diversification	1—Over 80% agricultural
Social heterogeneity	2—Mostly homogeneous
Social stratification	3—Average
Sex discrimination	2—Low
Social discrimination	2—Low
Literacy	2—20–40% literacy
Partisanship	4—High
Group patterns	4—High
Community norms	4—Participatory
Societal norms	4—Participatory
Government orientation	3—Indifference
Bureaucratic capacity	3—Indifference

Performance

Agricultural production/income	3—Some gains
Income distribution	3—Some gains
Access to public services	2—Few gains
Water	
Participation in government decision-making	2—No improvement
Participation in community decision-making	2—No improvement

Problems

Resistance: competition with Farm Systems Development Corporation
Internal divisions: mistrust
Malpractices: water stealing a big problem

Roles

Membership: 55 poor small farmers organized to get NIA assistance, elected leaders, and were responsible for maintenance of system, especially large-scale repairs; lost interest when NIA only offered a loan
Leadership: initiative for organization from barrio captain; officials elected by members
Government staff: NIA offered a loan

Bani Awwam Local Development Association, Yemen.

Type: Local Development Association.
Performance: 3—Good.

Background

This LDA was formed in response to a national decree. It is a part of the Confederation of Yemeni Development Associations and follows its guidelines for organization and structure. It elects members to the General Assembly and Board of Assembly but does not hold regular meetings. Despite the influx of cash from members of the community working in another country, the LDA has been unable to tap these resources. Limited finances make it difficult to undertake many projects. Residents are very suspicious of mismanagement and corruption among LDA officers, and lack of accountability and inaccessability exacerbate this mistrust. The LDA has built a road and a clinic despite many obstacles.

Source

Jon Swanson, *Local Government and Development in Bani Awwam, Hajja Province*. Working note no. 11. Ithaca, N.Y.: Rural Development Committee, 1981.

Tasks

Planning/goal-setting	3—Average
Conflict management	4—Effective
Resource mobilization	3—Average
Local tax revenue and individual assessments	
Resource management	2—Ineffective
Poor planning, design and construction; possible corruption	
Provision of services	3—Average
Integration of services	2—Ineffective
Control of bureaucracy	3—Average
Claim-making	4—Effective

Structure

Functions	4—Multiple
Formalization	3—Quasi-formal
Government linkage	4—High
Vertical linkage	5—National federation
Horizontal linkage	2—Sporadic
Some cooperation, often ineffective	
Initiative	3—Shared
Economic composition	5—Heterogeneous
Social composition	3—Moderate
Sex composition	1—All male
Incentives	2—Quasi-compulsory
Decision-making structure	2—Executive plus committee
General Assembly elected, administrative board makes most decisions	
Normative orientation	3—Neutral
Size	5—Very large

Environment

Topography	2—Unfavorable
Resource endowment	2—Unfavorable
Infrastructure	2—Unfavorable
Income level	2—Low
Income distribution	2—Modest inequality
Diversification	1—Over 80% agricultural
Settlement patterns	2—Unfavorable
Social heterogeneity	1—Homogeneity
Social stratification	3—Average
Sex discrimination	5—Very high
Social discrimination	2—Low

Literacy	1—Under 20% literacy
Partisanship	4—High
Group patterns	2—Kinship
Community norms	4—Participatory
Societal norms	3—Neutral
Government orientation	5—Active support
Bureaucratic capacity	3—Indifference

Performance

Education	3—Some gains
Health	3—Some gains
Clinic	
Water supply	3—Some gains
Transportation	4—Identifiable gains
Roads	
Access to public services	4—Relative gains
Sex discrimination	2—No improvement
Social discrimination	2—No improvement
Participation in government decision-making	3—Some increase
Participation in community decision-making	2—No improvement

Problems

Resistance: lack of confidence in LDA officers' handling of money

Internal divisions: intervillage/interregional conflict over land, water, and other resources

Roles

Membership: all villagers are members and elect General Assembly and Administrative Board; no move to change leadership, though they are dissatisfied; pay taxes and assessments

Leadership: members of General Assembly and Administrative Board are elected; they initiate projects and collect money

Government staff: CYDA initiated the LDA, supervised elections, and established organizational and structural guidelines; no other assistance

Banki Water Supply Project Committees, India.

Type: Interest Association.

Performance: 4—Very Good.

Background

This was one of three projects in a larger program. In Banki, it was set up in 1962 in seven villages. Heads of local *panchayats* requested the project from the Planning, Research, and Action Institute of Lucknow. The aim was to (a) develop a piped water supply, help the people to own it, and establish sound management practices; (b) demonstrate the benefits of piped water supply; (c) demonstrate the impact of its provision on health. The state government contributed funds for tube wells and an overhead tank; the remainder came from WHO/UNICEF. Initial resistance of local people was overcome, demonstrated by the fact that by 1973, 350 out of 836 families had piped water, and the rest used 42 public stand-posts. A Waterworks Executive Committee was set up in consultation with heads of *panchayats,* consisting of one member from each village. Community involvement was through local surveys and informal "evening sittings." The committee evolved into a body recognized by the government, appointed by the people, and fully responsible for the operation of the system. Through user fees, the system became self-sufficient, as the outside actors withdrew. Its success is demonstrated by high levels of participation and effective management practices.

Source

K. K. Misra, "Safe Water in Rural Areas: An Experiment in Promoting Community Participation in India," *International Journal of Health Education,* 18:53–59.

Task

Planning/goal-setting	4—Effective
Community given complete control over system	
Conflict management	3—Average
Resource mobilization	4—Effective
User fees covered maintenance costs plus savings for emergency fund	
Resource management	4—Effective
Provision of services	5—Quite effective
Control of bureaucracy	3—Average
Claim-making	3—Average

Structure

Functions	1—Single
Piped water supply	
Formalization	5—Highly formal
Elected officers, bylaws	
Government linkage	3—Moderate
Vertical linkage	2—Two tiers
Horizontal linkage	2—Sporadic
Initiative	3—Shared
Economic composition	5—Heterogeneous
Social composition	5—Heterogeneous
53% Hindus, 47% Muslims	
Sex composition	3—Mixed
Incentives	4—Quasi-voluntary
Membership voluntary, but must pay user fees	
Decision-making	3—Executive committee
Normative orientation	4—Egalitarian
Size	2—Small

Environment

Resource endowment	4—Favorable
Infrastructure	3—Neutral
Income level	2—Low
Income distribution	4—Substantial inequality
Diversification	3—40–60% agricultural

26.7% farmers, 27.3% in petty grain business, 26% white collar, 24.8% crafts, 16.9% agricultural labor, 2.8% other

Settlement patterns	4—Favorable
Social heterogeneity	5—Heterogeneous
Social stratification	5—Extreme
Sex discrimination	5—Very high
Social discrimination	5—Very high
Literacy	2—20–40% literacy
70% illiteracy	
Partisanship	3—Moderate
Group patterns	1—Caste
Community norms	3—Neutral
Societal norms	4—Participatory
Government orientation	4—Support
Bureaucratic capacity	5—Active support

Performance

Health	5—Significant gain
Lowered mortality	
Water supply	5—Significant gain
Participation in community decision-making	4—Definite increase

Problems

Resistance: due to misconceptions about costs, water quality, and traditional use of wells; overcome through house-to-house surveys, informal "evening sittings" to explain health benefits, and education program.

Roles

Membership: open to all who pay user fees; members elect the Waterworks Executive Committee

Leadership: *Panchayat* leaders requested the project, participated in planning and managing it; Waterworks Executive Committee primarily responsible fo management, hiring staff, etc.

Government staff: financial contributions for tube wells and overhead tank; vital education program carried out by government paraprofessionals; Planning, Research, and Action Institute of Lucknow and Central Health Education Bureau in India set up project with intent to turn it over to community

Outside agency: WHO/UNICEF donated much of the funding, though eventually withdrew

Centers for Social and Economic Development, Bolivia.

Type: Local Development Association.
Performance: 5—Outstanding.

Background

DESEC, the Center for Social and Economic Development, initiated in 1963, is a private sector program assisting campesinos. The base institution is the "Center," a voluntary association of resident peasants who solve problems collectively. When there are several centers, each with an elected board of directors, they elect representatives to the regional ARADO federation. Each has a leadership council, which sends representatives to the national level. Specific projects at ARADO level are undertaken by special committees, usually for such things as milk production, livestock, etc. DESEC provides technical services, home construction, health services, and agricultural services.

Source

Elliott Morss, et al., *Strategies for Small Farmer Development,* vol. 2, G1–G15. Boulder, Colo.: Westview Press, 1976.

Tasks

Planning/goal-setting	4—Effective
Range of activities for local development	
Conflict management	4—Effective
Resource mobilization	5—Quite effective
Voluntary contributions by peasants and by outside organizations	
Resource management	5—Quite effective
Provision of services	5—Quite effective
Integration of services	5—Quite effective
Claim-making	4—Effective

Structure

Functions	5—Comprehensive
Agricultural, livestock, artisan crafts, housing, education, health services	
Formalization	4—Formal
Government linkage	4—High
Committees linked to agencies providing particular services	
Vertical linkage	5—National federation
200 base groups, 8 regional groups, one national body	
Initiative	5—Outside
Private sector program	
Economic composition	3—Moderate
Social composition	2—Low heterogeneity
Sex composition	3—Mixed
Incentives	5—Voluntary
Decision-making structure	3—Executive committee
Normative orientation	4—Egalitarian
Size	1—Very small
15 members/group	

Environment

Topography	2—Unfavorable
Resource endowment	2—Unfavorable
Infrastructure	2—Unfavorable
Income level	1—Very low
Income distribution	3—Definite inequality
Diversification	1—Over 80% agricultural

Social heterogeneity	2—Mostly homogeneous
Social stratification	4—Relative
Sex discrimination	3—Moderate
Social discrimination	4—High
Literacy	1—Under 20% literacy
Partisanship	3—Moderate
Group patterns	4—Patron-client
Community norms	4—Participatory
Societal norms	3—Neutral
Government orientation	4—Support
Bureaucratic capacity	4—Support

Performance

Agricultural production/income Raised income, demonstration effect of projects	5—Significant gains
Nonagricultural production/income Artisan development	3—Some gains
Education	4—Identifiable gains
Health	4—Identifiable gains
Income distribution	4—Relative gains
Access to public services	4—Relative gains
Social discrimination	4—Definite reduction
Participation in government decision-making	2—No improvement
Participation in community decision-making	5—Significant increase

Problems

Resistance: Catholic Action Group and Christian Democratic Party at first resisted DESEC leader's focus on economic rather than religious activities; able to convince peasants to support DESEC

Roles

Membership: small farmers and landless contributed resources and participated enthusiastically

Leadership: conceived, organized, and operated by Juan Demeure; leaders of local organizations are elected; participate in committees to implement projects

Government staff: government has recognized DESEC and gives financial grants

Outside catalyst: OXFAM funded a potato project; Inter-American Foundation supported a self-help housing project; a commercial bank loan helped finance a rice mill

Farmers' Associations, Taiwan.

Type: Interest Association.
Performance: 5—Outstanding.

Background

Farmers' Associations originally established when Taiwan was a colony of Japan. It is like a profit-making corporation owned by most of the farmers in the township. It monopolizes or dominates such activities as grain processing and storing, marketing, and credit. Profits are redirected to the local economy, especially through agricultural extension. Most activities of the FA are controlled by central government.

Source

Benedict Stavis, *Rural Local Governance and Agricultural Development in Taiwan.* Ithaca, N.Y.: Rural Development Committee, 1974.

Tasks

Planning/goal-setting	4—Effective
Conflict management	4—Effective
Resource mobilization	5—Quite effective
Resource management	5—Quite effective
Provision of services	5—Quite effective
Integration of services	5—Quite effective
Control of bureaucracy	4—Effective
Claim-making	4—Effective

Structure

Functions	4—Multiple
Formalization	5—Highly formal
Government linkage	4—High
Vertical linkage	5—National federation
Horizontal linkage	2—Sporadic
Initiative	4—Government
Economic composition	5—Heterogeneous
Social composition	3—Moderate
Sex composition	2—Mostly male
Incentives	4—Quasi-voluntary
Decision-making structure	2—Executive plus committee
Normative orientation	4—Egalitarian
Size	5—Very large
Averages 4,000–6,000 (though has small agricultural units within it)	

Environment

Infrastructure	4—Favorable
Income level	3—Middle
Income distribution	2—Modest inequality
Diversification	2—60–80% agricultural
Social heterogeneity	1—Homogeneous
Social stratification	3—Average
Sex discrimination	3—Moderate
Social discrimination	2—Low
Literacy	3—40–60% literacy
Partisanship	3—Moderate
Group patterns	3—Familism
Community norms	4—Participatory
Societal norms	4—Participatory
Government orientation	5—Active support
Bureaucratic capacity	5—Active support

Performance

Agricultural production/income	5—Significant gains
Nonagricultural production/income	4—Relative gains
Lending for diversification	
Public facilities	3—Some gains
Income distribution	3—Some gains
Asset distribution	4—Relative gains
Reinforces land reform	
Access to public services	3—Some gains
Government participation	3—Some increase
Community participation	4—Definite increase

Problems

Subordination: manipulated by the state

Roles

Membership: legally, members supposed to control FA; all purchase shares and pay dues

Leadership: competition for elected positions in FA for the status; little real authority; supervisory bodies make most decisions for FA, especially Provincial Farmers' Association

Government staff: Nationalist Party involved in selection of directors and general managers at every level

Federacion Campesina de Venezuela, Venezuela.

Type: Interest Association.
Performance: 5—Outstanding.

Background

The FCV is a national organization of state and local peasant union groups. It is run by an 11-member National Executive Committee, elected every three years by a Campesino Congress. Local unions are organized with a Junta Directiva of officers, standing for election annually. There are 3,500 local unions ranging in size from 40 to several hundred members. The local, state, and national leaders have extensive formal and informal linkages with other parts of the political system, especially political parties. Many leaders hold party positions. Formal linkage with parts of the political system, such as representation of boards and committees, is mandated by law.

Source

John Duncan Powell, *The Role of Federacion Campesina in the Venezuelan Agrarian Reform Process,* Ph.D. diss., published as *Political Mobilization of the Venezuelan Peasant*. Cambridge, Mass.: Harvard University Press, 1971.

Tasks

Planning/goal-setting	4—Effective
Strategy for representing occupational interests	
Conflict management	4—Effective
Kept partisan conflict to a minimum	

Resource mobilization	4—Effective
Political mobilization	
Provision of services	5—Quite effective
Services for land reform beneficiaries	
Integration of services	4—Effective
Control of bureaucracy	5—Quite effective
Due to political ties	
Claim-making	5—Quite effective
Structure	
Functions	3—Multiple
Agrarian reform, political mobilization, credit, marketing	
Formalization	5—Highly formal
Government linkage	4—High
Vertical linkage	5—National federation
Horizontal linkage	2—Sporadic
Initiative	5—Outside
Political party behind FCV	
Economic composition	4—Some heterogeneity
Social composition	4—Some heterogeneity
Sex composition	2—Mostly male
Incentives	5—Voluntary
Decision-making structure	4—Assembly plus committee
Normative orientation	4—Egalitarian
Size	1—Very small
Environment	
Income level	2—Low
Income distribution	4—Substantial inequality
Diversification	3—40–60% agricultural
Social heterogeneity	4—Mostly heterogeneous
Social stratification	3—Average
Sex discrimination	4—High
Social discrimination	2—Low
Literacy	2—20–40% literacy
Partisanship	5—Very high
Group patterns	4—Patron-client
Community norms	4—Participatory
Societal norms	4—Participatory
Government orientation	5—Active support
Bureaucratic capacity	4—Support
Performance	
Agricultural production/income	4—Identifiable gains
Education	5—Significant gains
Vocational schools	
Income distribution	4—Relative gains
Asset distribution	5—Absolute gains

Access to public services — 4—Relative gains
Participation in government decision-making — 5—Significant increase
Participation in community decision-making — 5—Significant increase

Problems

Internal divisions: Potential partisan conflict held back through allocation of seats on executive boards in proportion to partisan support

Roles

Membership: campesinos provide feedback on activities, attend meetings, and participate in FCV activities

Leadership: leaders generally better educated

Government staff: National Agrarian Institute, Banco Agricola, and Ministry of Agriculture linked to FCV

Morningside/Delightful Buying Club, Jamaica.

Type: Cooperative.
Performance: 1—Very Poor.

Background

Two approaches were involved in the formation of this buying club: (1) the Jamaica Agricultural Society (JAS) as nominal sponsor wanted members to buy shares in a farm input buying club, operating as a JAS farmers' store; (2) farmers wanted to purchase shares in a commercial grouping aimed at maximizing dividends. Members would call it a preco-op for tax purposes, would buy from JAS farmers' store, and would learn co-op methods. In 1974, the JAS branch organizer presented the idea, which was initially rejected by the farmers. Eventually, one man set it up, operating out of his home from 1974 to 1977. Another man purchased 5,000 shares in a takeover, put his family in control, and diverted profits. The club split in half over religious issues, and Unity Buying Club was set up by Seventh-Day Adventists, who objected to M/D Buying Club operating on Saturdays. Both clubs are financially solvent, but not contributing much to improving the farmers' lives.

Source

David D. Gow et al., *Local Organizations and Rural Development: A Comparative Reappraisal,* vol. 2, 327–39. Washington: Development Alternatives, 1979.

Tasks

Planning/goal-setting — 2—Ineffective
Conflict management — 1—Quite ineffective
Split into two clubs
Resource mobilization — 3—Average
Raised funds, but no continuing mobilization after establishment
Resource management — 2—Ineffective
Funds go to large wages for dominant family
Provision of services — 3—Average

Structure

Functions	1—Single
Buys and sells farming inputs	
Formalization	5—Highly formal
Government linkage	3—Moderate
Linked to JAS, though little control	
Vertical linkage	1—None
Horizontal linkage	1—None
Initiative	2—Leaders
Economic composition	4—Some heterogeneity
Social composition	2—Low heterogeneity
Sex composition	3—Mixed
Incentives	5—Voluntary
Decision-making structure	1—Executive
Normative orientation	2—Inegalitarian
Voting by number of shares	
Size	2—Small

Environment

Diversification	2—60–80% agricultural
Social heterogeneity	3—Moderate heterogeneity
Social stratification	3—Average
Sex discrimination	2—Low
Social discrimination	2—Low
Literacy	3—40–60% literacy
Partisanship	5—Very high
Group patterns	3—Familism
Community norms	3—Neutral
Societal norms	4—Participatory
Government orientation	3—Indifference
Bureaucratic capacity	3—Indifference

Performance

Income distribution	2—Few gains
Access to public services	2—Few gains
Participation in government decision-making	2—No improvement
Participation in community decision-making	2—No improvement

Problems

Subordination: manipulated by one family; competition with JAS over control of LO

Internal divisions: conflict between resident and nonresident families, and between Seventh-Day Adventists and those of other religions; led to formation of separate club

Malpractices: exploitation by dominant family, whose members were paid huge salaries

Differential participation: no participation by the poor

Roles

Membership: farmers buy shares in the club; they participate little in management, though they can remove an incompetent manager; benefits are the return farmers get on their shares

Leadership: first managing director set up the club, operated it from his home; secretary of JAS branch society was buying chairman; bought out by outside family, operating for personal profit

Outside agency: JAS helped set up club; had representatives within it

Mothers' Club of Oryu Li, South Korea.

Type: Interest Association.

Performance: 5—Outstanding.

Background

This Mothers' Club was founded in 1968 with the purpose of reducing fertility and improving the life of the members' children. Activities expanded beyond family planning to goodwill/productive activities such as a chestnut tree nursery, credit union, day care nursery, sponsoring village dinners, etc. Its main purpose remained the improvement of the lives of their children. The original members were the 10 wealthiest women of the village, but a change in leadership led to opening membership to all. However, present members have accumulated such a large joint fund that it is difficult for new members to deposit the amount of money required; they are allowed to be quasi-members, and can attend meetings and join the credit union. By 1974, the club had 50 full members, and 112 quasi-members. It is run by the leader in a very strict executive manner. The village has had a dramatic increase in economic well-being thanks to the club's activities; 100% of members practice contraception; and the power of women in village affairs has increased.

Source

Dr. Lawrence Kincaid, Hyung-Jong Park, Kyung-Kyoon Chung, and Chin-Chuan Lee, *Mothers' Clubs and Family Planning in Rural Korea: The Case of Oryu Li.* Honolulu, Hawaii: East-West Communication Institute, 1975; also Frances F. Korten and Sarah Young, "The Mothers Clubs of Korea," in *Managing Community-Based Population Programmes.* Kuala Lumpur: International Committee for the Management of Population Programmes, 1978.

Tasks

Planning/goal-setting	5—Quite effective
Credit union, sewing, school, nursery, day care, etc.	
Conflict management	4—Effective
Tempestuous but effective change in leadership	
Resource mobilization	5—Quite effective
Resource management	4—Effective
All loan monies utilized with good payback	
Provision of services	5—Quite effective
Diversified its services	

Integration of services	4—Effective
Control of bureaucracy	3—Average
Unable to prevent railroad scheme from tearing out their bean plants	
Claim-making	4—Effective
Structure	
Functions	4—Multiple
Family planning, income generating activities, community service	
Formalization	4—Formal
Government linkage	4—High
Vertical linkage	1—None
Horizontal linkage	1—None
Initiative	5—Outside
Planned Parenthood Federation of Korea	
Economic composition	2—Low heterogeneity
Social composition	2—Low heterogeneity
Sex composition	5—All female
Incentives	4—Quasi-voluntary
Must pay share of savings fund to become member	
Decision-making structure	1—Executive
Normative orientation	3—Neutral
Size	1—Very small
Environment	
Resource endowment	1—Very unfavorable
Very poor village	
Infrastructure	2—Unfavorable
Income level	1—Very low
Income distribution	2—Modest inequality
Diversification	1—Over 80% agricultural
Settlement patterns	4—Favorable
Social heterogeneity	2—Mostly homogeneous
Social stratification	3—Average
Sex discrimination	4—High
One woman beaten to death for using family planning	
Partisanship	3—Moderate
Group patterns	2—Kinship
Tight, clan-dominated village	
Community norms	2—Nonparticipatory
Societal norms	3—Neutral
Government orientation	4—Support
Bureaucratic capacity	4—Support
Performance	
Agricultural production/income	5—Significant gains
Cash crops	

Nonagricultural production/income Brought in factory	3—Some gains
Education Sewing school, day care, student study room, classes for club	5—Significant gains
Health 100% use contraceptives	5—Significant gains
Nutrition	3—Some gains
Income distribution	3—Some gains
Access to public services Day care, classes	4—Relative gains
Sex discrimination	4—Definite reduction
Social discrimination	3—Some reduction
Participation in government decision-making	2—No improvement
Participation in community decision-making	4—Definite increase

Problems

Resistance: initial resistance by men, parents-in-law and husbands of members especially; when they became convinced this was not women's liberation but family improvement, they became supportive

Malpractices: good leadership avoids corruption; leader takes pains to avoid appearance of using club for own interests

Roles

Membership: originally limited to 10 leading women in village, later expanded; members contribute to savings fund; each member is expected to attend meetings and work parties, and is charged for not doing so; little control of leadership

Leadership: first leader was wife of village chief; she was able to overcome initial resistance to club; second leader was her assistant, a new woman in village, better educated, politically clever; she expanded the club's activities; strong leader, enjoying the confidence of members and officials, receiving no compensation

Government staff: the Myon secretary originally met with the village chief to request him to appoint a leader to form a club; no continued support, however

Outside agency: Planned Parenthood Federation of Korea initiated club and provided continued support

Puebla Farmer Committees, Mexico.

Type: Interest Association.
Performance: 4—Very Good.

Background

The Puebla Project was a large rural development program concentrating on rapidly increasing production and net income. It covered 32 *municipios* (counties), about 43,300 farmers. Technical assistance began in 1968, when new maize recommendations were developed by CIMMYT (the international maize and wheat research center in Mexico). The new technology required greater fertilizer use, higher plant densities, and strict weed control. Local organizations were formed to disseminate the new methods, and to help farmers obtain fertilizer and credit. The project was sponsored by CIMMYT and the Mexican government.

Source

CIMMYT, *The Puebla Project: Seven Years of Experiments, 1968–73*. El Bataan, Mexico: CIMMYT, 1974.

Tasks

Planning/goal-setting	4—Effective
Conflict management	4—Effective
Able to resolve disputes	
Resource management	4—Effective
Provision of services	5—Quite effective
Credit, fertilizer, extension	
Integration of services	5—Quite effective
Control of bureaucracy	4—Effective
Got favorable decisions from government in dispute over fertilizer distribution	
Claim-making	3—Average

Structure

Functions	1—Single
Formalization	3—Quasi-formal
Most not legally constituted	
Government linkage	4—High
Vertical linkage	1—None
Horizontal linkage	1—None
Initiative	4—Government
Economic composition	2—Low heterogeneity
Social composition	2—Low heterogeneity
Sex composition	2—Mostly male
Incentives	4—Quasi-voluntary
Decision-making structure	2—Executive plus committee
General coordinator elected at village level, committees administer	
Normative orientation	3—Neutral
Size	1—Very small

Environment

Resource endowment	2—Unfavorable
Income level	2—Low
Income distribution	3—Definite inequality
Diversification	1—Over 80% agricultural
Social heterogeneity	2—Mostly homogeneous
Social stratification	4—Relative
Sex discrimination	4—High
Social discrimination	4—High
Literacy	1—Under 20% literacy
Group patterns	4—Patron-client
Community norms	4—Participatory

Societal norms	3—Neutral
Government orientation	4—Support
Bureaucratic capacity	5—Active support

Performance

Agricultural production/income	5—Significant gain
23% increase in real income in three years	
Income distribution	4—Relative gains
Access to public services	4—Relative gains
Participation in government decision-making	2—No improvement
Participation in community decision-making	2—No improvement

Problems

Resistance: some resistance by local elites, especially fertilizer dealers. Clear government support helped overcome resistance

Internal divisions: some individuals failed to repay loans, which led the other members to seize property or use legal means of redress

Roles

Membership: at the beginning of the project in 1967, few farmers felt any desire to join; difficulty in buying fertilizer and obtaining credit led to the formation of LOs; commitment consists in collective responsibilities and liabilities

Leadership: a general representative/coordinator is elected by members; because LOs are so small, members participate in decision-making

Government staff: project staff initiated project and gained support of municipal authorities; Ministry of Agriculture and State of Puebla participated a little; government-funded credit sources cooperated with LOs

Outside agency: project staff from CIMMYT provided initiative and worked closely with LOs

San Jose Credit Cooperative, Philippines.

Type: Cooperative.
Performance: 4—Very Good.

Background

Co-op was initiated in 1967 by the local Catholic priest. The membership and assets grew rapidly. By 1975, however, mismanagement was evident. Members organized to confront the managers and forced two audits on them. Eventually, they replaced the management.

Source

Mary Hollnsteiner, *Mobilizing the Rural Poor through Community Organization.* Rome: Rural Organization Action Project, FAO, 1979.

Tasks

Planning/goal-setting	4—Effective
Members effectively planned and organized to regain control of co-op	
Resource mobilization	4—Effective
Funds raised by members for audits	
Provision of services	4—Effective

Control of bureaucracy	4—Effective
Able to force audit and change in management	
Conflict management	4—Effective
Able to unify members against the management	
Structure	
Functions	1—Single
Credit only	
Formalization	4—Formal
Registered	
Government linkage	2—Low
Vertical linkage	1—None
Horizontal linkage	1—None
Initiative	5—Outside
Economic composition	2—Low heterogeneity
Social composition	2—Low heterogeneity
Sex composition	3—Mixed
Incentives	5—Voluntary
Decision-making	3—Committee
Originally centralized and executive, but changed to committee management	
Normative orientation	4—Egalitarian
Size	2—Small
Environment	
Income level	2—Low
Income distribution	3—Definite inequality
Diversification	1—All agricultural
Social heterogeneity	3—Moderate
Social stratification	3—Average
Sex discrimination	2—Low
Social discrimination	1—Very low
Literacy	3—40–60% literacy
Partisanship	4—High
Group patterns	4—Patron-client
Community norms	4—Participatory
Societal norms	4—Participatory
Government orientation	2—Non-support
Bureaucratic capacity	1—Opposition
Performance	
Nonagricultural production/income	3—Some gains
Small businesses set up through loans	
Education	3—Some gains
School loans	
Health	3—Some gains
Loans for emergency medical care	
Income distribution	3—Some gains

Participation in government decision-making	2—Small increase
Faced down government opposition to confrontational tactics	
Participation in community decision-making	4—Identifiable increase

Problems

Resistance: local officials resisted efforts of members to control co-op by refusing them use of town auditorium, declaring the meetings to be illegal, and spreading rumors that the military would arrest them; members met anyway and forced reforms

Malpractices: mismanagement and corruption; members forced audits and dismissal of corrupt managers

Roles

Membership: members are of varied economic means, including the landless; after overthrow of corrupt officials, members took over leadership and elected new management

Leadership: original leaders were corrupt; new leaders nominated by members and elected by secret ballot; none had a high school education, but were active in the mobilization of members against corrupt managers

Outside agency: co-op initiated in 1967 by local Catholic priest in response to his discovery that 80% of parishioners had their land mortgaged to usurers at high interest rates; source mentions members calling in "community organizers," but no details given

San Luis Reconstruction Committees, El Salvador.

Type: Local Development Association.
Performance: 2—Poor.

Background

The American Friends Service Committee sponsored projects in San Luis, particularly the construction of a drinking water pipeline, and road improvement between San Luis and another town. Initiative and assistance came from the "Administracion del Valle de la Esperanza," a government agency. Local organizations called *Comites Pro-Recontruccion* were formed. Efforts were hampered by top-down bureaucracies, authoritarian attitudes of officials, and distrust of villagers.

Source

Gerritt Huizer, "A Community Development Experience in a Central American Village," *International Review of Community Development,* 12 (1963):161–86.

Tasks

Planning/goal-setting	3—Average
Resource mobilization	4—Effective
Labor given by villagers	
Resource management	3—Average
Control of bureaucracy	2—Ineffective
Claim-making	2—Ineffective

Structure

Functions	2—Dual
Formalization	3—Quasi-formal
Government linkage	4—High
Vertical linkage	1—None
Horizontal linkage	1—None
Initiative	4—Government
Economic composition	3—Moderate
Social composition	3—Moderate
Sex composition	2—Mostly male
Incentives	5—Voluntary
Decision-making structure	3—Committee
Normative orientation	3—Neutral
Size	3—Medium

Environment

Income level	2—Low
Income distribution	5—High inequality
Diversification	1—Over 80% agricultural
Social heterogeneity	3—Moderate
Social stratification	5—Extreme
Sex discrimination	4—High
Social discrimination	4—High
Literacy	1—Under 20% literacy
Group patterns	4—Patron-client
Community norms	4—Participatory
Societal norms	2—Nonparticipatory
Government orientation	3—Indifference
Bureaucratic capacity	2—Nonsupport

Performance

Water supply	4—Identifiable gains
Transportation	4—Identifiable gains
Access to public services	3—Some gains
Participation in government decision-making	2—No improvement
Participation in community decision-making	3—Some increase

Problems

Resistance: villagers distrusted government officials, and maintained passive resistance

Subordination: top-down bureaucratic style suffocated LOs

Malpractices: corruption in handling money, no one willing to take case to higher authorities; threat of investigation finally solved problem

Roles

Membership: members are poor campesinos who contributed labor to projects; they had no control over leaders

Government staff: government agency responsible for reconstruction and development of the region set up LOs and provided inputs

Outside catalyst: American Friends Service Committee mediated between villages and local authorities

Self-Help Water Supply Committees, Malawi.
Type: Interest Association.
Performance: 5—Outstanding.

Background

In 10 years, 900 miles of piping have been laid and 2,000 village taps installed, serving 400,000 people in rural areas. The project is sponsored by the Department of Community Development. The government supplies pipes and materials; people provide self-help labor to dig trenches and lay piping. The District Development Council requests the project of the government and holds a public meeting of traditional and administrative authorities. If there is sufficient interest, a main committee is formed and also section (village) committees which organize self-help labor, work schedules, and distribution of labor among villages. Sixteen branch line systems operate from each storage tank along a main line, with each branch line committee responsible for maintenance of its branch line. Project implementation depends heavily upon cooperation and active support of village and area chiefs and party leaders; it acts through existing local leadership.

Source

Lindsey Robertson, *The Development of Self-Help Gravity Piped Water Projects in Malawi.* Lilongwe: Ministry of Community Development, 1978. Also J. Gus Liebenow, "Malawi: Clean Water for the Rural Poor," *American Universities Field Staff Reports, Africa,* no. 40.

Tasks

Planning/goal-setting	4—Effective
Conflict management	4—Effective
Resource mobilization	5—Quite effective
Tremendous mobilization of labor	
Resource management	4—Effective
Provision of services	5—Quite effective
Water to 400,000 people	
Integration of services	3—Average
Control of bureaucracy	3—Average
Claim-making	3—Average

Structure

Functions	1—Single
Water supply	
Formalization	3—Quasi-formal
Government linkage	4—High
Vertical linkage	3—Three tiers
Horizontal linkage	1—No linkage
Initiative	3—Shared
Economic composition	2—Low heterogeneity
Social composition	2—Low heterogeneity
Sex composition	2—Mostly male
Incentives	4—Quasi-voluntary
Strong social sanctions to contribute labor	
Decision-making structure	2—Committees plus executive

Main committee and section committees directed by chief or party leader

Normative orientation	3—Neutral
Size	3—Medium
Environment	
Resource endowment	2—Unfavorable
Infrastructure	2—Unfavorable
Income level	1—Very low
Income distribution	2—Modest inequality
Diversification	1—Over 80% agricultural
Social heterogeneity	2—Mostly homogeneous
Social stratification	2—Little
Literacy	1—Under 20% literacy
Partisanship	1—Very low
Group patterns	2—Kinship
Community norms	4—Participatory
Societal norms	3—Neutral
Government orientation	5—Active support
Bureaucratic capacity	5—Active support
Performance	
Water supply	5—Significant gains
Access to public services	4—Relative gains
Participation in government decision-making	2—No improvement
Participation in community decision-making	3—Some increase

Roles

Membership: not clear who from within community make up the committees, but virtually all in community participate in construction

Leadership: traditional leaders and party leaders direct village-level operations

Government staff: government field assistants (technical paraprofessionals) live in villages and provide technical knowledge to assist implementation; Rural Water Engineer coordinates field assistants, coordinates project between villages and supply sources, and provides technical expertise

Sidamo Mahabar Associations, Ethiopia.

Type: Local Development Association.
Performance: 4—Very Good.

Background

Mahabar voluntary self-help associations were usually initiated by local converted Christians, and most are assisted by the Ministry of Community Development with technology and some capital. They are committee organizations with elected officers, a membership fee, and ability to sanction members through expulsion. The *mahabar* are involved in community development and planning; setting up co-ops for marketing and credit; raising capital for projects; and dispute settlement. They encourage members to use *mahabar* courts to avoid the costs of national courts. The associations include members of all generations and various lineages and ethnic groups.

Source

John H. Hamer, "Prerequisites and Limitations of Voluntary Self-Help Associations: A Case Study and Comparison," *Anthropological Quarterly,* 19(2):107–34; "Preference, Principle, and Precedent: Dispute Settlement and Changing Norms in Sidamo Associations," *Ethnology,* 19(1):89–109.

Tasks	
Planning/goal-setting	4—Effective
Conflict management	5—Quite effective
Members must use local courts of elders before using official courts; especially successful at settlements when community interests are involved	
Resource mobilization	5—Quite effective
Contract out labor to raise capital	
Provision of services	4—Effective
Control of bureaucracy	4—Effective
Handled conflict, kept it out of courts	
Structure	
Functions	4—Multiple
Community planning, co-ops, capital formation, large projects, dispute settlement	
Formalization	4—Formal
Elected officers, recognized by government	
Government linkage	3—Moderate
Community Development officers assist	
Vertical linkage	1—No linkage
Horizontal linkage	2—Sporadic
Initiative	2—Leaders
Economic composition	3—Moderate
Social composition	3—Moderate
Sex composition	3—Mixed
Incentives	5—Voluntary
Decision-making	4—Assembly plus committee
Normative orientation	4—Egalitarian
Size	2—Small
Environment	
Resource endowment	4—Favorable
Infrastructure	2—Unfavorable
Income level	3—Middle
Diversification	2—60–80% agricultural
Social heterogeneity	3—Moderate heterogeneity
Various religious groups	
Social stratification	2—Little
Sex discrimination	3—Moderate
Partisanship	3—Moderate
Clan competition	
Group patterns	2—Kinship
Community norms	5—High participatory

Societal norms	3—Neutral
Government orientation	4—Support
Bureaucratic capacity	4—Support
Performance	
Agricultural production/income	3—Some gains
New crops; co-ops; marketing and credit	
Transportation	4—Identifiable gains
Access to public services	4—Relative gains
Courts open to all	
Reduced social discrimination	3—Some reduction
Government participation	2—No improvement
Community participation	5—Significant increase

Problems

Resistance: local elite (government and merchants) made accusations against *mahabar*; died out with success of LO

Subordination: government attempt to impose its notion of legality; elections, fairness, active participation, and sanctions reduced likelihood of capture

Internal divisions: between villages and clan; overcome through legitimacy and fairness of LO

Ineffectiveness: due to lack of state support, or competition, or ex-members competing for short-term gains

Roles

Membership: anyone may join for a fee; initiative originally from Christian converts, who were well-educated, willing to take risks; members contribute money, labor, and elect leaders

Leadership: led by young, modern men and traditional elders; elders did not set up LO, though are crucial for conflict management; network of leaders of different *mahabar,* accountable through elections; members can ostracize corrupt leaders, while leaders can enforce sanctions on misbehaving members

Government staff: Ministry of Community Development has field workers to handle large projects and supply technical assistance; intermediary with government; not enough development officers available

Small Farmer Cooperatives, Ecuador.

Type: Cooperative.
Performance: 5—Outstanding.

Background

This organization was formed in 1972 to reduce the exploitation of campesinos through supply stores, marketing assistance, and credit. It has achieved these ends through consumer stores, a rotating loan fund, and educational courses. The organization is governed through a General Assembly composed of two representatives from each of the associated coops; officers are elected. The original twelve coops expanded to 120 by 1978, with each local organization serving about fifty families. Most of the programs started by the Organization are fairly self-reliant, and offer quality services at low prices. It has gained some legitimacy in government circles.

Sources

Eugene J. Meehan, *In Partnership with People: An Alternative Development Strategy*. Washington: Inter-American Foundation, 1978, 134–137.

Tasks

Planning/goal-setting	4—Effective
Conflict management	4—Effective
Able to expand services	
Resource mobilization	4—Effective
Interest payments have added $35,000 to loan fund	
Resource management	4—Effective
Provision of services	5—Quite effective
Sell staples 5–10% below market price, high quality, honest weight, marketing services	
Integration of services	5—Quite effective
Practical education complements services	
Claim-making	5—Quite effective
Representation, advocacy	

Structure

Functions	4—Four plus
Supply, marketing, credit, consumer stores, education	
Formalization	3—Quasi-formal
Government linkage	3—Moderate
Vertical linkage	4—Loose federation
Horizontal linkage	2—Sporadic
Initiative	2—Leaders
Economic composition	3—Moderate heterogeneity
Social composition	2—Low heterogeneity
Sex composition	3—Mixed
Incentives	5—Voluntary
Decision-making	4—Assembly plus committees
General Assembly composed of two representatives from each co-op, with officers elected	
Normative orientation	4—Egalitarian
Size	1—Very small

Environment

Topography	2—Unfavorable
Resource endowment	2—Unfavorable
Infrastructure	2—Unfavorable
Income level	2—Low
Income distribution	3—Definite inequality
Diversification	1—Over 80% agricultural
Stratification	5—Extreme

Sex discrimination — 4—High
Social discrimintion — 5—Very high
Literacy — 1—Under 20% literacy
Group patterns — 4—Patron-client
Community norms — 4—Participatory
Societal norms — 2—Nonparticipatory
Government orientation — 3—Neutral
Bureaucratic capacity — 3—Neutral

Performance

Agricultural production/income — 3—Some gains
Nonagricultural production/income — 3—Some gains
Education — 5—Significant gains
Income distribution — 4—Relative gains
Asset distribution — 3—Some gains
Some action on land and water disputes
Access to public services — 4—Relative gains
Literacy in particular
Reduction in social discrimination — 4—Identifiable
Reduced exploitation of peasants by provision of credit, marketing and education
Participation in government decision-making — 4—Identifiable increase
Participation in community decision-making — 5—Major increase

Roles

Membership: mostly poor Indians; initiative from local persons subsequently supported by IAF; control by members through elections
Leadership: elected by members
Government staff: some support for LO once it was set up and functioning well
Outside agency: Inter-American Foundation, a nonprofit government corporation set up by U.S. Congress in 1969, helps to support indigenous private and semiautonomous institutions in Latin America and Caribbean; provided credits from 1977

Subak Irrigation Association, Indonesia.

Type: Interest Association.
Performance: 5—Outstanding.

Background

A *subak* association manages the water for the *subak,* which is a complex of rice fields obtaining water from one conduit. Some small *subaks* share leadership. All *subaks* automatically belong to a union of several *subaks*. At the district level, water is allocated on a mutually advantageous schedule. Water available per *subak* is directly proportional to water need, measured in terms of land productivity. *Subak* management is well-established. Members make decisions as a group.

Source

Aubrey Birkelbach, Jr., "The Subak Association," in *Indonesia.* Ithaca, N.Y.: Southeast Asia Program, Cornell University, 1973, 153–169; also Clifford Geertz,

"Tihingan: A Balinese Village," in *Villages in Indonesia,* ed. Koentjaraningrat. Ithaca, N.Y.: Cornell University Press, 1967.

Tasks

Planning/goal-setting	5—Quite effective
Manages large system collectively	
Conflict management	5—Quite effective
Resource mobilization	5—Quite effective
Fees, labor contribution	
Resource management	5—Quite effective
Provision of services	5—Quite effective
Integration of services	4—Effective

Structure

Functions	1—Single
Water management	
Formalization	3—Quasi-formal
Government linkage	1—Autonomous
Vertical linkage	3—Three tiers
Horizontal linkage	3—Regular
Plans water distribution	
Initiative	1—Members
Economic composition	5—Heterogeneous
Social composition	2—Low heterogeneity
Sex composition	2—Mostly male
Incentives	1—Compulsory
Decision-making structure	4—Assembly plus committee
Most decisions by assembly at large	
Normative orientation	5—High egalitarianism
Size	2—Small

Environment

Income level	1—Very low
Income distribution	4—Substantial inequality
Diversification	1—Over 80% agricultural
Social heterogeneity	2—Mostly homogeneous
Social stratification	4—Relative
Sex discrimination	4—High
Social discrimination	2—Low
Literacy	1—Under 20% literacy
Partisanship	3—Moderate
Group patterns	3—Familism
Community norms	4—Participatory
Societal norms	3—Neutral
Government orientation	3—Indifference
Bureaucratic capacity	3—Indifference

Performance

Agricultural production/income	5—Significant gains
Water supply	3—Some gains
Access to public services	3—Some gains

Participation in government decision-making	2—No improvement
Participation in community government decision-making	5—Significant increase

Problems

Subordination: prohibition against government officials holding office in association limits outside influence

Malpractices: corruption controlled by holding open public tax assessment meetings, so head of association can't collect too much in taxes

Ineffectiveness: traditional leader poorly educated, may not be "progressive"

Roles

Membership: farmers with over 1/10 hectare of land belong; others may contract for water; some members provide physical or financial support, proportionate to use of system; can purchase exemption; laws codify admission, taxation, responsibilities, and rights

Leadership: titular leader is *Klian subak,* who is elected. Most decisions made by assembly as a whole. *Klian* is an administrator for the group

Sukuma Cotton Cooperative, Tanzania.

Type: Cooperative.
Performance: 5—Outstanding.

Background

In 1952, a Sukuma bookkeeper noted that the Asians who controlled cotton marketing were cheating the Sukumas. Through his initiative, 32 co-ops were established, despite opposition from the colonial government. Eventually, they were recognized and registered. In 1954 the co-ops consolidated into unions, and in 1955 the Victoria Federation of Cooperative Unions was established. By 1961 the federation had expanded into owning and operating cotton ginneries and an office building. The cotton co-op led to a doubling of output and an accumulation of cash. Local decision-making groups ran the co-ops in a way compatible with Sukuma culture. The co-op was seen by members in the context of the independence movement and Africanization. There is no information on whether the co-op exists today.

Source

Gottfried O. Lang, Warren J. Roth, and Martha B. Lang, "Sukumaland Cooperatives as Mechanisms of Change," in *The Anthropology of Development in Sub-Saharan Africa,* ed. David Brokensha and Marion Pearsall. Lexington: Kentucky Society for Applied Anthropology, monograph no. 10, 1969.

Tasks

Planning/goal-setting Able to expand to 324 co-ops in 1959	4—Effective
Resource mobilization Raised capital for transport and marketing, got fees and labor from members	4—Effective
Resource management Encouraged greater production	4—Effective

Provision of services	4—Effective
Scales, trucks, ginnery, all bypassed middlemen	
Control of bureaucracy	4—Effective
Able to resist government opposition	
Claim-making	4—Effective
Forced government to repeal an assessment on their cotton	
Structure	
Functions	2—Dual
Cotton marketing and processing, owned office building and hotel	
Formalization	3—Quasi-formal
Government linkage	1—Autonomous
Vertical linkage	3—Three tiers
Co-ops, unions, and federation	
Horizontal linkage	2—Sporadic
Initiative	2—Leader
One person instrumental, but helped by Young Men's Societies	
Economic composition	3—Moderate
Social composition	1—Homogeneous
Incentives	4—Quasi-voluntary
Ostracism as method of social control. Established virtual monopoly of cotton buying	
Decision-making structure	3—Committee
Had officers and committee. Similar operation to traditional village work societies	
Size	1—Very small
Environment	
Resource endowment	2—Unfavorable
Infrastructure	2—Unfavorable
Income level	1—Very low
Income distribution	1—Relative equality
Diversification	1—Over 80% agricultural
Settlement patterns	2—Unfavorable
Social heterogeneity	1—Homogeneous
Social stratification	1—Very little
Sex discrimination	3—Moderate
Social discrimination	3—Moderate
Literacy	1—Under 20% literacy
Partisanship	2—Low
Group patterns	5—Associational
Community norms	4—Participatory
Societal norms	4—Participatory
Government orientation	2—Nonsupport
Bureaucratic capacity	2—Nonsupport

Performance

Agricultural production/income	5—Significant gains
1954–62 doubled production	
Income distribution	3—Some gains
Access to public services	3—Some gains
Participation in government decision-making	3—Some increase
Participation in community decision-making	3—Some increase

Problems

Resistance: colonial officials opposed to co-op; sided with processors in disputes; group solidarity prevailed, along with threats to withhold cotton

Roles

Membership: one Sukuma bookkeeper set up co-op; members vote for committeemen and officers; all donate time and labor

Leadership: original leader set up structure, dealt with government, was good at face-to-face communication; leaders stressed contribution of co-op to nationalist movement; set up committee system to involve members in decision-making

Tiv Farmers' Associations, Nigeria.

Type: Interest Association.
Performance: 5—Outstanding.

Background

A Tiv agricultural officer familiar with "bams" suggested to senior tribal leadership the formation of a broadly based farmers' association to increase production through improving extension. In 1966, a senior council was established, consisting of 60 representatives of Tiv areas. In 1968, the association decided to organize permanent groups at the village level, with a monthly system of dues. The village branch elects a representative to the district organization, which elects representatives to the division organization. Representatives from three divisional groups form the Council of the Farmers' Association. The farmers' associations have helped increase Tiv production and developed an institutional base in the area.

Source

Elliott Morss, et al., *Strategies for Small Farmer Development,* vol. 2, 231–41. Boulder, Colo.: Westview Press, 1976.

Tasks

Planning/goal-setting	4—Effective
Expansion of activities over time	
Conflict management	4—Effective
Resource mobilization	5—Quite effective
Entirely self-sufficient, with 3,300 members	
Resource management	5—Quite effective
Operates on own budget	
Provision of services	5—Quite effective
Integration of services	4—Effective
Claim-making	4—Effective
Makes use of government services	

Structure	
Functions	3—Multiple
Extension, marketing, roads, creation of agro-businesses	
Formalization	4—Formal
Government linkage	3—Moderate
Vertical linkage	3—Federation
5 tiers, large and complex	
Initiative	2—Leaders
Economic composition	4—Some heterogeneity
Social composition	1—Homogeneous
Sex composition	3—Mixed
Incentives	5—Voluntary
Most pay fees	
Decision-making structure	4—Assembly plus committee
Councils, with evaluation sessions by full membership	
Normative orientation	4—Egalitarian
Size	1—Very small
Environment	
Income level	1—Very low
Diversification	1—Over 80% agricultural
Social heterogeneity	1—Homogeneous
Social stratification	2—Little
Sex discrimination	3—Moderate
Social discrimination	2—Low
Literacy	1—Under 20% literacy
Partisanship	4—High
Group patterns	2—Kinship
Community norms	5—High participatory
Societal norms	4—Participatory
Government orientation	4—Support
Bureaucratic capacity	4—Support
Performance	
Agricultural production/income	5—Significant gains
Increased 72% in 10 years	
Transportation	3—Some gains
Roads	
Public facilities	3—Some gains
Markets	
Income distribution	3—Some gains
Access to public services	3—Some gains
Participation in government decision-making	3—Some increase
Participation in community decision-making	4—Definite increase

Problems

Ineffectiveness: lack of extension officers to carry out program as desired

Roles

Membership: pays fees, assists in LO activities

Leadership: leaders are generally "progressive" farmers; traditional leaders also active

Government staff: extension officer initiated the LO and took part in its implementation; government contributed money and appointed Tiv extension officers in the area

Village Health Care Teams, Niger.

Type: Interest Association.

Performance: 1—Very Poor.

Background

The village health team approach began with several pilot projects initiated and conducted by expatriate doctors. Two national services, Health and Animation Rurale, became involved in a national program of village health teams in the 1960s. A third service, Animation Feminine, was also involved in several areas. The health team was to comprise voluntary health workers, a secretary, and president. Teams would supervise health treatments and the pharmacy, mobilize workers for local activities, and resolve financial and organizational problems encountered by the program. Villages which had previously been "animated" by the Animation Rurale program were selected on the assumption that they had the structure for broadly based village activity. Team managers were to be given three days' training. The teams themselves would include both men and women. The concept was, however, inapplicable because of the lack of a structure for participation among the local people. The village health care program failed.

Source

Robert Charlick, Animation and Village Health Care: The Pilot Project of Matameye, 1968–71. (Case study on Animation Rurale written for Rural Development Committee, Cornell University.)

Tasks

Planning/goal-setting	2—Ineffective
Conflict management	1—Quite ineffective
Resource mobilization	2—Ineffective
Few resources raised to pay for the village health care team	
Resource management	2—Ineffective
Provision of services	2—Ineffective
Integration of services	1—Quite ineffective
Control of bureaucracy	2—Ineffective
Claim-making	1—Quite ineffective

Structure

Functions	1—Single
Health care	
Formalization	2—Informal
Elected officers but operated under traditional norms	

Government linkage	5—Directed
Vertical linkage	1—No linkage
Horizontal linkage	1—No linkage
Initiative	4—Government
Economic composition	2—Low heterogeneity
Social composition	2—Low heterogeneity
Sex composition	3—Mixed
Animation Feminine involved in some areas	
Incentives	5—Voluntary
No enforcement of payment requirements	
Decision-making structure	1—Executive
Animation Rurale or headman dominated	
Normative orientation	3—Neutral
Size	1—Very small
Environment	
Topography	2—Unfavorable
Resource endowment	1—Very unfavorable
Infrastructure	2—Unfavorable
Income level	1—Very low
Income distribution	1—Relative equalities
Diversification	1—Over 80% agricultural
Settlement patterns	3—Neutral
Social heterogeneity	1—Homogeneous
Social stratification	2—Little
Local elites, headmen and merchants slightly better off	
Sex discrimination	4—High
Women excluded from decision-making in most village affairs	
Literacy	1—Under 20% literacy
Partisanship	2—Low
Group patterns	3—Families
Community norms	4—Participatory
Societal norms	3—Neutral
Government orientation	3—Indifference
Bureaucratic capacity	3—Indifference
Performance	
Health	4—Identifiable gains
Access to public services	3—Some gains
Participation in government decision-making	2—No improvement
Participation in community decision-making	2—No improvement

Problems

Resistance: passive resistance by villagers, who accepted control by headman

Subordination: politics in Hausa village dominated by headman and his supporters, who retained control over village health workers

Malpractices: some corruption, misappropriation of funds, etc.; VHW usually selected from among headman's client group

Ineffectiveness: lack of realism on the part of program planners

Differential participation: women rarely involved, though teams of animators were to include women

Roles

Membership: no data on how team members were selected; though program envisaged broad participation, no sign of this

Leadership: members of team took little initiative in setting up LO; health management committee largely ignored by villagers; VHW accountable to headman and Animation Rurale supervisors

Government staff: Animation Rurale staff played major role in organization and supervision; health service involved in initial training of VHW, though no in-service training provided; Animation Rurale supervised VHW in authoritarian manner

APPENDIX D

Small Farmers' Groups in Nepal

The Small Farmers' Development Program (SFDP) is based on an approach devised by FAO/UNDP in 1973 under the Asian Survey for Agrarian Reform and Rural Development (ASARRD). In Nepal it was introduced in 1975 first as a pilot program in Dhanusha and Nuwakot with the Agricultural Development Bank of Nepal (ADBN) as the implementing agency. The program was preceded by, and emerged out of, field-level workshops and a final workshop in Kathmandu, in all of which small farmers and landless laborers themselves participated.

The pilot program sought to motivate small farmers and landless rural labor to form organizations of their own below the level of cooperatives—small (15 to 20 members), homogeneous, multifunction groups around a common-nucleus income-raising activity based on group work plans and group action, supported by an integrated program of supervised credit, extension, and technical backstopping. A group-organizer/action-research-fellow (GO/ARF) was stationed in each district to play the vital catalytic and monitoring role and link with the ADBN and the various line agencies. A communication coordination network for all levels from central government to the farmer was established, using a committee structure that already existed—from the National Coordinating Committee at the center to the Subproject Implementation Committee in the field.

It was conceived that the GO would guide the groups to achieve self-reliance, so that after two or three years he could himself withdraw. Eventually, the small farmers' groups were to federate into an association with all group plans aggregating into an "area plan."

The groups were to receive credit for individual members, as well as for the groups as a whole on the basis of group demand, under group liability and with credible income-raising actions plans only, without any other

This account is based on a report by Dharam Ghai and Anisur Rahman, written for the International Fund for Agricultural Development (IFAD) mission to Nepal in 1978. It was published in *Development: Seeds of Change (SID)*, 1981:1.

collateral being required. The group themselves were to develop their own group savings fund with member contributions for providing consumption, distress and emergency loans to their members.

In three and a half years, from September 1975 to February 1979, 45 groups were formed in Dhanusha, comprising 510 farm families. Of these, three are landless groups and five are women's groups. In Nuwakot, where the project started in March 1976, 49 groups were formed in three years, comprising 755 families with nine women's groups.

Expansion of SFDP

Encouraged by the success of the pilot program in Dhanusha and Nuwakot, the government decided to expand the program in other areas. As of May 1979 as many as 24 projects have been launched in 24 districts, servicing a total of 370 groups comprising 3,992 small farmers' landless households in 37 *panchayats*. The total loan disbursed in all the projects taken together was 5.6 million rupees.

A recent report on the SFDP sums up its achievements thus:

> Through the formation of small groups not only is the economic status of the small farmers inproving, but also their social and political status. The "receiving mechanisms" of the small farmers are being created and strengthened. On the basis of their group strength, the peasants are gradually coming out of their so-called "culture of silence." They have a group voice now to demand various services from the different line agencies and local institutions. And they are becoming members of local cooperatives in ever-increasing numbers. Already as a result of group action, forty-five small farmers have been elected to local Panchayat bodies. Slowly but surely all this is increasing the strength of the poorest peasants vis-à-vis the big landholders and moneylenders.
>
> In the frequent meetings which take place in this area now between people, from different government departments, from FAO and the small farmers, there are certain distinct changes that have become noticeable. No longer do the small farmers and landless peasants just sit passively at the outskirts of meetings. They are very much in the center of things, now, participating in meetings actively and confidently. And no longer are they afraid, as they used to be, to air their views before government officers. They even reel out statistics about their groups without faltering and without the aid of written reports.
>
> As they put it in their own words, "Because we are a group now and we stick to each other, we have suddenly become more powerful. The moneylenders are afraid to exploit us now. The government officials speak to us, they even speak nicely. We are also no more afraid to enter the bank of the office of the cooperative society." [Kamla Bhasin, *Breaking barriers: a South Asian Experience of Training for Participatory Development*. Report of the Freedom-from-

Hunger Campaign/Action for Development Regional Change Agents' Programme, Bangkok, FAO, March–May 1978]

Some Illustrative Glimpses

In the following pages are presented the information and impressions gathered from a visit to three SFDP projects in May 1979.

Ramnagar project. There are 206 small farm families in the program, out of a total of 529 small farm families in the Ramnagar Panchayat. The Panchayat has in all 711 farm families. Small farmers as defined in this project are those holding up to 3 *bighas* of nonirrigated or 2 *bighas* of irrigated land (1 *bigha* = 0.67 hectares).

The 206 small farm families in the program have formed themselves into 22 groups, starting with two groups two years ago. The groups are mostly formed among relations, friends, and neighbors. Some of the groups are along caste lines, but others are mixed.

The activities include paddy and wheat cultivation, fish-breeding, buffalo-raising, piggery, road construction and maintenance, compost pit, construction of common meeting hall, latrine construction, adult education, family planning, training and visits to demonstration farms, visits to other SFDP groups, and cultural functions.

Group formation is voluntary. Members pay 5 rupees each per month into the group savings fund.

Groups deliberate on demands for credit, for individual and group projects. After these are passed in the group, the demands are submitted to the project chief (GO), who may ask groups to modify demands which he considers technically or otherwise unsound. The demands are thereafter placed before the Loan Committee. The Loan Committee consists of the project chief, the chairman of the local Sajha cooperative, and an ADBN official. The cooperative advances credit on the basis of the Loan Committee's recommendations. A loan is given to each group against collective liability at an interest rate of 11–14%.

A 90% recovery rate has so far resulted. The defaults are stated to be really hard cases and the groups are content to wait for another year for their recovery and even write off the most difficult cases. In the latter case they would repay the cooperative from out of the group savings fund.

Incomes of the farmers in the groups have increased significantly as a result of participation in the program. It was reported that many of the families previously had no assured food beyond five to six months, while now they have difficulty for two to three months only. One small farmer is now a member of the cooperative, while previously there was none. Three are members of the *panchayat,* a new phenomenon in the village. The groups expect to send more members to the *panchayat* in next year's election.

The GO has worked as junior technician for five years. He has one assistant with him. Together, according to the GO, they can handle the needs of 45 groups.

In the opinion of the GO, out of 200 agricultural diploma-holders with whom he was trained, about 15 would be able to handle the task of working as catalyst for such small farmers' group formations and to assist them in their various activities. More can be trained. Training methods, according to the GO, should include visits to SFDP groups and working with experienced GOs.

One group collectively breeds fish in a *panchayat*-owned pond. Distribution of the harvest is according to labor.

Two groups took a collective lease on land for group farming simultaneously with individual farming in individual plots. One group failed in this effort, because of the attention that had to be given to individual plots. The other group is maintaining collective farming, with a good production record.

Interview with a farmer belonging to one of the groups. The group was formed with seven members when the project was initiated in the *panchayat,* and the village leaders told the small farmers about the facilities to be provided in the program. It expanded gradually to 15 members. There is one landless member in the group, to whom the group leader has given part of his land at a cheaper price. He also got a buffalo on credit and profited by raising and selling it. He is now raising another buffalo. Two more of the landless want to join the group and are likely to be taken in.

Before joining the program the interviewee had 13 *kathas* of land and was doing a priest's job to supplement his work on the land. Now he has 22 *kathas* of land (1 *katha* = .03 hectare). He still works for a wage sometimes—about 20 days last year.

During a visit to a village under the Ramnagar project [we found] a fairly large tank was under construction by the voluntary labor of the members of one 13-member group formed only four months before. In reply to questions the members said that they had selected their leader by consensus and because he was wise and could keep accounts. The group leader provides leadership for whatever is decided in group meetings, which are held once every week. Although they possess land, the group members were previously not always in a position to cultivate it as no credit was available, so they used to cut and sell wood. This was the situation with most members in 1978. But now they are in a position to cultivate their own land. They still have a problem to obtain bullocks for plowing—they have received one only through the bank, and all the members need it at the same time.

The visit also revealed, for example, one nonmember small farmer who seemed very poor; he did not want to join a group for fear of collective liability for credit, whereby he would be responsible for loans taken by others (the contrast this represents underlines the significance of other poor farmers' forming groups and accepting collective liability for individual credit).

In another case, a wheat and rice mill was bought one month earlier by another group—as a group-owned project—and installed in a shed constructed by the voluntary labor of the group members. The group has hired one *mistry* (technician) for one month to train them to operate the mill, and every member was learning from him and operating the machine by turn. After one month the group would take over the operation completely. The project thus had a self-reliant training component built into it.

Manigram project (Anandban Panchayat). Launched two years earlier, this project showed that farmers were initially suspicious of such new ideas, thinking it was like the land-reform movement, which brought them little tangible benefit. There was much moneylending in the area by the village rich, who also opposed the project. Under the initiative of the GO, the first group was organized secretly at midnight with five members. They were given bullocks, one to each on credit under group liability. They sold the bullocks after harvest and made some money. The Red Cross people also helped in the initial motivational work.

In the beginning there was a threat to the GO's personal security: there was an attempt to poison him, and his house was raided and all his belongings stolen. He had to be protected by two security guards for a while. Now the program has gathered strength as more groups have been formed.

Today there are 35 groups in the project—25 men's and ten women's, 387 members in all with 150 women.

The project chief (GO) was asked about mistakes he had committed and the lessons he had learned. He said he had initially misunderstood his task—he had thought he would be a bureaucrat with office and power. He saw his proper role as a motivator rather than a doer, after reading *Small Is Beautiful,* and was also inspired by the visit of a delegation from the National Institute of Bank Management in India (Bombay), which has launched an "Other India" program of motivating and assisting small farmers in self-reliant group action through cadres grouped into "spearhead teams."

It was also his mistake initially, the GO said, to make a lot of decisions for the farmers. Now the decision-making was increasingly in the hands of the groups.

The farmers require paddy for consumption, and a paddy loan was initially tied to the pig or goat harvest, not the paddy harvest. This was also a mistake, and the repayment schedule was not kept.

The project gradually picked up. There were landless laborers, sharecroppers, contract farmers, and also slave labor bonded against loans. In some cases debt and bondage were inherited. Eleven bonded laborers belonging to a group were liberated—the group borrowed one bullock, and the members also made personal contributions to raise the money to pay off the masters.

The GO reported on excess land in the possession of the rich and about 37 *bighas* of land were redistributed.

Group savings is being used to provide distress loans to members—16,629 rupees have been accumulated so far. Individual saving is also rising: previously [money] used to be wasted on gambling; now members have opened bank accounts.

In marriages, all members contribute personally and, in addition, loans are made from the group savings fund at the 14% interest rate. Generally, marriage expenditure has been cut down from around 5,000 rupees to something on the order of 3,000 rupees. In one instance, however, marriage expenditure was still as high as 56,000 rupees.

The groups got 100 buffaloes. They did not find a good market for milk and so are making butter. Poultry and calf-raising are some other activities, and there is a group vegetable farm.

Women's groups are taking training in nutritional diet, family planning, sweater knitting, and hosiery. They are also making table mats.

Two local high schools have been involved in this project. They have started two first aid centers.

Three National Development Service (NDS) students are also in the village, helping in adult education. They joined the SFDP farmers in constructing a bridge and the school building. There is a manpower shortage, and the GO thought that it would be helpful if the NDS students were formally attached to his project.

Asked about the nature of training required for the tasks of project chief, the GO suggested the following: more practical training and less theory; maximum field training; familiarity with village conditions and local language and development of a rapport with local people; manuals developed by field workers.

We visited a training center for weaving where 35 girls belonging to four or five groups were taking training in a rented room in which the looms were also rented. Asked what they would do after training, the girls said they would like to have credit to purchase the looms, but they were afraid to borrow individually and take the looms to their respective houses, as they were not sure whether they could repay the loans, whether they would be able to sell their products and where, and so on. They would therefore prefer to have all looms purchased under group ownership and installed in a room like their present one, where they would come and work together, and they would like to have help in marketing their products (this meant a project possibly to build a low-cost building to keep the looms, credit for 35 looms and the raw materials, provision of a storage facility, and marketing assistance).

Nuwakot subproject in Tupche Panchayat. Part of one of the two pilot projects, this subproject stood out as one of the best visited. The SFDP office itself was a model—an ordinary-looking low-cost structure with mud walls and mat floors. It has cost only 10,000 rupees to construct, built with the free labor of farmers in the village, and was an open house for the villagers. The project chief was a person with about 20 years' experience in the ADBN, who had visibly won the affection of the villagers and was even able to obtain the support of the village rich in the project.

The cultivation looked well cared for, and the maize fields were particularly impressive. A number of drinking-water projects had been completed with the contributory labor of the farmers, and pipes obtained on group credit. Big farmers were charged twice as much as small farmers in allocating contributions to the cost of construction in these public basic-need projects, built and managed by small farmers for use by all.

A group project for a handloom had been completed and was in operation for a year; the cost of building materials had been paid for from group savings funds, construction done by voluntary labor, and 14,000 rupees borrowed for the equipment. Group members work in the factory on a piece-wage basis, receiving 300–400 rupees a month each. They work there all the year round except for June to August, when they work in the fields. Unemployment thus has been wiped out. In one year 4,000 rupees of the loan had been repaid. After the full loan is repaid, this surplus will go to the group savings fund, with which the group can undertake further projects.

We saw an ingenious fish pond, constructed by the voluntary labor of a group to retain water running down the hill. It was owned by the whole group. The income, the member farmers told us, will not be distributed individually but will be kept in the group savings account, from which individuals can have loans—a group decision.

One landless group took possession of some uncultivated land for cultivation. Everyone supported this, though this was illegal. Later on, all orchards in the vicinity were registered in the name of landless groups.

Last year in the *panchayat* election in Tupche, two-thirds of the members were elected from SFDP groups, an event of far-reaching significance for the area.

Self-reliance

While the groups were taking a lot of decisions themselves, they still wanted the GO to stay, three years after the launching of the project. The GO considered full self-reliance in less than six to seven years difficult, and listed the following factors as important to promote self-reliance: Financial self-reliance: reliance on the group savings fund more than on outside loans, hence the importance of building up the group savings fund; regular meetings to raise group decision-making capability; election of their own representatives in the *panchayat;* direct links of the groups with the line agencies of the government.

Bibliography

Abedin, Md. Zainul. 1979. Two Case Studies from the Village of Ballovpur under the Comilla Sub-Project of the Small Farmer and Landless Labourers Development Project. Bangkok, Thailand: Food and Agriculture Organization, Regional Office for Asia and the Far East.

Abel, M. E. 1975. Irrigation Systems in Taiwan: Management of a Decentralized Public Enterprise. Staff paper. St. Paul: Department of Agricultural Economics, University of Minnesota.

Adelman, Irma, and George Dalton, 1971. Factor Analysis of Modernization in Village India. In *The Modernization of Village Communities,* ed. G. Dalton, 504–13. Garden City, N.Y.: Natural History Press.

Adelman, Irma, Cynthia Taft Morris, and Sherman Robinson, 1976. Policies for Equitable Growth. *World Development,* 4:(7),561–82.

Ahmed, Manzoor. 1980. BRAC: Building Human Infrastructure to Serve the Rural Poor. In Coombs (1980:362–468).

Alam, Shamsul, ed. 1979. *Small Farmers and Landless Labourers Development Project: Action Research Project on the Development of Small Farmers and Landless Labourers.* Dacca, Bangladesh: Abco Press.

Alberti, Giorgio. 1976. Persistence and Change in Structure and Values in the Sugar Plantations of Northern Peru. In Nash et al. (1976).

Alexander, K. C. 1980. *Rural Organizations in South India: The Dynamics of Laborer and Tenant Unions and Farmer Associations in Kerala and Tamil Nadu.* Ithaca, N.Y.: Rural Development Committee, Cornell University.

Almond, Gabriel A. 1973. Approaches to Developmental Causation. In Almond et al. (1973:1–30).

Almond, Gabriel A., Robert Mundt and Scott Flanagan, eds. 1973. *Crisis, Choice, and Change: Historical Studies of Political Development.* Boston: Little, Brown.

American Institutes for Research. 1973. *Village-Level Disposing Conditions for Development Impact.* Bangkok, Thailand: Asia/Pacific Office, AIR.

American Public Health Association. 1981. Revolving Funds: Trials and Errors in the Search for Self-Sufficiency. *Salubritas,* June. Washington: APHA.

Anderson, Peggy. 1968. New System in Niger. *Africa Report,* 13(8):12–17.

Antrobus, Peggy, with Barbara Rogers. 1980. *Hanover Street: An Experiment to Train Women in Welding and Carpentry.* SEEDS pamphlet series. New York: Population Council, Carnegie Corporation, and Ford Foundation.

APROSC. 1980. *Evaluation Study of Participatory Small-Scale Irrigation Projects.* Kathmandu, Nepal: Agricultural Projects Service Centre.

Apter, David. 1965. *The Politics of Modernization.* Chicago: University of Chicago Press.

——. 1968. *Some Conceptual Approaches to the Study of Modernization.* Englewood Cliffs, N.J.: Prentice-Hall.

Aqua, Ronald. 1974. *Local Institutions and Rural Development in Japan.* Ithaca, N.Y.: Rural Development Committee, Cornell University. Revised version in Uphoff (1982–83:11,328–93).

Arocena, Leopoldo M. 1979. *Barangay General Luna, Llanera, Nueva Ecija, Philippines. Group No. 3 (Success) Group No. 1 (Failure).* Case Studies on Small Farmer Development. Manila: Small Farmer Development Program, Ministry of Agrarian Reform.

Ashford, Douglas E. 1967. *National Development and Local Reform: Political Participation in Morocco, Tunisia, and Pakistan.* Princeton, N.J.: Princeton University Press.

Asian Development Bank. 1977. *Rural Asia: Challenge and Opportunity.* New York: Praeger.

Barkan, Joel. 1981. Self-Help Organization, State, and Society: Linkage and Development in Rural Kenya. Paper prepared for Conference on Local Political Organization and Rural Development Policy, University of Iowa, Iowa City, 14–16 September.

Barkan, Joel, Frank Holmquist, David Gachuki, and Shem Migot-Adholla. 1979. *Is Small Beautiful? The Organizational Conditions for Effective Small-Scale Self-Help Development Projects in Rural Kenya.* Iowa City: Comparative Legislative Research Center, University of Iowa.

Barkan, Joel, and John Okumu, eds. 1979. *Politics and Public Policy in Kenya and Tanzania.* New York: Praeger.

Barraclough, Solon. 1971. Farmers' Organizations in Planning and Implementing Rural Development. In Weitz (1971b:364–90).

——. 1974. *Agrarian Structure in Latin America.* Lexington, Mass.: Lexington Books.

Bates, Robert. 1981. *Markets and States in Tropical Africa: The Political Basis of Agricultural Policies.* Berkeley: University of California Press.

Beckford, George. 1972. *Persistent Poverty: Underdevelopment in Plantation Economies of the Third World.* New York: Oxford University Press.

Belloncle, Guy. 1979. *Quel développement rural pour l'Afrique noire?* Dakar, Senegal: Nouvelles Editions Africaines.

Belloncle, Guy, and Dominque Gentil. 1968. Pedagogic de l'implantation de mouvement cooperatif au Niger. *Archives internationales de sociologie de la cooperation,* 23 (Jan.–June):50–71.

Benor, Daniel, and James Harrison. 1977. *Agricultural Extension: The Training and Visit System.* Washington: World Bank.

Bentley, Arthur F. 1908. *The Process of Government.* Chicago: University of Chicago Press.

Berger, Peter. 1977. *To Empower People: The Role of Mediating Structures in Public Policy.* Washington: American Enterprise Institute.

Bergmann, Herbert. 1974. Les notable villageois: Chef de village et imam face à la

cooperative rurale dans une région de Senégal. *Bulletin de l'Ifan,* 25(2, series B):283–322. Abstracted in Münkner (1979).
Bhatty, K. M. 1979. *Social Determinants of Water Management in Daudzai.* Peshawar: Pakistan Academy for Rural Development.
Bienen, Henry. 1967. *Tanzania: Party Transformation and Economic Development.* Princeton, N.J.: Princeton University Press.
——. 1974. *Kenya: The Politics of Participation and Control.* Cambridge, Mass.: Harvard University Press.
Binder, Leonard, et al. 1971. *Crises and Sequences in Political Development.* Princeton, N.J.: Princeton University Press.
Birkelbach, Aubrey, Jr. 1973. The Subak Association. In *Indonesia,* 153–69. Ithaca, N.Y.: Southeast Asia Program, Cornell University.
Blair, Harry W. 1974. *The Elusiveness of Equity: Institutional Approaches to Rural Development in Bangladesh.* Ithaca, N.Y.: Rural Development Committee, Cornell University. Revised version in Uphoff (1982–83:1,387–478).
——. 1978. Rural Development, Class Structure, and Bureaucracy in Bangladesh. *World Development,* 6:1,65–83.
——. 1982. *The Political Economy of Participation in Local Development Programs: Short-Term Impasse and Long-Term Change in South Asia and the United States from the 1950s to the 1970s.* Ithaca, N.Y.: Rural Development Committee, Cornell University.
Blustain, Harvey. 1982a. *Resource Management and Agricultural Development in Jamaica: Lessons for Participatory Development.* Ithaca, N.Y.: Rural Development Committee, Cornell University.
——. 1982b. Clientelism and Local Organizations. In *Strategies for Organization of Small-Farm Agriculture in Jamaica,* ed. H. Blustain and E. LeFranc, 192–210. Ithaca, N.Y.: Rural Development Committee, Cornell University; Kingston, Jamaica: Institute of Social and Economic Research, University of the West Indies.
Blustain, Harvey, and Arthur Goldsmith. 1979. Farmers' Organizations and Local Institutions in the Two Meetings and Pindars River Watersheds. Mimeo. Kingston: USAID Mission.
Boer, Leen. 1982. The Variety of Local Contexts of Participatory Development Projects: An Attempt at Systematization. In Galjart and Buijs (1982:126–83).
Bottomore, T. B. 1964. *Elites and Society.* Harmondsworth, U.K.: Penguin.
Bottrall, Anthony. 1977. Evolution of Irrigation Associations in Taiwan. *Agricultural Administration,* 4(4):245–50.
——. 1980. Planning Local Organisations: In Search of a Method. Mimeo. London: Overseas Development Institute.
Boulding, Kenneth E. 1963. Toward a Pure Theory of Threat Systems. *American Economic Review,* June, 424–34.
Boumann, F. J. A. 1977. Indigenous Savings and Credit Societies in the Third World: A Message? *Savings and Development Quarterly Review,* 4,181–218.
Breslin, Patrick. 1982. The Technology of Self-Respect: Cultural Projects among Aymara and Quechua Indians. *Grassroots Development,* 6(1):33–37.
Brinkerhoff, Derick W. 1980. Participation and Rural Development Effectiveness: An Organizational Analysis of Four Cases. Ph.D. diss., School of Education, Harvard University.

Brokensha, David, and Marion Pearsall. 1969. *The Anthropology of Development in Sub-Saharan Africa.* Monograph no. 10. Lexington, Ky.: Society for Applied Anthropology.

Brown, Chris. 1982. Locally-Initiated Voluntary Organizations: The Burial Societies of Botswana. *Rural Development Participation Review,* 3(3):11–15.

Brown, Chris, et al. 1982. *Rural Local Institutions in Botswana: Four Village Surveys and Analysis for Kgatleng District.* Ithaca, N.Y.: Rural Development Committee, Cornell University.

Brown, Lester. 1970. *Seeds of Change: The Green Revolution and Development in the 1970s.* New York: Praeger.

Bruce, Judith. 1980. *Market Women's Cooperatives: Giving Women Credit.* SEEDS pamphlet series. New York: Population Council, Carnegie Corporation, and Ford Foundation.

Bryant, Coralie, and Louise G. White. 1982. *Managing Development in the Third World.* Boulder, Colo.: Westview Press.

Buijs, Dieke. 1982a. The Participation Process: When It Starts. In Galjart and Buijs (1982:50–75).

———. 1982b. On Admittance, Access, Cooperation, and Participation: The Basic Concepts of the "Access and Participation" Research. In Galjart and Buijs (1982:8–20).

———. 1982c. Traditional and Modern Organizations. In Galjart and Buijs (1982: 76–85).

Bunting, A. H., ed. 1970. *Change in Agriculture.* London: Gerald Duckworth.

Cardoso, F. H. 1972. Dependency and Development in Latin America. *New Left Review,* July–Aug., 83–95.

Carroll, Thomas. 1971. Peasant Cooperation in Latin America. In Worsley (1971: 218–25).

Cernea, Michael. 1979. *Measuring Project Impact: Monitoring and Evaluation in the PIDER Rural Development Project—Mexico.* Staff working paper no. 332. Washington: World Bank.

———. 1982. Modernization and Development Potential of Traditional Grassroots Peasant Organizations. In *Directions of Change: Modernization Theory, Research and Realities,* ed. M. O. Attir, B. Holzner, and Z. Suda. Boulder, Colo.: Westview Press.

———. 1983. *A Social Methodology for Community Participation in Local Investments: The Experience of Mexico's PIDER Program.* Staff working paper no. 598. Washington: World Bank.

Chambers, Robert. 1974. *Managing Rural Development: Ideas and Experience from East Africa.* Uppsala, Sweden: Scandinavian Institute of African Studies.

———. 1975. *Water Management and Paddy Production in the Dry Zone of Sri Lanka.* Colombo, Sri Lanka: Agrarian Research and Training Institute.

———. 1978. Project Selection for Poverty-Focused Rural Development: Simple Is Optimal. *World Development,* 6(Feb.):209–19.

———. 1980. *Rural Poverty Unperceived: Problems and Remedies.* Staff working paper no. 400. Washington: World Bank.

———. 1983. *Rural Development: Putting the Last First.* London: Longmans.

Charlick, Robert. 1980. Animation Rurale: Experience with "Participatory" Development in Four West African Nations. *Rural Development Participation Review,* 1(Winter):1–6.

——. 1981. Animation Rurale: Participatory Approaches to Rural Development in Five Countries. Paper prepared for Conference on Local Political Organization and Rural Development Policy, University of Iowa, Iowa City, 14–16 September.

——. 1984. *Animation Rurale Revisited: Participatory Techniques for Improving Agriculture and Social Services in Five Francophone Nations.* Report to Office of Multi-Sectoral Development, Bureau of Science and Technology. Washington: U.S. Agency for International Development. (To be published by Rural Development Committee, Cornell University.)

Chenery, Hollis, and Alan Strout. 1966. Foreign Assistance and Economic Development. *American Economic Review,* Sept., 679–739.

Chowdhury, A. H. 1979. *Case Studies of Marketing (Success) and Landless Fishermen (Failure) Groups, Fakirakanda Village, Mymensingh District, Bangladesh.* Mymensingh: Small Farmer and Landless Labourers Development Project, Bangladesh Agricultural University.

CIMMYT. 1974. *The Puebla Project: Seven Years of Experiments, 1967–73.* El Bataan, Mexico: CIMMYT.

Clark, G. C., M. M. Crowley, M. E. Janz, and B. N. de los Reyes. 1979. *Report on the Third Annual Evaluation Workshop and Recommendations for Follow-up Action on the Small Farmers Development Project in Nepal.* Bangkok, Thailand: FAO Regional Office for Asia and the Far East.

Cline, William R. 1975. Distribution and Development: A Survey of Literature. *Journal of Development Economics,* Feb., 359–40.

Cohen, John M., and Norman Uphoff. 1977. *Rural Development Participation: Concepts and Measures for Project Design, Implementation and Evaluation.* Ithaca, N.Y.: Rural Development Committee, Cornell University.

Cohen, John M., Mary Hebert, David B. Lewis, and Jon C. Swanson. 1981. Development from Below: Local Development Associations in the Yemen Arab Republic. *World Development,* 9(11–12):1039–61.

Colburn, Forrest. 1981. *Guatemala's Rural Health Professionals.* Ithaca, N.Y.: Rural Development Committee, Cornell University.

Compton, J. Lin. 1982. Evaluating the Training and Visit System: A Recent Experience. Mimeo. Ithaca, N.Y.: Department of Extension and Adult Education, Cornell University.

Conrad, Charles, and Joyce Conrad. 1975. *Fifty Years: North Dakota Farmers Union.* Bismark, N.D.: NDFU.

Coombs, Philip. 1980. *Meeting the Basic Needs of the Rural Poor.* London: Pergamon Press.

Cornelius, Wayne. 1973. Nation Building, Participation and Distribution: The Politics of Social Reform under Cardenas. In Almond et al. (1973:392–498).

Coser, Lewis. 1956. *The Functions of Social Conflict.* New York: Free Press.

Coward, E. Walter, Jr. 1972. Irrigation and Organization: Research in Progress. In *View From the Paddy: Empirical Studies of Philippine Rice Farming and Tenancy,* ed. Frank Lynch. Quezon City: Institute of Philippine Culture.

——. 1979a. Participation in Irrigation Development: A Philippine Example. *Rural Development Participation Review,* 1(1):10–12.

———. 1979b. Principles of Social Organization in an Indigenous Irrigation System. *Human Organization,* 38(1):28–36.

Crampton, John A. 1965. *The National Farmers' Union.* Lincoln: University of Nebraska Press.

Cummings, Ralph W., Jr. 1982. *Improving the Lot of the People Left Behind.* Unpublished book manuscript.

Curtis, Donald. 1980. *Appropriate Village-Level Institutions: Some Generalizations from the Case of Lesotho's Village Water Supplies.* Agricultural Administration Network paper no. 8. London: Overseas Development Institute.

De, Nitish R. 1979. Organising and Mobilising: Some Building Blocks of Rural Work Organisations. *Human Features,* Winter, 32–64.

Delancey, Mark, and Virginia Delancey. 1982. Savings and Credit Alternatives for Rural Development in Anglophone Cameroon. Case study written for Rural Development Committee, Cornell University.

de los Reyes, Romana P., and Ma. Francisca P. Viado. 1979. Profiles of Two Communal Gravity Systems. *Philippine Agricultural Engineering Journal,* 10(2):14–18.

de Silva, G. V. S., et al. 1979. Bhoomi Sena: A Struggle for People's Power. *Development Dialogue* (Uppsala), 2, 3–70.

de Silva, N. G. R. 1981. Farmer Participation in Water Management: The Minipe Project in Sri Lanka. *Rural Development Participation Review,* 3(1):16–19.

Devine, Dennis. 1981. An Aborted Democracy. *Worldview,* May, 19–20.

Devitt, Paul. 1977. Notes on Poverty-Oriented Rural Development. In *Extension, Planning and the Poor,* ed. G. Hunter. London: Overseas Development Institute.

Dobyns, Henry F., Paul Doughty, and Harold Lasswell. 1971. *Peasants, Power, and Applied Social Change: Vicos as a Model.* Beverly Hills, Calif.: Sage.

Doherty, Victor S. 1980. Human Nature and the Design of Agricultural Technology. In *Proceedings of the International Workshop on Socioeconomic Constraints to the Development of Semi-Arid Agriculture in the Tropics,* ed. J. G. Ryan and H. L. Thompson. Hyderabad, India: ICRISAT.

Doherty, Victor S., and N. S. Jodha. 1979. Conditions for Group Action Among Farmers. In Wong (1979:207–23).

Dore, Ronald F. 1971. Modern Cooperatives in Traditional Communities. In Worsley (1971:43–60).

Dorner, Peter, ed. 1977. *Coops and Communes.* Madison: University of Wisconsin Press.

Dorsey, George. 1978. Milk and Justice. *Ceres,* Nov.-Dec., 31–38.

Downing, T. E. and McGuire Gibson, eds. 1974. *Irrigation's Impact on Society.* Tucson: University of Arizona Press.

Duewel, John. 1984. *Peasant Strategies for Cultivating Water User Associations: Two Case Studies from Central Java.* Ithaca, N.Y.: Cornell Irrigation Studies Papers, Cornell University.

Dye, Thomas R. 1973. *Politics in States and Communities.* Englewood Cliffs, N.J.: Prentice-Hall.

Eckstein, Harry, and Ted Robert Gurr. 1975. *Patterns of Authority: A Structural Basis for Political Inquiry.* New York: Wiley.

Eckstein, Schlomo. 1971. Land Reform and Cooperative Farming: An Evaluation of the Mexican Experience. In Weitz (1971b:294–309).

Edel, Matthew D. 1969. The Colombian Community Action Program: Costs and Benefits. *Yale Economics Essays,* 9(2):3–55.

Eicher, Carl K., and Doyle C. Baker. 1982. *Research on Agricultural Development in Sub-Saharan Africa: A Critical Survey.* East Lansing: Department of Agricultural Economics, Michigan State University.

Eicher, Carl, and Lawrence Witt, eds. 1964. *Agriculture in Economic Development.* New York: McGraw-Hill.

Elliott, Charles. 1975. *Patterns of Poverty in the Third World.* New York: Praeger.

ESCAP/FAO. 1979. *Learning from Rural Women: Village Level Success Cases of Rural Women's Group Income-Raising Activities.* Bangkok, Thailand: Economic and Social Commission for Asia and Pacific/Food and Agriculture Organization.

Esman, Milton J. 1972. The Elements of Institution Building. In *Institution Building and Development: From Concepts to Application,* ed. Joseph Eaton, 19–39. Beverly Hills, Calif.: Sage.

——. 1978a. *Landlessness and Near-Landlessness in Developing Countries.* Ithaca, N.Y.: Rural Development Committee, Cornell University.

——. 1978b. Development Administration and Constituency Organizations. *Public Administration Review,* 38(2):166–72.

Esman, Milton J., and John D. Montgomery. 1980. The Administration of Human Development. In Knight (1980:183–234).

Esman, Milton J., and Norman Uphoff. 1982. *Local Organization and Rural Development: The State of the Art.* Ithaca, N.Y.: Rural Development Committee, Cornell University.

Esman, Milton, et al. 1980. *Paraprofessionals in Rural Development.* Ithaca, N.Y.: Rural Development Committee, Cornell University.

Espiritu, Caesar Augusto, and Clarence J. Dias. 1982. Project Sarilakas: Towards a Partnership of Equals. In *Third World Legal Studies: Law in Alternative Strategies of Rural Development,* 262–83. New York: International Center for Law in Development.

Etzioni, Amitai. 1960. *A Comparative Analysis of Complex Organizations.* New York: Free Press.

Fals Borda, Orlando. 1976. The Crisis of Rural Cooperatives: Problems in Africa, Asia and Latin America. In Nash et al. (1976:439–56).

FAO. 1978–79. *Field Action for Small Farmers, Small Fishermen and Peasants. Vol. 1: The Field Workshop: A Methodology for Planning, Training and Evaluation of Programmes for Small Farmers/Fishermen and Landless Agricultural Labourers. Vol. 2: Small Farmer Development Manual.* Bangkok, Thailand: Food and Agriculture Organization.

——. 1979. *Report of World Conference on Agrarian Reform and Rural Development,* Rome, 12–20 July.

Favier, J. 1970. Aspects of the Cooperative Movement in Dahomey. *Cooperative Information,* 2(44).

Feder, Ernest. 1971. *The Rape of the Peasantry: Latin America's Landholding System.* New York: Doubleday.

Fernando, Tissa. 1977. *Cooperative Societies in Developing Nations: Some Problems at the Grassroots as Seen in Four Sri Lanka Villages.* Geneva: United Nations Research Institute for Social Development.

Festinger, Leon. 1957. *A Theory of Cognitive Dissonance*. Evanston, Ill.: Row, Peterson.

Field, John Osgood. 1980. Development at the Grassroots: The Organizational Imperative. *Fletcher Forum* (Tufts University), 4(2):145–65.

Fite, Gilbert C. 1965. *Farm to Factory: A History of the Consumers Cooperative Association*. Columbia: University of Missouri Press.

Flores, Xavier. 1970. *Agricultural Organization and Development*. Geneva: International Labour Office.

Fortmann, Louise. 1980. *Peasants, Officials and Participation in Rural Tanzania: Experience with Villagization and Decentralization*. Ithaca, N.Y.: Rural Development Committee, Cornell University.

Fortmann, Louise, and Emery Roe. 1981. *The Water Points Survey*. Gaborone, Botswana: Ministry of Agriculture.

Fountain, D. E. 1973. Programme of Rural Public Health: Vanga Hospital, Republic of Zaire. *Contact*, 13. Geneva: Christian Medical Commission.

Fox, Roger. 1979. Potentials and Pitfalls of Product Marketing through Group Action by Small Farmers. *Agricultural Administration*, 6(4):305–16.

Franda, Marcus. 1979. *India's Rural Development: An Assessment of Alternatives*. Bloomington: Indiana University Press.

——. 1981. Conservation, Water and Human Development at Sukhomajri. *American Universities Field Staff Reports*, Asia, no. 13.

Frank, A. G. 1968. *Capitalism and Underdevelopment in Latin America*. New York: Monthly Review.

Freire, Paolo. 1970. *The Pedagogy of the Oppressed*. New York: Herder & Herder.

Fremerey, Michael. 1981. *The Motivator Concept: Thoughts on a "Small Strategy" for Rural Development in the Third World*. Frankfurt am Main: Deutsches Institut für Internationale Pädagogische Forschung.

Fresson, Silvianne. 1979. Public Participation on Village Level Irrigation Perimeters in the Matam Region of Senegal. In D. Miller (1979:96–137).

Friedmann, John, and Clyde Weaver. 1979. *Territory and Function: The Evolution of Regional Planning*. London: Edward Arnold.

Galaj, Dyzma. 1973. The Polish Peasant Movement in Politics: 1895–1969. In Landsberger (1973a:316–47).

Galjart, Benno. 1967. *Itaguai: Old Habits and New Patterns in a Brazilian Land Settlement*. Wageningen: Center for Agricultural Publishing and Documentation.

——. 1981a. Participatory Development Projects: Some Conclusions from Research. *Sociologia Ruralis*, 21,142–59.

——. 1981b. Cooperation as Pooling. Paper presented to IUAES Symposium on Traditional Cooperation and Social Organisation in Relation to Modern Cooperative Organisation and Enterprise, Amsterdam, 23–24 April.

Galjart, Benno, and Dieke Buijs, eds. 1982. *Participation of the Poor in Development: Contributions to a Seminar*. Leiden, Netherlands: Institute of Cultural and Social Studies, University of Leiden.

Galjart, Benno, et al. 1982. *Participatie, toegang tot ontwikkeling: Verslag van een onderzoek naar de mogelijkheden van participate in ontwikkelingsprojekten*. Leiden, Netherlands: Institute of Cultural and Social Studies, University of Leiden.

Ganewatte, Piyasena. 1971–72. *A Study of Rural Leadership: Weliyawa Village Leadership Study; Patterns of Leadership in Colonisation Schemes*. Mimeo. Colombo, Sri

Lanka: Mahaweli Development Board and Land Commissioner's Department.

Garzon, Jose M. 1981. Small-Scale Public Works, Decentralization and Linkages. In *Linkages to Decentralized Units,* ed. D. Leonard and D. R. Marshall. Berkeley: Institute of International Studies, University of California.

Geertz, Clifford. 1962. The Rotating Credit Association: A "Middle Rung' in Development. *Economic Development and Cultural Change,* 10(3):241–63.

——. 1967. Tihingan: A Balinese Village. In *Villages in Indonesia,* ed. Koentjaraningrat. Ithaca, N.Y.: Cornell University Press.

Geertz, Hildred. 1963. Indonesian Cultures and Communities. In *Indonesia,* ed. Ruth McVey, 24–96. New Haven: Yale University Press.

Gerschenkron, Alexander. 1966. *Bread and Democracy in Germany.* New York: Harold Fertig.

Ghai, Dharam, and Anisur Rahman. 1979. *Rural Poverty and the Small Farmers' Development Programme in Nepal.* Geneva: Rural Employment Policies Branch, International Labour Office.

——. 1981. The Small Farmers' Groups in Nepal. *Development,* 1981, no. 1, 23–28.

Glennie, Colin E. R. 1979. The Rural Piped Water Programme in Malawi: A Case Study in Community Participation. M.S. thesis, Department of Civil Engineering, Imperial College of Science and Technology, University of London.

——. 1982. *A Model for the Development of a Self-Help Water Supply Program.* Technology advisory group working paper no. 1. Washington: World Bank.

Goldsmith, Arthur A. 1981. Popular Participation and Rural Leadership in the Saemaul Movement. In *Toward a New Community Life: Reports of International Research Seminar on the Saemaul Movement,* ed. Man-Gap Lee, 427–57. Seoul, South Korea: Institute of Saemaul Undong Studies, Seoul National University.

Goldsmith, Arthur, and Harvey Blustain. 1980. *Local Organization and Participation in Integrated Rural Development in Jamaica.* Ithaca, N.Y.: Rural Development Committee, Cornell University.

Golladay, Frederick. 1983. Meeting the Needs of the Poor for Water Supply and Waste Disposal. In *Appropriate Technology for Water Supply and Sanitation.* Washington: World Bank.

Gooch, Toby. 1979. *An Experiment with Group Ranches in Upper Volta.* Pastoral Network paper 9b. London: Overseas Development Institute.

Goodell, Grace. 1980. From Status to Contract: The Significance of Agrarian Relations of Production in the West, Japan and in "Asiatic" Persia. *European Journal of Sociology,* 21, 285–325.

——. 1982. Conservatism and Foreign Aid. *Policy Review,* Winter, 111–31.

Goulet, Denis. 1971. *The Cruel Choice: A New Concept in the Theory of Development.* New York: Atheneum.

Government of India. 1978. *Draft Five-Year Plan.* New Delhi: Government of India Press.

Gow, David, and Jerry Van Sant. 1981. *Beyond the Rhetoric of Rural Development Participation: Can It Be Done?* IRD working paper no. 9. Washington: Development Alternatives, Inc.

Gow, David D., et al. 1979. *Local Organizations and Rural Development: A Comparative Reappraisal,* 2 volumes. Washington: Development Alternatives, Inc.

Gran, Guy. 1983. *Development by People: Citizen Construction of a Just World.* New York: Praeger.

Grant, James P. 1973. Development: An End of the Trickle Down? *Foreign Policy,* Fall, 43–65.

Green, James W. 1975. *Local Initiative in Yemen: Exploratory Studies of Four Local Development Associations.* Washington: Office of Technical Support, USAID.

Greenwood, Davydd. 1973. *The Political Economy of Peasant Family Farming: Some Anthropological Perspectives on Rationality and Adaptation.* Ithaca, N.Y.: Rural Development Committee, Cornell University.

Griffin, Keith. 1976. *Land Concentration and Rural Poverty.* New York: Holmes & Meier.

Griffin, Keith, and Ajit Kumar Ghose. 1979. Growth and Impoverishment in the Rural Areas of Asia. *World Development,* 7(2):361–83.

Grijpstra, Bouwe. 1982a. Initiating and Supervising Agencies: Group Approaches in Rural Development. In Galjart and Buijs (1982:199–219).

——. 1982b. Approaches to Initiating and Supervising Groups for Rural Development. *Rural Development Participation Review:* 3(2):1–7.

Grindle, Merilee. 1977. *Bureaucrats, Politicians and Peasants in Mexico: A Case Study in Public Policy.* Berkeley: University of California Press.

——, ed. 1980. *Politics and Policy Implementation in the Third World.* Princeton, N.J.: Princeton University Press.

Gross, Bertram. 1964. *The Managing of Organizations.* New York: Free Press.

Gunawardena, A. M. T., and A. Chandrasiri. 1981. *Training and Visit System of Extension.* Colombo, Sri Lanka: Agrarian Research and Training Institute.

Hadden, Susan G. 1974. *Decentralization and Rural Electrification in Rajasthan, India.* Ithaca, N.Y.: Rural Development Committee, Cornell University. Revised version in Uphoff (1982–83:1,225–88).

Haeri, M. Hossein, and M. Taghi Farvar. 1980. *Traditional Rural Institutions and Their Implications For Development Planning: Studies from Hamadan Province of Iran.* Tokyo: United Nations University.

Hafid, Anwar, and Yujiro Hayami. 1976. Mobilizing Local Resources for Irrigation Development: The Subsidi Desa Case of Indonesia. Agricultural Economics Department paper 76-18. Los Baños, Philippines: International Rice Research Institute.

Hamer, John H. 1976. Prerequisites and Limitations in the Development of Voluntary Self-Help Associations: A Case Study and Comparison. *Anthropological Quarterly,* 19(2):107–34.

——. 1980. Preference, Principle and Precedent: Dispute Settlement and Changing Norms in Sidamo Associations. *Ethnology,* 19(1):89–109.

Hansen, Roger D. 1971. *The Politics of Mexican Development.* Baltimore: Johns Hopkins University Press.

Haragopal, G. 1980. *Administrative Leadership and Rural Development in India.* New Delhi: Light and Life Publishers.

Harbeson, John W. 1972. Cooperative Societies, Local Politics and Development: A Kenya Case Study. Paper presented at 15th annual meeting of the African Studies Association, Philadelphia, 8–11 November.

Harik, Iliya. 1974. *The Political Mobilization of Peasants: Change in an Egyptian Village.* Bloomington: Indiana University Press.

Hartfiel, Ann. 1982. Two Women's Production Cooperatives. *Grassroots Development,* 6(1):38–41. Washington: Inter-American Foundation.

Hayami, Yujiro, and Vernon W. Ruttan. 1971. *Agricultural Development: An International Perspective.* Baltimore: Johns Hopkins University Press.

Hayter, Theresa. 1971. *Aid as Imperialism.* Harmondsworth, U.K.: Penguin.

Healy, Kevin. 1980. Innovative Approaches to Development Participation in Rural Bolivia. *Rural Development Participation Review,* 1(3):15–18.

Heim, Franz G., Akin Rabibhadana, and Chirmsak Pinthong. 1983. *How to Work With Farmers: A Manual for Field Workers.* Khon Kaen, Thailand: Research and Development Institute, Khon Kaen University.

Herring, E. Pendleton. 1936. *Public Administration and the Public Interest.* New York: McGraw-Hill.

Hickey, Gerald C., and Robert A. Flammang. 1977. *The Rural Poor Majority in the Philippines: Their Present and Future Status as Beneficiaries of A.I.D. Programs.* Manila: USAID.

Hicks, John D. 1931. *The Populist Revolt.* Minneapolis: University of Minnesota Press.

Hicks, Ursala K. 1961. *Development from Below.* Oxford: Clarendon Press.

Hilton, Rodney H. 1973. Peasant Society, Peasant Movements and Feudalism in Medieval Europe. In Landsberger (1973a:67–94).

Hirschman, Albert O. 1967. *Development Projects Observed.* Washington: Brookings Institution.

——. 1970. *Exit, Voice and Loyalty.* Cambridge, Mass.: Harvard University Press.

——. 1981. *Essays in Trespassing: Economics to Politics and Beyond.* Cambridge: Cambridge University Press.

Hodsdon, Dennis. 1979. *The Federation of Free Farmers in the Philippines.* Geneva: International Labour Office.

Holdcroft, Lane E. 1978. *The Rise and Fall of Community Development in Developing Countries: A Critical Analysis and an Annotated Bibliography.* MSU rural development paper no. 2. East Lansing: Department of Agricultural Economics, Michigan State University.

Hollnsteiner, Mary Racelis. 1963. *The Dynamics of Power in a Philippine Municipality.* Quezon City: Community Development Research Council, University of the Philippines.

——. 1977. Local Initiatives and Modes of Participation in Asian Cities. *Assignment Children* (UNICEF), 40,11–48.

——. 1979. Mobilizing the Rural Poor through Community Organization. Mimeo. Rome: Rural Organization Action Project, FAO.

Holmquist, Frank. 1979. Class Structure, Peasant Participation and Rural Self-Help. In Barkan and Okumu (1979:129–153).

Honadle, George. 1979. Beneficiary Involvement in Project Implementation: Experience in the Bicol. *Rural Development Participation Review,* 1(1):12–13.

Honadle, George, and Rudi Klauss. 1979. *International Development Administration: Implementation Analysis for Development Projects.* New York: Praeger.

Hoselitz, Bert F. 1957. Non-economic Factors in Economic Development. *American Economic Review,* May, 28–42.

Hossain, Mahbub, Raisal Awal, Mahmood Qazi, and Kholiquzzaman Ahmad. 1979.

Participatory Development Efforts in Rural Bangladesh: A Case Study. World Employment Programme research working papers. Geneva: International Labour Office.

Howell, John. 1982. *Managing Agricultural Extension: The T and V System in Practice*. Agricultural Administration Unit discussion paper 8. London: Overseas Development Institute.

Howse, C. J. 1974. Agricultural Development without Credit. *Agricultural Administration,* 1(4):259–62.

Huizer, Gerrit. 1963. A Community Development Experience in a Central American Village. *International Review of Community Development,* 12,161–86.

Hunt, E. McCauley. 1974. A Smallholder Milk Cooperative in Gujarat, India. *Development Digest,* 12(2):45–52.

Hunt, Eva, and Robert C. Hunt. 1974. Irrigation, Conflict and Politics: A Mexican Case. In Downing and Gibson, (1974:129–57).

Hunter, Guy. 1969. *Modernizing Peasant Societies*. New York: Oxford University Press.

———. 1976. Organizations and Institutions. In Hunter, Bunting, and Bottrall (1976:197–207).

———. 1978. *Agricultural Development and the Rural Poor*. London: Overseas Development Institute.

———. 1980a. Criteria for Choice of Institutional Forms. Mimeo. London: Overseas Development Institute.

———. 1980b. *Enlisting the Small Farmer: The Range of Requirements*. AAU/ODI occasional paper no. 4. London: Overseas Development Institute.

Hunter, Guy, and Janice Jiggins. 1977. Farmer and Community Groups. Mimeo. London: Agricultural Administration Unit, Overseas Development Institute.

Hunter, Guy, A. H. Bunting, and Anthony Bottrall, eds. 1976. *Policy and Practice in Rural Development*. London: Croom Helm with the Overseas Development Institute.

Huntington, Richard. 1980. Councils' Inaction: The Triple Marginality of Local Development Committees in Africa. Mimeo. Cambridge, Mass.: Department of Anthropology, Harvard University.

Huntington, Samuel P. 1968. *Political Order in Changing Societies*. New Haven: Yale University Press.

Hutupea, R., et al. 1978. The Organization of Farm-Level Irrigation in Indonesia. In *Irrigation Policy and Management in Southeast Asia,* 167–74. Los Baños, Philippines: International Rice Research Institute.

Hyden, Goran. 1970. Cooperatives and Their Socio-Political Environment. In Widstrand (1970:61–80).

———. 1973. *Efficiency vs. Distribution in East African Cooperatives: A Study in Organizational Conflicts*. Management and Administration series no. 1. Nairobi, Kenya: East Africa Literature Bureau.

———. 1978–79. Cooperatives and Local Leadership Patterns. *Rural Africana,* 3(Winter):49–59.

———. 1980. Cooperatives and the Poor: Comparing European and Third World Experience. *Rural Development Participation Review,* 2(1):9–12.

——. 1981a. *Beyond Ujamaa in Tanzania: Underdevelopment and an Uncaptured Peasantry.* Berkeley: University of California Press.

——. 1981b. Modern Cooperatives and the Economy of Affection in Sub-Saharan Africa. Paper presented at IUAES Symposium on Traditional Cooperation and Social Organization in Relation to Modern Cooperative Organization and Enterprise, Amsterdam, 23–24 April.

——. 1983. *No Shortcuts to Progress: African Development Management in Perspective.* Berkeley: University of California Press.

IAF. 1982. *Inter-American Foundation Annual Report 1981.* Washington: Inter-American Foundation.

ICA. 1978. *Co-operatives and the Poor:* London: International Cooperative Alliance.

Ilchman, Warren F., and Norman Uphoff. 1969. *The Political Economy of Change.* Berkeley: University of California Press.

ILO. 1974. *Sharing in Development: A Programme of Employment, Equity and Growth for the Philippines.* Geneva: International Labour Office.

——. 1975. *Time for Transition: A Mid-term Review of the Second United Nations Development Decade.* Geneva: International Labour Office.

——. 1977. *Employment, Growth and Basic Needs: A One-World Problem.* New York: Praeger.

Inayatullah. 1972. *Cooperatives and Development in Asia: A Study of Cooperatives in Fourteen Rural Communities of Iran, Pakistan and Ceylon.* Geneva: United Nations Research Institute for Social Development.

Isely, Raymond B. 1979. Reflections on an Experience in Community Participation in Cameroon. *Ann. Soc. Belge Med. Trip,* 59, suppl., 103–15.

Isely, Raymond B., and Craig R. Hafner. 1982. Facilitation of Community Organization. *Water Supply and Management,* 6(5):431–42.

Isely, Raymond B., and Jean F. Martin. 1977. The Village Health Committee: Starting Point for Rural Development. *WHO Cronicle,* 31:307–15.

Isely, Raymond B., L. L. Sanwogou, and Jean F. Martin. 1979. Community Organization as an Approach to Health Education in Rural Africa. *International Journal of Health Education,* 22(July–Sept.):3–9.

Islam, Md. Nazrul. 1979. *Case Studies of Krishi Group Ka (Success) and Rickshaw Pullers' Group (Failure), Boyra Village, Mymensingh District, Bangladesh.* Mymensingh: Small Farmers and Landless Labourers Development Programme, Bangladesh Agricultural University.

Isles, Carlos, and M. Collado. 1979. Farmer Participation in Communal Irrigation Development: Lessons from Laur. *Philippines Agricultural Engineering Journal,* 10(2):3–4.

Jackson, George D., Jr. 1973. Peasant Political Movements in Eastern Europe. In Landsberger (1973a:259–315).

Janzen, John M. 1969. The Cooperative in Lower Congo Economic Development. In Brokensha and Pearsall (1969:70–76).

Jedlicka, Allen D. 1977. *Organization for Rural Development: Risk Taking and Appropriate Technology.* New York: Praeger.

Johnson, E. A. G. 1970. *The Organization of Space in Developing Countries.* Cambridge, Mass.: Harvard University Press.

Johnston, Bruce F., and William C. Clark. 1982. *Redesigning Rural Development: A Strategic Perspective.* Baltimore: Johns Hopkins University Press.

Johnston, Bruce F., and Peter Kilby. 1975. *Agriculture and Structural Transformation: Economic Strategies in Late-Developing Countries.* New York: Oxford University Press.

Joshi, Janardan. 1980. *Background Paper on Small Farmers Development Programme: Nepal.* Regional workshop on organizing small farmer groups for income generating activities. Kathmandu: His Majesty's Government and FAO/ESCAP, 17–27 November.

Kahl, Joseph. 1976. *Modernization, Exploitation and Dependency in Latin America.* New Brunswick, N.J.: Transaction Books.

Kale, Pratima, and Philip Coombs. 1980. Social Work and Research Center: An Integrated Team Approach in India. In Coombs (1980:289–361).

Kanhare, Vijay P. 1980. The Struggle in Dhulia: A Women's Movement in India. In *Participation in Research: Case Studies of Participatory Research in Adult Education,* ed. Helen Callaway. Amersfoort, Netherlands: Studiecentrum ncvo.

Karanja, Edward. 1972. The Problem of Amalgamating Cooperative Societies: The Case of Northern Tetu. In Widstrand (1972:105–33).

Kasfir, Nelson. 1977. *The Shrinking Political Arena: Participation and Ethnicity in African Politics, with a Case Study of Uganda.* Berkeley: University of California Press.

Killick, A. J. 1978. *Development Economics in Action: A Study of Economic Policies in Ghana.* New York: St. Martin's Press.

Kimble, David. 1963. *A Political History of Ghana: The Rise of Gold Coast Nationalism, 1850–1928.* Oxford: Clarendon Press.

Kincaid, D. Lawrence, Hyung-Jong Park, Kyung-Kyoon Chung, and Chin-Chuan Lee. 1976. *Mothers' Clubs and Family Planning in Rural Korea: The Case of Oryu Li.* Honolulu: East-West Communication Institute.

Kindleberger, Charles. 1958. *Economic Development.* New York: McGraw-Hill.

King, Roger. 1975. Experience in Administration of Cooperative Credit and Marketing Societies in Northern Nigeria. *Agricultural Administration,* 2(3):195–207.

———. 1976. Farmers' Cooperatives in Northern Nigeria: A Case Study Used to Illustrate the Relationship between Economic Development and Institutional Change. Ph.D. diss., Department of Agricultural Economics, Ahmadu Bello University, Nigeria, and Department of Agricultural Economics, University of Reading, U.K.

———. 1981. Cooperative Policy and Village Development in Northern Nigeria. In *Rural Development in Tropical Africa,* ed. J. Heyer, P. Roberts, and G. Williams, 259–80. New York: St. Martin's Press.

Kloppenburg, Jack. 1983. Group Development in Botswana: The Principles of Collective Farmer Action. *Research in Economic Anthropology,* vol. 5, ed. G. Dalton, 311–33. Greenwich, Conn.: JAI Press.

Knapp, Joseph G. 1969. *The Rise of American Cooperative Enterprise: 1620–1920.* Danville, Ill.: Interstate Printers.

Kneerim, Jill. 1980. *Village Women Organize: The Mraru Bus Service.* SEEDS pamphlet series. New York: Population Council, Carnegie Corporation, and Ford Foundation.

Knight, Peter, ed. 1980. *Implementing Programs of Human Development*. Staff working paper no. 403. Washington: World Bank.

Korten, David C. 1980. Community Organization and Rural Development: A Learning Process Approach. *Public Administrative Review*, 40(5):480–511.

Korten, David C., and Felipe B. Alfonso, eds. 1981. *Bureaucracy and the Poor: Closing the Gap*. Singapore: McGraw-Hill.

Korten, David C., and Norman Uphoff. 1981. *Bureaucratic Reorientation for Participatory Development*. NASPAA working paper no. 1. Washington: National Association of Schools of Public Affairs and Administration.

Korten, Frances F. 1982. *Building National Capacity to Develop Water Users' Associations: Experience from the Philippines*. Staff working paper no. 528. Washington: World Bank.

Korten, Frances F., and Sarah Young. 1978. The Mothers Club of Korea. In *Managing Community-Based Population Programmes*. Kuala Lumpur, Malaysia: International Committee for the Management of Population Programmes.

Kwoh, Min Hioh. 1964. *Farmers Associations and Their Contributions toward Agricultural and Rural Development in Taiwan*. Bangkok, Thailand: FAO Regional Office for Asia and the Far East.

Labovitz, S. 1970. The Assignment of Numbers to Rank Order Categories. *American Sociological Review*, 35,515–24.

Lamb, Geoff. 1974. *Peasant Politics: Conflict and Development in Murang'a*. New York: St. Martin's Press.

Landsberger, Henry. 1971. *Latin American Peasant Movements*. Ithaca, N.Y.: Cornell University Press.

——. 1973a. *Rural Protest: Peasant Movements and Social Change*. New York: Barnes & Noble.

——. 1973b. Peasant Unrest: Themes and Variations. In Landsberger (1973a:1–64).

Landsberger, Henry, and Cynthia Hewitt de Alcantara. 1971. From Violence to Pressure-Group Politics and Cooperation: A Mexican Case Study. In Worsley (1971:293–346).

Landau, Y. H., et al. 1976. *Rural Communities, Intercooperation and Development*. New York: Praeger.

Lang, Gottfried, Warren J. Roth, and Martha B. Lang. 1969. Sukumaland Cooperatives as Mechanisms of Change. In Brokensha and Pearsall (1969:48–63).

Lassen, Cheryl A. 1980. *Reaching the Assetless Poor: Projects and Strategies for Their Self-Reliant Development*. Ithaca, N.Y.: Rural Development Committee, Cornell University.

Lee, Man-Gap. 1979. *Participation in Saemaul Movement*. Seoul, South Korea: Institute of Saemaul Studies, Seoul National University.

Lele, Uma. 1975. *The Design of Rural Development: Lessons from Africa*. Baltimore: Johns Hopkins University Press.

——. 1981. Co-operatives and the Poor: A Comparative Perspective. *World Development*, 9(1):55–72.

Lemarchand, René. 1972. Political Clientelism and Ethnicity in Tropical Africa. *American Political Science Review*, 66(1):68–90.

Leonard, David K. 1977. *Reaching the Peasant Farmer: Organization Theory and Practice in Kenya*. Chicago: University of Chicago Press.

———. 1982a. Analyzing the Organizational Requirements for Serving the Rural Poor. In Leonard and Marshall (1982:1–39).

———. 1982b. Choosing among Forms of Decentralization and Linkage. In Leonard and Marshall (1982:193–226).

Leonard, David K., and Dale Rogers Marshall, eds. 1982. *Institutions of Rural Development for the Poor: Decentralization and Organizational Linkages*. Berkeley: Institute of International Studies, University of California.

Lewis, Henry T. 1971. *Ilocano Rice Farmers*. Honolulu: University Press of Hawaii.

Lewis, W. Arthur. 1955. *A Theory of Economic Growth*. Homewood, Ill.: Irwin.

Liebenow, J. Gus. 1981. Malawi: Clean Water for the Rural Poor. *American Universities Field Staff Reports*, Africa, no. 40.

Lipset, S. M., James Coleman, and Martin Trow. 1962. *Union Democracy*. Garden City, N.Y.: Doubleday.

Lipton, Michael. 1977. *Why Poor People Stay Poor: Urban Bias in World Development*. Cambridge, Mass.: Harvard University Press.

Lopez, Alberto V. 1979. *Barangay Liberty, Pantabangan, Nueva Ecija, Philippines: Group #4 (Success) and Group #1 (Failure)*. Case studies on small farmer development. Manila: Small Farmer Development Program, Ministry of Agrarian Reform.

Lowdermilk, Max K., A. C. Early, and D. M. Freeman. 1978. *Irrigation Water Management in Pakistan: Constraints and Farmer Responses, a Comprehensive Survey*. Technical report no. 46, vol. 4. Fort Collins: Colorado State University.

Lynch, Barbara D. 1982. *The Vicos Experiment: A Study of the Impacts of the Cornell-Peru Project in a Highland Community*. AID special evaluation study no. 7. Washington: USAID.

McClintock, Cynthia. 1981. *Peasant Cooperatives and Political Change in Peru*. Princeton, N.J.: Princeton University Press.

McGrath, Mary Jean, ed. 1978. *Cooperatives, Small Farmers and Rural Development*. Madison: University Center for Cooperatives, University of Wisconsin.

Maddick, Henry. 1963. *Democracy, Decentralization and Development*. Bombay: Asia Publishing House.

Maeda, Justin H. J. 1981. Creating National Structures for People Centered Agrarian Development. In Korten and Alfonso (1981:136–62).

Maguire, Robert. 1979. *Bottom-up Development in Haiti*. Washington: Inter-American Foundation.

Manor, James. 1977. Structural Change in Karnataka Politics. *Economic and Political Weekly*, 29 Oct., 1865–1869.

———. 1982. The Dynamics of Political Integration and Disintegration. In *The States of South Asia: Problems of National Integration*, ed. A. J. Wilson and D. Dalton. Honolulu: University Press of Hawaii.

March, James G., and Herbert A. Simon. 1958. *Organizations*. New York: Wiley.

March, Kathryn, and Rachel Taqqu. 1982. *Women's Informal Associations and the Organizational Capacity for Development*. Ithaca, N.Y.: Rural Development Committee, Cornell University.

Marshall, Ray, and Lamond Godwin. 1971. *Cooperatives and Rural Poverty in the South*. Baltimore: Johns Hopkins University Press.

Meehan, Eugene J. 1978. *In Partnership with People: An Alternative Development Strategy*. Washington: Inter-American Foundation.

Meister, Albert. 1966. *Développement économique des pays de l'Est African*. Paris: Presses Universitaires de France.

——. 1969. *Participation, animation et développement*. Paris: Editions Anthropos.

——. 1970. Développement communautaire et animation rurale en Afrique. *L'Homme et al Société*, 19.

Mellor, John W. 1966. *The Economics of Agricultural Development*. Ithaca, N.Y.: Cornell University Press.

Mensah, Moise. 1977. An Experience of Group Farming in Dahomey: The Rural Development Cooperatives. In Dorner (1977:277–86).

Michels, Robert. 1915. *Political Parties*. Reprint. Glencoe, Ill.: Free Press, 1959.

Migdal, Joel. 1974. *Peasants, Politics and Revolution*. Princeton, N.J.: Princeton University Press.

Miller, Barbara D. 1980. *Local Social Organizations and Local Project Capacity*. Syracuse, N.Y.: Maxwell School, Syracuse University.

Miller, Duncan. 1979. *Self-Help and Popular Participation in Rural Water Systems*. Paris: Development Centre, Organization for Economic Cooperation and Development.

Miller, Robert F. 1977. Group Farming Practices in Yugoslavia. In Dorner (1977: 163–98).

Millikan, Max, and David Hapgood. 1967. *No Easy Harvest: The Dilemma of Agriculture in Underdeveloped Countries*. Boston: Little, Brown.

Mintz, Sidney. 1979. The Rural Proletariat and the Problem of Rural Proletarian Consciousness. In *Peasants and Proletarians: The Struggle of Third World Workers*, ed. Robin Cohen et al., 173–97. New York: Monthly Review Press.

Misra, K. K. 1975. Safe Water in Rural Areas: An Experiment in Promoting Community Participation in India. *International Journal of Health Education*, 18,53–59.

Mitchell, Robert. 1965. Survey Materials Collected in the Developing Countries: Sampling, Measurement and Interviewing Obstacles to Intra- and Inter-National Comparisons. *International Social Science Journal*, 17, 665–85.

Montgomery, John D. 1972. The Allocation of Authority in Land Reform Programs: A Comparative Study of Administrative Process and Outputs. *Administrative Science Quarterly*, 17(1):62–75.

Montgomery, John, and Milton Esman. 1971. Popular Participation in Development Administration. *Journal of Comparative Administration*, 3(3):358–82.

Montiel, Christina. 1980. *Rural Organizations in the Philippines*, ed. Marie S. Fernandez. Manila: Institute of Philippine Culture, Ateneo de Manila.

Moore, Cynthia. 1981. *Paraprofessionals in Village-Level Development in Sri Lanka: The Sarvodaya Shramadana Movement*. Ithaca, N.Y.: Rural Development Committee, Cornell University.

Moore, M. E. 1979. Social Structure and Institutional Performance: Local Farmers' Organizations in Sri Lanka. *Journal of Administration Overseas*, 18(4):240–49.

——. 1983. Irrigation Management in Taiwan. Unpublished paper. Sussex, U.K.: Institute of Development Studies.

Morell, David, and Chaianan Samudavaniya. 1981. *Political Conflict in Thailand*. Cambridge, Mass.: Oelgeschlager, Gunn & Hain.

Moris, Jon. 1981. *Managing Induced Rural Development*. Bloomington: Program of Advanced Studies in Institution Building and Technical Assistance Methodology (PASITAM), Indiana University.

Morland, Robert C. 1955. *Political Prairie Fire*. Minneapolis: University of Minnesota Press.

Morss, Elliott R., John K. Hatch, Donald R. Mickelwait, and Charles F. Sweet. 1976. *Strategies for Small Farmer Development*, 2 vols. Boulder, Colo.: Westview Press.

Mosher, Arthur T. 1966. *Getting Agriculture Moving: Essentials for Development and Modernization*. New York: Praeger.

———. 1969. *Creating a Progressive Rural Structure to Serve a Modern Agriculture*. New York: Agricultural Development Council.

———. 1971. *To Create a Modern Agriculture: Organization and Philosophy*. New York: Agricultural Development Council.

Moulik, T. M. 1980. Action Research on Rural Development for Rural Poor: The Dharampur and Jawaja Projects. Mimeo. Ahmedabad: Indian Institute of Management.

———. 1981. Taiwan Farmers' Organisations: Some Observations. Mimeo. Ahmedabad: Indian Institute of Management.

Münkner, Hans. 1976. *Cooperatives: For the Rich or for the Poor?* Marburg: Institute for Cooperation in Developing Countries.

———. 1979. *Cooperatives and Rural Poverty*. Marburg: Institute for Cooperation in Developing Countries.

———. 1981. Possibilities and Problems of Transformation of Local Village Groups into Pre-Cooperatives: Experiences in French-Speaking Countries of West Africa. Paper presented at IUAES Symposium on Traditional Cooperation and Social Organization in Relation to Modern Cooperative Organization and Enterprise, Amsterdam, 23–24 April.

Myrdal, Gunnar. 1968. *Asian Drama: An Inquiry into the Poverty of Nations*, 3 vols. New York: Pantheon.

Naipaul, V. S. 1977. *India: A Wounded Civilization*. Harmondsworth, U.K.: Penguin.

Nambodiri, N. K., Lewis F. Carter, and Hubert M. Blalock, Jr. 1975. *Applied Multivariate Analysis and Experimental Design*. New York: McGraw-Hill.

Nash, June, Jorge Dandler, and Nicholas Hopkins, eds. 1976. *Popular Participation in Social Change: Cooperatives, Collectives and Nationalized Industry*. Chicago: Aldine.

Nelson, Joan. 1979. *Access to Power: Politics and the Urban Poor in Developing Nations*. Princeton, N.J.: Princeton University Press.

Nerfin, Marc, ed. 1977. *Another Development: Approaches and Strategies*. Stockholm: Dag Hammarskuöld Foundation.

Nicholson, Norman K. 1973. *Panchayati Raj, Rural Development and the Political Economy of Village India*. Ithaca, N.Y.: Rural Development Committee, Cornell University.

NIPA. 1976. *Organization for Participation in Rural Development in Zambia*. Lusaka: Administration for Rural Development Project, National Institute of Public Administration, Zambia.

Oakley, P. 1980. Can Jorge Do It? *RRDC Bulletin*, no. 9. Reading, U.K.: University of Reading.

O'Brien, Robert M. 1979. The Use of Pearson's R with Ordinal Data. *American Sociological Review*, 44,851–57.

ODM. 1975. *Overseas Development: The Changing Emphasis in British Aid Policies.* Cmnd. 6270. London: Her Majesty's Stationery Office.
Okonjo, Kamene. 1979. Rural Women's Credit Systems: A Nigerian Example. In *Learning about Rural Women: Studies in Family Planning,* ed. S. Zeidenstein, cited in Cernea (1982).
Olson, Mancur. 1965. *The Logic of Collective Action.* Cambridge, Mass.: Harvard University Press.
Owens, Edgar, and Robert Shaw. 1972. *Development Reconsidered: Bridging the Gap between Government and People.* Lexington, Mass.: D. C. Heath.
Oxby, Clare, 1980. How to Benefit the Poor via Local Organisations. Mimeo. London: Overseas Development Institute.
Paddock, William, and Elizabeth Paddock. 1973. *We Don't Know How.* Ames: Iowa State University Press.
Paige, Jeffrey. 1975. *Agrarian Revolution: Social Movements and Export Agriculture in the Underdeveloped World.* New York: Free Press.
Pandey, S. M. 1975. *Agricultural Workers' Union in Panipat: A Case Study.* In Seth (1975).
Parras, N. 1976. *Operativos Campesinos.* Tegucigalpa, Honduras: PROCCARA.
Paul, Samuel. 1982. *Managing Development Programs: The Lessons of Success.* Boulder, Colo.: Westview Press.
Peterson, Stephen B. 1982a. Government, Cooperatives, and the Private Sector in Peasant Agriculture. In Leonard and Marshall (1982:73–124).
——. 1982b. Alternative Local Organizations Supporting the Agricultural Development of the Poor. In Leonard and Marshall (1982:125–50).
Petras, James F., and Robert LaPorte. 1971. *Cultivating Revolution: The United States and Agrarian Reform in Latin America.* New York: Random House.
Poats, Rutherford M. 1972. *Technology for Developing Nations.* Washington: Brookings Institution.
Popkin, Samuel. 1979. *The Rational Peasant: The Political Economy of Rural Society in Vietnam.* Berkeley: University of California Press.
——. 1981. Public Choice and Rural Development: Free Riders, Lemons and Institutional Design. In Russell and Nicholson (1981:43–80).
Powell, John Duncan. 1971. *Political Mobilization of the Venezuelan Peasant.* Cambridge, Mass.: Harvard University Press.
Pradhan, Bharat B. 1980. Strategy of the Development of Small Farmers through Group Approach: The Case of Nepal. Paper for workshop on small farmer development and credit policy. Kathmandu, Nepal: Agricultural Development Bank of Nepal and Department of Agricultural Economics and Rural Sociology, Ohio State University.
Pradhan, Prachanda P. 1980. *Local Institutions and People's Participation in Rural Public Works in Nepal.* Ithaca, N.Y.: Rural Development Committee, Cornell University.
Rabibhadana, Akin. 1983. The Transformation of Tambon Yokkrabat, Changwat Samut Sakorn. *Thai Journal of Development Administration,* 22(1):73–104.
Raha, Shankar Kumar. 1979. *Case Studies of Two Groups: Rickshaw Pulling Group (Success) and Paddy Processing Group (Failure), Digharkanda Village.* Mymensingh: Small Farmer and Landless Labourers Development Project, Bangladesh Agricultural University.

Rahman, Anisur. 1981. *Some Dimensions of People's Participation in the Bhoomi Sena Movement.* Report no. 81.2. Geneva: United Nations Research Institute for Social Development.

———. 1983. *SARILAKAS: A Pilot Project for Stimulating Grassroots Development in the Philippines.* Geneva: ILO.

Ralston, Lenore, James Anderson, and Elizabeth Colson. 1981. *Voluntary Efforts in Decentralized Management.* Berkeley: Institute of International Studies, University of California.

Raper, Arthur. 1970. *Rural Development in Action: The Comprehensive Experiment at Comilla, East Pakistan.* Ithaca, N.Y.: Cornell University Press.

Ratnapala, Nandasena. 1980. The Sarvodaya Movement: Self-Help Rural Development in Sri Lanka. In Coombs (1980:469–523).

Redfield, Robert. 1967. *The Little Community and Peasant Society and Culture.* Chicago: University of Chicago Press.

Riggs, Fred. 1964. *Administration in Developing Countries.* Boston: Houghton Mifflin.

Roberts, Bryan. 1973. *Organizing Strangers: Poor Families in Guatemala City.* Austin: University of Texas Press.

Robertson, Lindsey. 1978. The Development of Self-Help Gravity Piped Water Projects in Malawi. Mimeo. Lilongwe: Ministry of Community Development.

Robinson, David M. 1982. Water Users' Organizations in Two Large Philippine Irrigation Systems: Constraints and Benefits of a Participatory Approach to Water Management. Ph.D. diss., Department of Government, Cornell University.

Robinson, Marguerite S. 1975. *Political Structure in a Changing Sinhalese Village.* Cambridge: Cambridge University Press.

———. 1982. The Law of the Fishes: The Politics of a Small Indian Village and Its Sub-District (Taluk) in Medak District, Andhra Pradesh (1957–1981). Unpublished book manuscript, Harvard Institute for International Development.

Roca, Santiago. 1980. *Participatory Processes and Action of the Rural Poor in Anta, Peru.* Geneva: ILO.

Rodney, Walter. 1972. *How Europe Underdeveloped Africa.* Washington: Howard University Press.

Roe, Emery, and Louise Fortmann. 1982. *Season and Strategy: The Changing Organization of the Rural Water Sector in Botswana.* Ithaca, N.Y.: Rural Development Committee, Cornell University.

Rondinelli, Dennis. 1981. Government Decentralization in Comparative Perspective: Theory and Practice in Developing Countries. *International Review of Administrative Sciences,* 47,133–45.

———. 1982. The Dilemma of Development Administration: Complexity and Uncertainty in Control-Oriented Bureaucracies. *World Politics,* 35(1):43–72.

Rostow, W. W. 1960. *The Stages of Economic Growth: A Non-Communist Manifesto.* New York: Cambridge University Press.

Russell, Clifford, and Norman Nicholson, eds. 1981. *Public Choice and Rural Development.* Baltimore: Johns Hopkins University Press.

Rustow, Dankwart. 1967. *A World of Nations.* Washington: Brookings Institution.

Saunders, Robert S. 1977. Traditional Cooperation, Indigenous Peasants' Groups and Rural Development: A Look at Possibilities and Experiences. Mimeo. Washington: World Bank.

Savino, Margaret. 1984. *Community Development Paraprofessionals in Bolivia: The NCDS Promotores in the Field.* Ithaca, N.Y.: Rural Development Committee, Cornell University.

Schiller, O. 1969. *Cooperation and Integration in Agricultural Production.* Bombay: Asia Publishing House.

Schultz, T. W. 1964. *Transforming Traditional Agriculture.* New Haven: Yale University Press.

Scitovsky, Tibor. 1954. Two Concepts of External Economies. *Journal of Political Economy,* 62(2):143–51.

Scott, James C. 1972. Patron-Client Politics and Political Change. *American Political Science Review,* 66(1):91–113.

———. 1976. *The Moral Economy of the Peasant.* New Haven: Yale University Press.

Seers, Dudley. 1969. The Meaning of Development. *International Development Review,* Dec., 2–6.

Seibel, Hans Dieter. 1981. Indigenous Self-Help Organizations and Rural Development: Some Liberian and Ghanaian Cases. *Rural Development Participation Review,* 3(1):11–16.

Seibel, Hans Dieter, and Andreas Massing. 1974. *Traditional Organizations and Economic Development: Studies of Indigenous Cooperatives in Liberia.* New York: Praeger.

Seligson, Mitchell. 1980. *Peasants of Costa Rica and the Development of Agrarian Capitalism.* Madison: University of Wisconsin Press.

———. 1982. *Peasant Participation in Costa Rica's Agrarian Reform: A View from Below,* Ithaca, N.Y.: Rural Development Committee, Cornell University.

Selznick, Philip. 1952. *The Organizational Weapon.* New York: McGraw-Hill.

Seth, A. 1975. *Survey of Peasant Organizations in India: A Summary of Case Studies.* Bangkok: FAO Regional Office.

Shadid, Wasif, Wil Prins, and Peter J. M. Nas. 1982. Access and Participation: A Theoretical Approach. In Galjart and Buijs (1982:21–49).

Sharpe, Kenneth Evans. 1977. *Peasant Politics: Struggle in a Dominican Village.* Baltimore: Johns Hopkins University Press.

Sheppard, Jim. 1981. Liberia: A Tale of Patience. *Salubritas,* 5(1).

Sheth, D. L., ed. 1975. *Citizens and Parties: Aspects of Competitive Politics in India.* Bombay: Allied Publishers.

Shrestha, Bihari K. 1980. Nuwakot District, Nepal. In *The Practice of Local-Level Planning: Case Studies in Selected Rural Areas in India, Nepal and Malaysia.* Bangkok, Thailand: U.N. Economic and Social Commission for Asia and Pacific.

Silberfein, Marilyn, 1982. A New Look at the Phenomenon of Clustered Settlements. Draft paper. Washington: Office of Multi-Sectoral Development, USAID.

Simpas, Santiago, Ledivina Cariño, and Arturo Pacho. 1974. *Local Government and Rural Development in the Philippines.* Ithaca, N.Y.: Rural Development Committee, Cornell University. Revised version in Uphoff (1982–83:111).

Siy, Robert Y., Jr. 1982. *Community Resource Management: Lessons from the Zanjera.* Quezon City, Philippines: University of the Philippines Press.

Skinner, G. William. 1964–65. Marketing and Social Structure in Rural China. *Journal of Asian Studies,* 24,3–42, 195–228.

——. 1976. Mobility Strategies in Late Imperial China: A Regional Systems Analysis. In Smith (1976:1,327–64).

Sklar, Richard. 1963. *Nigerian Political Parties: Power in an Emergent African Nation.* Princeton, N.J.: Princeton University Press.

Smith, Carol A., ed. 1976. *Regional Analysis. Vol. I: Economic Systems; Vol. II: Social Systems.* New York: Academic Press.

Smith, G. A., and S. J. Dock. 1981. The Savings Development Movement in Rural Areas: A Popular People's Program. In *Case Studies in Nonformal Education,* B1–15. Harare: Institute of Adult Education, University of Zimbabwe.

Smith, William E., Francis J. Lethem, and Ben A. Thoolen. 1980. *The Design of Organizations for Rural Development Projects: A Progress Report.* Staff working paper no. 375. Washington: World Bank.

Smock, Audrey. 1971. *Ibo Politics.* Cambridge, Mass.: Harvard University Press.

Snyder, Francis G. 1978. Legal Innovation and Social Change in a Peasant Community: A Senegalese Village Police. *Africa,* 48(3):231–47.

Sobhan, Iqbal. 1976. The Planning and Implementation of Rural Development Projects: An Empirical Analysis. Report for AID under contract AID/CM/ta-147-533, Dec. Washington: USAID.

Soenarjono, Danoewidjojo. 1980. IPPA Youth Projects: An Indonesian Experiment in Population Education. In Coombs (1980:635–97).

Solomon, D. D. 1972. Characteristics of Local Organizations and Service Agencies. *Sociologia Ruralis,* 12(3/4):334–60.

Somjee, A. H., and Geeta Somjee. 1978. Cooperative Dairying and the Profiles of Social Change in India. *Economic Development and Social Change,* 26(3):577–90.

Southworth, Herman, and Bruce F. Johnston, eds. 1967. *Agricultural Development and Economic Growth.* Ithaca, N.Y.: Cornell University Press.

Ståhl, Michael. 1974. *Ethiopia: Political Contradictions in Agricultural Development.* Uppsala, Sweden: Raben and Sjögren.

Stavis, Benedict. 1974a. *Rural Local Governance and Agricultural Development in Taiwan.* Ithaca, N.Y.: Rural Development Committee, Cornell University. Revised version in Uphoff (1982–83:11,166–271).

——. 1974b. *Peoples' Communes and Rural Development in China.* Ithaca, N.Y.: Rural Development Committee, Cornell University. Revised version in Uphoff (1982–83:11,13–165).

Steeves, Jeffrey B. 1975. The Politics and Administration of Agricultural Development in Kenya: The Kenya Tea Development Authority. Ph.D. diss., Department of Political Economy, University of Toronto.

——. 1978a. Class Analysis in Rural Africa: The Kenya Tea Development Authority. *Journal of Modern African Studies,* 16(1):123–32.

——. 1978b. The Structure of Participation in Agricultural Development: The Kenya Tea Development Authority. Unpublished paper, Department of Economics and Political Science, University of Saskatchewan.

Stiller, Ludwig, and Ram P. Yadav. 1979. *Planning for People: A Study of Nepal's Planning Experience.* Kathmandu: Research Centre for Nepal and Asian Studies, Tribhuvan University.

Stöhr, Walter B., and D. R. Fraser Taylor, eds. 1981. *Development from Above or Below? The Dialectics of Regional Planning in Developing Countries.* New York: Wiley.

Sussman, Gerald. 1981. The Pilot Project and the Design of Implementing Strategies: Community Development in India. In Grindle (1980:103–22).

Swanson, Jon C., and Mary Hebert. 1982. *Rural Society and Participatory Development: Case Studies of Two Villages in the Yemen Arab Republic.* Ithaca, N.Y.: Rural Development Committee, Cornell University.

Sweet, Charles F., and Peter F. Weisel. 1979. Process versus Blueprint Models for Designing Rural Development Projects. In Honadle and Klauss (1979:127–45).

Talagune, A. B. 1982. "Change Agents" to Promote Participatory Village Development in Sri Lanka. *Rural Development Participation Review,* 3(3):21–24.

Tendler, Judith. 1976. *Inter-Country Evaluation of Small Farmer Organizations in Ecuador and Honduras: Final Report.* Program evaluation study. Washington: USAID, Latin America Bureau, Office of Development Programs.

——. 1979. *New Directions for Rural Roads.* Washington: Office of Evaluation, USAID.

Tendler, Judith, with Kevin Healy and Carol M. O'Laughlin. 1983. *What to Think about Cooperatives: A Guide from Bolivia.* Washington: Inter-American Foundation.

Texier, J. M. 1976. The Promotion of Cooperatives in Traditional Societies. In Hunter, Bunting, and Bottrall (1976:215–22).

Thai Khadi Research Institute. 1980. *A Self-Help Organization in Rural Thailand: The Question of Appropriate Policy Inputs:* Bangkok: Thai Khadi Research Institute, Thammasat University.

Thomas, Barbara. 1980. The *Harambee* Self-Help Experience in Kenya. *Rural Development Participation Review,* 1(3):1–5.

Thomas, Fraser M. 1968. *Culture and Change in India: The Barpali Experiment.* Boston: University of Massachusetts Press.

Tilakaratna, S. 1982. *Grassroots Self-Reliance in Sri Lanka: Organization of Betel and Coir Yarn Producers.* Working paper no. 24. Geneva: World Employment Program, International Labour Office.

Tocqueville, Alexis de. 1835. *Democracy in America.* Reprint. New York: Vintage, 1954.

Toth, C., and T. J. Cotter. 1978. Learning from Failure. *Focus,* 3,25–28.

Truman, David. 1951. *The Governmental Process.* New York: Knopf.

Tufte, Edward R. 1969. Improving Data Analysis in Political Science. *World Politics,* 21,641–54.

United Nations. 1962. *Decentralization for National and Local Government.* New York: United Nations.

UNRISD. 1975. *Rural Cooperatives as Agents for Change: A Research Report and a Debate.* Vol. 8. Geneva: U.N. Research Institute for Social Development.

Uphoff, Norman. 1979. Assessing the Possibilities for Organized "Development from Below" in Sri Lanka. Paper for Ceylon Studies Seminar, 1978–79 series. University of Peradeniya. Mimeo.

——. 1980. Political Considerations in Human Development. In Knight (1980:3–108).

———. 1981. The Institutional-Organizer (IO) Programme in the Field after Three Months: A Report on Trip to Ampare/Gal Oya, June 17–20, 1981. Mimeo. Colombo: Agrarian Research and Training Institute.

———. 1982a. The Institutional-Organizer (IO) Programme in the Field after Ten Months: A Report on Trip to Ampare/Gal Oya, Sri Lanka, Jan. 14–17, 1982. Mimeo. Colombo: Agrarian Research and Training Institute.

———. 1982b. Contrasting Approaches to Water Management Development in Sri Lanka. In *Third World Legal Studies: Law in Alternative Strategies of Rural Development*. New York: International Center for Law in Development, 202–49.

———. 1982c. The Institutional-Organizer (IO) Programme in the Field after Sixteen Months: A Report on Trip to Ampare/Gal Oya, Sri Lanka, June 1982. Mimeo. Colombo: Agrarian Research and Training Institute.

———, ed. 1982–83. *Rural Development and Local Organization in Asia,* 3 vols. New Delhi: Macmillan.

———. 1983a. The Institutional-Organizer (IO) Programme in the Field after Twenty-One Months: A Report on Trip to Ampare/Gal Oya, Sri Lanka, Jan. 7–14, 1983. Mimeo. Colombo: Agrarian Research and Training Institute.

———. 1983b. Rural Development Strategy: The Central Role of Local Organizations, and Changing "Supply-Side" Bureaucratics. *Studies on Agrarian Reform and Rural Poverty*. Rome: FAO.

Uphoff, Norman T., John M. Cohen and Arthur A. Goldsmith. 1979. *Feasibility and Application of Rural Development Participation: A State of the Art Paper.* Ithaca, N.Y.: Rural Development Committee, Cornell University.

Uphoff, Norman T., and Milton J. Esman. 1974. *Local Organization for Rural Development: Analysis of Asian Experience.* Ithaca, N.Y.: Rural Development Committee, Cornell University. Updated version in Uphoff (1982–83:111).

Uphoff, Norman, and Warren Ilchman, eds. 1972. *The Political Economy of Development.* Berkeley: University of California Press.

Uphoff, Norman, and R. D. Wanigaratne. 1982. Rural Development and Local Organization in Sri Lanka. In Uphoff (1982–83:1,479–549).

Uphoff, Norman, M. L. Wickramasinghe, and C. M. Wijayaratne. 1981. "Optimum" Participation in Water Management: Issues and Evidence from Sri Lanka. Paper prepared for Rural Development Committee, Cornell University, and Agrarian Research and Training Institute, Colombo.

USAID. 1975. *Implementation of "New Directions" in Development Assistance.* Report prepared by U.S. Agency for International Development for Committee on International Relations on Implementation of the Foreign Assistance Act of 1973, 94th Con., 1st Sess., 22 July.

van Heck, Bernard. 1979. *Participation of the Poor in Rural Organizations: A Consolidated Report on the Studies in Selected Countries of Asia, Near East and Africa.* Rome: FAO.

Van Sant, Jerry, and Peter F. Weisel. 1979. *Community Based Integrated Rural Development (CBIRD) in the Special Territory of Aceh, Indonesia.* Washington: Development Alternatives, Inc., and Research Triangle Park, N.C.: Research Triangle Institute.

Vengroff, Richard. 1974. Popular Participation and the Administration of Rural Development: The Case of Botswana. *Human Organization,* 33,303–9.

Vincent, Joan. 1976. Rural Competition and the Cooperative Monopoly: A Ugandan Case Study. In Nash, Dandler, and Hopkins (1976).

Volken, Henry, Ajoy Kumar, and Sara Kaithathara. 1982. *Learning from the Rural Poor: Shared Experiences of the Mobile Orientation and Training Team*. New Delhi: Indian Social Institute.

Wade, Robert. 1982a. Group Action for Irrigation. *Economic and Political Weekly* 17(39):103–6.

———. 1982b. The System of Administrative and Political Corruption: Canal Irrigation in South India. *Journal of Development Studies,* 18(3):287–328.

Wallman, Sandra. 1969. The Farmech Mechanization Project, Basutoland (Lesotho). In Brokensha and Pearsall (1969:14–21).

Wasserstrom, Robert. 1982. La Libertad: A Women's Cooperative in Highland Bolivia. *Grassroots Development,* 6(1):7–12, Washington: Inter-American Foundation.

Weber, Eugen. 1976. *Peasants into Frenchmen*. Stanford, Calif.: Stanford University Press.

Weitz, Raanan. 1971a. *From Peasant to Farmer: A Revolutionary Strategy for Development*. New York: Columbia University Press.

———. 1971b. *Rural Development in a Changing World*. Cambridge, Mass.: M.I.T. Press.

Westney, Eleanor. 1983. Organizational Development and Social Change in Meiji Japan. Unpublished book manuscript, School of Management, Massachusetts Institute of Technology.

Wharton, Clifton, Jr., ed. 1969. *Subsistence Agriculture and Economic Development*. Chicago: Aldine.

Whyte, William F. 1975. Conflict and Cooperation in Andean Communities. *American Ethnologist,* 2(2):373–92.

———. 1981. *Participatory Approaches to Agricultural Research and Development*. Ithaca, N.Y.: Rural Development Committee, Cornell University.

Whyte, William F., and Giorgio Alberti. 1976. *Power, Politics and Progress: Social Change in Rural Peru*. New York: Elsevier.

Whyte, William F., and Damon Boynton. 1983. *Higher-Yielding Human Systems for Agriculture*. Ithaca, N.Y.: Cornell University Press.

Widstrand, C. G., ed. 1970. *Cooperatives and Rural Development in East Africa*. New York: Africana Publishers.

———, ed. 1972. *African Cooperatives and Efficiency*. Uppsala, Sweden: Scandinavian Institute of African Studies.

Willett, A. B. J. 1981. *Agricultural Group Development in Botswana,* 4 vols. Gaborone, Botswana: USAID.

Williams, Arthur. 1981. *Measuring Local Government Performance: Assessing Management, Decentralization and Participation*. Ithaca, N.Y.: Rural Development Committee, Cornell University.

Williamson, John. 1982. Case Studies of Conflicts in the Context of Development in Rural Nepal. Draft. Kathmandu: United Mission to Nepal.

Winans, Edgar V., and Angelique Haugerud. 1977. Rural Self-Help in Kenya: The Harambee Movement. *Human Organization,* 36(4):334–51.

Winder, David. 1979. Mexico: Carrizo Valley Settlement Scheme. *RRDC Bulletin*, no. 7. Reading, U.K.: University of Reading.
Wolf, Eric. 1966. *Peasants*. Englewood Cliffs, N.J.: Prentice-Hall.
——. 1969. *Peasant Wars of the Twentieth Century*. New York: Harper & Row.
Womack, John. 1968. *Zapata and the Mexican Revolution*. New York: Vintage.
Wong, John, ed. 1979. *Group Farming in Asia*. Singapore: Singapore University Press.
World Bank. 1975. *Rural Development: Sector Policy Paper*. Washington: World Bank.
——. 1976. *Village Water Supply: A World Bank Paper*. Washington: World Bank.
Worsley, Peter, ed. 1971. *Two Blades of Grass: Rural Cooperatives in Agricultural Modernization*. Manchester, U.K.: Manchester University Press.
Wortman, Sterling, and Ralph W. Cummings, Jr. 1978. *To Feed This World: The Challenge and the Strategy*. Baltimore: Johns Hopkins University Press.
Young, Frank W. 1980. Participation and Project Success: A Reanalysis of the Development Alternatives Study. *Rural Development Participation Review*, 1(3):10–15.
Young, Frank W., Mary Hebert, and Jon Swanson. 1981. *The Ecological Context of Local Development Participation in Yemen*. Yemen research program, working note no. 13. Ithaca, N.Y.: Rural Development Committee, Cornell University.
Zandstra, Hubert, Kenneth Swanberg, Carlos Zulberti, and Barry Nestel. 1979. *Caqueza: Living Rural Development*. Ottawa: International Development Research Centre.

Author Index

Case Index

Asterisked cases are those not included in statistical analysis, but reported on from literature references and observations. Boldface references are to Appendices C and D.

Subject Index

Library of Congress Cataloging in Publication Data

Esmann, Milton J., 1918–
Local organizations.

"April 1983."
Bibliography: p.
Includes index.
1. Rural development—Developing countries—Management. 2. Agriculture, Cooperative—Developing countries. I. Uphoff, Norman Thomas. II. Title.
HD1417.E83 1984 334'.683'091724 83-73340
ISBN 0-8014-1665-5 (alk. paper)